Geopolitics and Development

Geopolitics and Development examines the historical emergence of development as a form of governmentality, from the end of empire to the Cold War and the War on Terror. It illustrates the various ways in which the meanings and relations of development as a discourse, an apparatus and an aspiration, have been geopolitically imagined and enframed.

The book traces some of the multiple historical associations between development and diplomacy and seeks to underline the centrality of questions of territory, security, statehood and sovereignty to the pursuit of development, along with its enrolment in various (b)ordering practices. In making a case for greater attention to the evolving nexus between geopolitics and development and with particular reference to Africa, the book explores the historical and contemporary geopolitics of foreign aid, the interconnections between development and counterinsurgency, the role of the state and social movements in (re)imagining development, the rise of (re)emerging donors like China, India and Brazil, and the growing significance of South–South flows of investment, trade and development cooperation. Drawing on post-colonial and post-development approaches and on some of the author's own original empirical research, this is an essential, critical and interdisciplinary analysis of the complex and dynamic political geographies of global development.

Primarily intended for scholars and post-graduate students in development studies, human geography, African studies and international relations, this book provides an engaging, invaluable and up-to-date resource for making sense of the complex entanglement between geopolitics and development, past and present.

Marcus Power is a Professor of Human Geography at Durham University. His research interests include critical geopolitics and the spatialities of (post)development; visuality and popular geopolitics; energy geographies and low-carbon transitions in the global South; and China–Africa relations and the role of (re)emerging development donors in South–South cooperation. He is author of *Rethinking Development Geographies* (2003) and co-author of *China's Resource Diplomacy in Africa: Powering Development?* (2012).

Geopolitics and Development

Marcus Power

Routledge
Taylor & Francis Group
LONDON AND NEW YORK

First published 2019
by Routledge
2 Park Square, Milton Park, Abingdon, Oxon OX14 4RN

and by Routledge
52 Vanderbilt Avenue, New York, NY 10017

Routledge is an imprint of the Taylor & Francis Group, an informa business

British Library Cataloguing-in-Publication Data
A catalogue record for this book is available from the British Library

Library of Congress Cataloging-in-Publication Data
Names: Power, Marcus, 1971- author.
Title: Geopolitics and development / Marcus Power.
Description: Abingdon, Oxon ; New York, NY : Routledge, 2019. | Includes bibliographical references and index.
Identifiers: LCCN 2018043343| ISBN 9780415519564 (hardback : alk. paper) | ISBN 9780415519571 (pbk. : alk. paper) | ISBN 9780203494424 (ebook)
Subjects: LCSH: Geopolitics--History--20th century. | Geopolitics--History--21st century. | Economic development--History--20th century. | Economic development--History--21st century. | Postcolonialism.
Classification: LCC JC319 .P68 2019 | DDC 338.9--dc23
LC record available at https://lccn.loc.gov/2018043343

ISBN: 978-0-415-51956-4 (hbk)
ISBN: 978-0-415-51957-1 (pbk)
ISBN: 978-0-203-49442-4 (ebk)

Typeset in Times New Roman
by Integra Software Services Pvt. Ltd.

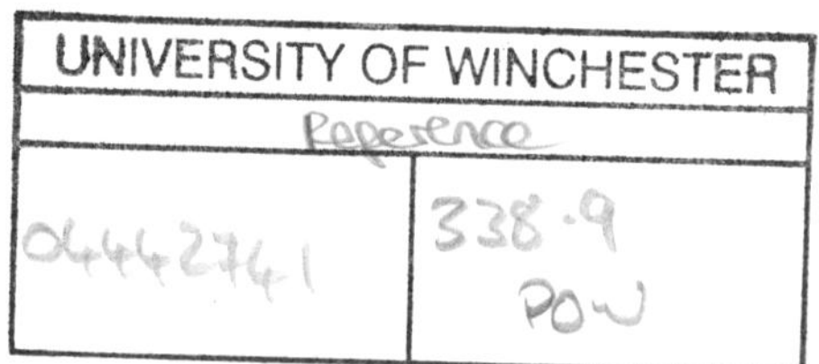

Printed and bound in Great Britain by
TJ International Ltd, Padstow, Cornwall

For Conor Ciarán Power

Contents

Figures

Acknowledgements

THERE are a number of people I would like to thank for the help and support they provided to me in writing this book. For their helpful comments, advice and feedback I would like to thank Joe Painter, Padraig Carmody and Andrew Brooks. Thanks also to James Sidaway, whose own work has been a huge inspiration to me, for his encouragement and support throughout my academic career but also for the insightful comments and helpful reading suggestions that he made as I wrote this book. I would also like to express my sincere gratitude to Andrew Mould at Routledge for all his continued patience, understanding and support throughout the writing process. This book has taken much longer to write than anticipated but Andrew's support for the project was consistent and unwavering throughout. I'd also like to thank Alaina Christensen, Faye Leerink and Egle Zigaite at Routledge for their help, support and editorial assistance and Drew Stanley for all his help with the copyediting process. Many thanks also to Chris Orton in the Design and Imaging Unit at Durham University for his help in preparing the map for figure 1.1.

There are also a number of people and organisations that I would like to thank for their help in securing permissions to use the images included within the book. Thanks to Yevgeniy Fiks for his help in accessing Soviet propaganda posters from the Wayland Rudd Collection and to Sergey Sheremet for his assistance with accessing photographic archives of Soviet military experts in Angola. Thanks also to the International Institute of Social History in Amsterdam for their help with access to Chinese propaganda posters held in the Stefan R Landsberger collection. I would also like to thank Kiluanji Kia Henda for permission to use the artwork included as figure 1.3 and Andy Singer for the permission to use the cartoon that appears as figure 6.1. Thanks also to the Imperial War Museum, the UN Photo Library, the US Peace Corps, the German Federal Archives (Bundesarchiv) and the Netherlands National Archive for helping with access to archival images. I would also like to thank staff at the John F Kennedy Presidential Library and Museum, at the US National Archives Still Picture Unit and at the Lyndon B Johnson Presidential Library for their assistance with accessing and reproducing a variety of archival images.

Of course, none of this would have been possible without the support of my friends and family. I would like to thank my mother Ann and my father Maurice for the inspiration and encouragement they have given me throughout my career. I would also like to thank my son Conor who has been a constant source of inspiration to me.

Finally, I would like to express my heartfelt thanks to Leanne Cornelius for all her love, patience, support and understanding. I could not have done this without you sweetheart and love you to the moon and back.

Abbreviations

AAPSO	Afro-Asian People's Solidarity Organisation
AAS	Association for Asian Studies
ACOTA	Africa Contingency Operations Training and Assistance
AfP	Alliance for Progress
AFRICOM	United States Africa Comnmand
AQIM	Al-Qaeda in the Islamic Mahgreb
AU	African Union
CCP	Communist Party of China
CERP	Commander's Emergency Response Program
CIA	Central Intelligence Agency
CIS	Center for International Studies
CJTF-HoA	Combined Joint Task Force-Horn of Africa
CL	Contingency locations
CMO	Civil–military operations
COMECON	Council for Mutual Economic Assistance
Comintern	Communist International
CORDS	Civil Operations and Revolutionary [later Rural] Development Support
CSL	Cooperative security locations
CIVETs	Colombia, Indonesia, Vietnam, Egypt, Turkey and South Africa
DAC	Development Assistance Committee
DoD	Department of Defence
DRC	Democratic Republic of the Congo
EAGLES	Emerging and growth-leading economies
ECA	Economic Cooperation Administration
ECOSOC	Economic and Social Council

FAF	Foreign Assistance Framework
FAO	Food and Agriculture Organization
FLN	Front de Libération National
FNLA	National Liberation Front of Angola
FOCAC	Forum on China–Africa Cooperation
FOL	Forward Operating Locations
FTA	Free Trade Area
GDP	Gross Domestic Product
GDR	German Democratic Republic
HCA	Humanitarian and Civic Assistance
HDI	The Human Development Index
HTS	Human Terrain System
HVP	Helmand Valley Project
IBSA	India–Brazil–South Africa Dialogue Forum
IFI	International Financial Institutions
ILO	International Labour Organization
IR	International Relations
ISGS	Islamic State in the Greater Sahel
ITEC	Indian Technical and Economic Cooperation scheme
Komsomol	All-Union Leninist Young Communist League
LDC	Less developed country
LPRP	Laotian People's Revolutionary Party
LRA	Lord's Resistance Army
MDG	Millennium Development Goals
Mercosur	Mercado Común del Sur
MINT	Mexico, Indonesia, Nigeria, Turkey
MIST	Mexico, Indonesia, South Korea, Turkey
MIT	Massachusetts Institute of Technology
MPLA	People's Movement for the Liberation of Angola
MSA	Mutual Security Agency
MST	Landless Rural Workers Movement (Brazil)
NAFTA	North American Free Trade Agreement
NAM	Non-Aligned Movement
NATO	North Atlantic Treaty Organization
NGO	Non-governmental organisation
NIEO	New International Economic Order
NLF	National Liberation Front
OBOR	One Belt, One Road
ODA	Overseas Development Assistance
OECD	Organisation for Economic Co-operation and Development
OEEC	Organisation for European Economic Co-operation

OPS	Office of Public Safety
OPSAAAL	Organization of Solidarity of the Peoples of Africa, Asia and Latin America
OSS	Office of Strategic Services
PEPFAR	President's Emergency Plan for AIDS Relief
PMSC	Private Military and Security Contractor
PRSP	Poverty Reduction Strategy Paper
RIDP	Rapti Integrated Development Project
SAP	Structural Adjustment Programme
SDC	Save Darfur Coalition
SDG	Sustainable Development Goals
SEZ	Special Economic Zone
SKSSAA	Soviet Afro-Asian Solidarity Committee
SOCAFRICA	Special Operations Command, Africa
SORO	Special Operations Research Office
SSDC	South–South Development Cooperation
SSRC	Social Science Research Council
SWAPO	South West African People's Organisation
TAZARA	Tanzania–Zambia Railway Authority
TCA	Technical Cooperation Administration
TCT	Three Cups of Tea
TPP	Trans-Pacific Partnership
TSCTP	Trans-Saharan Counterterrorism Partnership
TVA	Tennessee Valley Authority
UAE	United Arab Emirates
UCI	University College Ibadan
UNCTAD	United Nations Conference on Trade and Development
UNDP	United Nations Development Programme
UNESCO	United Nations Educational, Scientific and Cultural Organisation
UNFPA	United Nations Population Fund
UNICEF	United Nations International Children's Emergency Fund
UNITA	National Union for the Total Independence of Angola
UNSC	United Nations Security Council
UK	United Kingdom
US	United States
USAF	United States Air Force
USAID	United States Agency for International Development
USDA	United States Department of Agriculture
USSR	Union of Soviet Socialist Republics

WHO	World Health Organization
WTO	World Trade Organization
ZAPU	Zimbabwe African People's Union

Chapter 1

Introduction: geopolitics and the assemblage of development

INTRODUCTION: THE ANTI-POLITICS OF DEVELOPMENT

> "development" has beyond doubt been widely used as a hard drug, addiction to which, legally tolerated or encouraged, may stimulate the blissful feelings that typify artificial paradises. (Rist, 2010: 19)

ALTHOUGH it has consistently been imagined and represented as a neutral and depoliticised technical or managerialist domain, questions of international relations, geopolitics and foreign policy have always been at the very centre of the theory and practice of development. With its in-built sense of design, development has always represented forms of mobilisation associated with order and security (Duffield, 2002) and has various "strategic" effects: in its depoliticisation of poverty, for example, or in the (inconsistent) expansion of state bureaucratic power (Ferguson, 1990). While different strategies and technologies have come and gone, development has consistently served as a kind of antidote to the unruly material politics created by the challenges of governing fluid and changeable spaces, attempting to reconcile the inevitable disruptions of progress with the need for order (Cowen and Shenton, 1995: 27–43) – an objective it has consistently failed to achieve (Li, 2007). This book seeks to further interrogate this sense of development as a set of ameliorative and compensatory technologies of security and argues that what has consistently been central to the very logic and fabric of development is a desire to counter crises of disorder, insecurity and insurgency. In particular, the book explores the repeated enrolment of development in projects of counterinsurgency and the long-standing idea of development as a form of (violent) pacification. To fully understand the interface between security and development, or its uses as a form of countering insurgency, it is necessary to understand the wider nexus between geopolitics and development. In particular, the book argues that development is fundamentally a form of governmentality, is very much about diplomacy and has historically had multiple associations with questions of territory, governance and sovereignty or been implicated in various (b)ordering practices. (In)security and insurgency are not incidental to theories and practices of development and neither is development's obsession with questions of order and security somehow "new" or a recent rediscovery – they have always been at the very heart of (and are constitutive of) the development enterprise and imaginary.

As an arena of theory and state practice, development came to prominence in an era when the legitimacy of colonial rule was in rapid decline, such that it quickly became a mechanism for trying to shape, manage and control socio-economic and political change in an era of formal sovereignty and to create particular kinds of states in the process. The primary threats development was meant to "neutralise" included decolonisation, communism and various crises within industrial capitalism and, most recently, terrorist insurgencies in the global war on terror. This book will argue that geopolitical knowledges, discourses and practices have *always* played a key role both in the construction of development and in its contestation and seeks to build on Slater's (1993) contention that *all* conceptualisations of development contain and express a geopolitical imagination which conditions and enframes its meanings and relations. Development is conceptualised here as simultaneously a discourse, an apparatus and an aspiration (Sidaway, 2011: 2792) and in what follows I argue that the political geography of development requires and rewards much further critical scrutiny. Development Studies has tended to avoid a direct and sustained engagement with questions of (geo)politics in the global South whilst much of the literature in (sub)disciplines like IR and Political Geography has ignored or misrepresented questions of development in the non-Western world and has been characterised by a certain degree of parochialism (Power, 2010). Seeking to adopt a more interdisciplinary approach to transcend sub-disciplinary boundaries in Geography, the book provides a synthesis of scholarship in critical geopolitics, critical IR and critical development studies and seeks both to re-theorise development as an apparatus or assemblage and to make the case for a post-colonial approach to understanding its complex and dynamic geopolitics.

The focus of the book is both historical and contemporary, exploring the geopolitical enframing and imagination of development at key historical junctures such as the end of empire, the Cold War and the War on Terror. In a variety of ways however, the landscape of international development has shifted quite dramatically in the last three decades and thus in addition to attending to the historical role of global hegemons like the US and the USSR in shaping international development theory and practice, the book seeks to make an original contribution by exploring what the rise of "new" state donors from the global South (such as China, Brazil and India) means for established modes of development cooperation and for the theory and practice of development more generally. In doing so, my analysis is informed by some of my own recent research on China–Africa relations and South–South development cooperation (SSDC). For much of the post-war period the drivers of the global economy and the trustees of international development were unproblematically seen as the wealthy countries of Europe and North America whilst historically much of the theory and practice of development has been focused around North–South relations and interactions. Yet over the past few decades the order of international development has fundamentally changed with (re)emerging or "rising" powers from the global South taking a greater role in the global economy and international politics. The ontological hierarchy of Northern donors and Southern recipients has been upset by the rise of the South and the growth of SSDC, by the global financial crisis and by wider changes in global geographies of poverty and wealth, which have collectively contributed to decentring the field of international development, in terms of both the agents authorised to play and the practices considered legitimate (Mawdsley, 2017). The book charts this gradual disintegration of established ways of categorising global development and the meta-geographies and geopolitical divides that have previously structured North–South relations. As the world is experiencing rapid geo-economic transformations, marked by a downturn of European and American hegemony and the rise of Southern economies, the book argues the case for a renewed approach to unpacking the emerging dynamics of accumulation, dispossession and poverty, one that puts questions of international relations and geopolitics front and centre in the study of development.

In addition to tracing the rise and fall of the "Third World" as a geopolitical category and the rise of the South and SSDC, the book examines the growing articulation between security policies and aid dynamics, or the creeping "securitisation of development" (and "developmentalisation of security"). Driven by the security imperatives of the post 9/11 era and the ensuing War on Terror, underdevelopment has once again been framed as highly "dangerous". This has been marked by a wave of Western humanitarian and peace interventionism that "empowers international institutions and actors to individuate, group and act upon Southern populations" (Duffield, 2006b: 16) but also by a growing collusion between military and humanitarian actors around the idea of development. As an assemblage of practices that connect "violence to order, force to persuasion, civil to military power" (Bell, 2015: 18), counterinsurgency has experienced a renaissance in recent years following military setbacks in Iraq and Afghanistan, placing human social and economic development, the protection of civilians, political solutions and the reform of state security sectors at the forefront of military doctrine and practice, further blurring the line between security and development. This incarnation of security seeks to reprioritise development criteria in relation to supporting intervention, reconstructing crisis states and, in order to stem terrorist recruitment, protecting livelihoods and promoting opportunity within strategically important areas of instability (Duffield, 2002). In the "security–development nexus", fragile statehood and concerns about unstable or stateless regions have been linked to a range of threats to international security, shifting the Western emphasis to the containment of risks emanating from "undergoverned" spaces. Along with important geographical shifts in aid disbursements and allocations, the War on Terror has fundamentally altered the nature of donor cooperation with developing countries as the subordination of foreign aid to military, foreign policy and economic interests has increasingly altered the context in which development aid is framed and implemented.

Any claim to be able to change the lives of others is ultimately a claim to power (Li, 2007), as is the wider idea of the "makeability" of society which has long been a feature of the theory and practice of development. Whether they seek to move people from one place to another, to better provide for their needs, to rationalise their use of land, or to educate and to modernise, programmes for development are all, as Li (2007: 1) points out, "implicated in contemporary sites of struggle". The book examines the ways in which development has enabled a series of projects that seek to govern, to control, to bring order, along with its long-standing obsession with (re)modelling states in the periphery. In reference to "high-modernist" approaches to development in the South, Scott (1998) writes of the "blindness" of the state and about legibility as a central problem of statecraft, the gradual resolution of which enables the state to get a better handle on its subjects and their environment through a more permanent visibility that assures the automatic functioning of power. The book explores this idea of development as a resolution to problems of statecraft and as enabling particular forms of state power and the "capturing" of populations by increasing the *legibility* of citizens and territories and considers the changing historical role of the state in development. It also examines the continuing hegemony of neoliberalism in development theory and practice and explores the role of agencies like USAID in its diffusion, along with the various spaces of insurgency that state-led neoliberal models of development have created in their wake across the South.

Neoliberalism is often depicted as a kind of "tidal wave", emanating from the dominant metropoles, that rolls across all places, but although it is clear that development has long played a role in "strategies of social and cultural domination" (Escobar, 2012: vii), it is important to recognise the various ways in which it has been contested and resisted. Neoliberalism is best understood "as an assemblage of technologies, techniques and practices that are appropriated *selectively*, that come into uncomfortable encounters with 'local' politics and cultures, and that are mobile and connective (rather than 'global')" (Clarke, 2008: 138, emphasis added). The book traces these "selective" appropriations of neoliberalism in development along with its "uncomfortable" encounters with local cultural and political spaces. Development can of course operate as much as a discourse of entitlement as a discourse of control (Cooper and Packard, 1997; Nilsen, 2016) and the global South has given rise to some of the world's most intense and advanced popular struggles against neoliberalism as across Latin America, Africa and Asia movements of indigenous peoples, workers, peasants, women and shanty-town dwellers have challenged the dispossession, exclusion and poverty that have followed in the wake of neoliberalisation. Through practices of resistance, these movements are beginning to transform the direction and meanings of post-colonial development and to reshape conceptions of politics, participation, statehood and citizenship.

A relational view of development is key because neoliberal forms of capitalism did not simply arise in the "core" and spread from there to the "periphery" (Hart, 2010) but are the products of processes of spatial *interconnection* and it is thus important to hold North and South in a relational view when thinking about the complex and shifting territories of poverty and development (Roy and Crane, 2015). Along with geopolitics, development also needs to be situated in relation to the larger nexus of relations within which it is embedded (e.g. social, economic, cultural and racial) and it is important to examine the different discourses, institutions, forms of management and circuits of capital that have shaped them (Roy and Crane, 2015). The significance of questions of "race" in development, in particular, has often been neglected (White, 2002) yet biopolitics is inherently about race, acquiring its powers as a form of governance by securing a population's "purity" and "safety" within the context of an imagined, alien, raced, internal or external threat (Foucault et al., 2003 [1975–76]; Macey, 2009). Expressions of biopolitical forms of power always contain within them racialised anxieties and fears (Domosh, 2018) and as Mitchell (2017: 358) has shown, "race was and remains central" to liberal, humanitarian forms of aid.

THEORISING (POST-)DEVELOPMENT

> The postdevelopment agenda is not, as we see it, anti-development. The challenge of postdevelopment is not to give up on development, nor to see all development practice – past, present and future, in wealthy and poor countries – as tainted, failed, retrograde; as though there were something necessarily problematic and destructive about deliberate attempts to increase social wellbeing through economic intervention; as though there were a space of purity beyond or outside development that we could access through renunciation. The challenge is to imagine and practice development differently. (Gibson-Graham, 2005: 6)

> postdevelopment was meant to convey the sense of an era in which development would no longer be a central organising principle of social life. This did not mean that postdevelopment was seen as a new historical period to which its proponents believed we had arrived. (Escobar, 2012: xiii)

A "toxic" word (Rist, 2010: 24) fraught with a Faustian ambiguity, development is regarded as something which happens simultaneously to individuals, communities, nations and regions but ultimately as a cumulative, organic, natural and inherently progressive process that is somehow above and beyond the realm of the political. The word *development* itself, Rist (2010: 22) observes, has become a "modern shibboleth, an unavoidable password", which comes to be used "to convey the idea that tomorrow things will be better, or that more is necessarily better". In this sense "development" has taken on a "quasi-mystical connotation" (Munck, 1999: 198). Yet the very taken-for-granted quality of "development" (and indeed much of the lexicon of development discourse) leaves much of what is actually *done* in its name unquestioned. Although its failures as a socio-economic endeavour have now been widely recognised, "development discourse still contaminates social reality" and "remains at the centre of a powerful but fragile semantic constellation" (Esteva, 2009: 1). The language of development essentially defines worlds-in-the making (Cornwall, 2010), animating and justifying intervention with fulsome promises of the possible. As Sachs (2010: 1) contends, "development is much more than just a socio-economic endeavour; it is a perception which models reality, a myth which comforts societies, and a fantasy which unleashes passions."

Development is also very much an "anti-politics machine" (Ferguson, 1990: 270) that "insistently repos[es] political questions of land, resources, jobs or wages as technical 'problems' responsive to the technical 'development' intervention". Lummis (1996: 46) even argues that economic development is essentially politics *camouflaged* or a way of concealing power arrangements. The anti-politics of development ensures that political issues are "rendered technical" (Li, 2007) or as best addressed by experts and hence the terms of any public debate are limited to trivial or technical matters, constituting the boundary between those positioned as trustees, with the capacity to diagnose deficiencies in others, and those subject to expert direction. Anti-politics is often subliminal and routine as the structure of political-economic relations is written out of the diagnoses and prescriptions produced by development "experts" and as the process of development is perpetually depoliticised. Here, the contestability of many of the words in its lexicon (e.g. civil society) is "flattened" (Chandhoke, 2010) and they become "consensual hurrah-words" that have warmly persuasive qualities (much like "development" itself), gaining their purchase and power through their vague and euphemistic qualities, their capacity to embrace a multitude of possible meanings and also their normative resonance, all of which places the sanctity of its goals beyond reproach and disables any possible critique of "development", since it was equated "almost with life itself" (Rist, 2010: 20). The related idea of poverty reduction similarly has a luminous obviousness to it, defying mere mortals to challenge its status as a moral imperative (Toye, 2010: 45). The moral unassailability of the development enterprise is similarly secured by copious references to that nebulous, but emotive, category: "the poor" (Cornwall and Brock, 2005). The very concept of "poverty" however "covers up the inequality wrought by capitalism" (Kapoor, 2012: 34) and tends to assume that being poor is a question of unfortunate circumstances, mystifying its structural causes wherein wealth in some parts of the world "is the historical result of the pauperization of others" (34).

Development has often been depicted by some of its critics as a singular and monolithic "project" in a common failure to recognise its multiplicities. In an attempt to move past this, Hart (2010: 10) differentiates between "Big D" Development or "the multiply scaled projects of intervention in the 'Third World' that emerged in the context of decolonisation struggles and the Cold War" and "Little d" development which refers "to the development of capitalism as geographically uneven but spatially interconnected processes of creation and destruction, dialectically interconnected with discourses and practices of Development" (Hart, 2010: 119). This also includes a recognition that D/development has historically come to be defined by a multiplicity of "developers" entrusted with the task (Cowen and Shenton, 1996: 4) but also that geographically D/development has been played out in a multiplicity of places and localities such that its geopolitical significance cannot simply be "read off" from any one vantage point or set of coordinates. This useful recognition that development is both a project and a process has also been accompanied by a growing recognition of the plural origins of development discourses: Chinese intellectuals used the language of modernity in the 1920s (Cullather, 2000) and for decades before independence Indian nationalists sought to articulate their own rival visions of national development (long before they had been debated by modernisation theorists in the US in the 1950s) in order to seize the development process and fashion a uniquely Indian modernity (Bose, 1997; Prakash, 1999). Similarly, there is a growing recognition of the need for more *global* histories of key ideas and projects within development thinking, such as modernisation (Engerman and Unger, 2009), and a refusal of the idea that modernisation was purely or even predominantly an American "export".

Although Hart (2010) does recognise the dialectical nature of D/development in practice these distinctions are far messier and more complex, particularly in relation to the contemporary dynamics of South–South cooperation. As a dynamic "field of meaning" (Williams, 1976) development is conceptualised in this book as operating simultaneously across three dimensions: as an immanent process of politico-economic change; as an intentional project of amelioration led by international and other aid agencies; and as a set of social experiences and outcomes (Cowen and Shenton, 1996). Alongside this, the book seeks to trace some of the ways in which development has historically served as a "dominant problematic or interpretive grid through which the impoverished regions of the world are known" (Ferguson, 1990: xiii). It is this sense that through development, as a problematic or an "interpretive grid", the cultures, peoples, places and geopolitics of the Third World or Global South become more "knowable" and intelligible that the book seeks to examine along with the interventions that this then gives rise to. "Development thinking" has various strands of course (Hettne, 1995) and this includes *development theories* (logical propositions about how the world is structured which explain past and future developments), *development strategies* (the practical paths adopted by actors and agents, ranging from communities to central states and international institutions) and *development ideologies* (the different economic goals and political agendas associated with its pursuit). This book seeks to engage with all three of these elements but before doing so it is necessary to map out my own approach to theorising "development".

In particular the book seeks to conceptualise development as a form of governmentality and to (re)engage with and enhance theorisations of post-development. Development practices can usefully be understood as forms of government "structured through a variety of technics and micropolitics of power (from the map, to the national statistics, to forms of surveillance), to accomplish or attempt to accomplish, stable rule through certain sorts of governable subjects and governable objects" (Watts, 2003: 12). Foucault's concept of governmentality provides a valuable framework through which to analyze the contemporary governance of development and how it functions not solely through states but through multiple tactics and means that regulate the conduct of individuals and institutions by setting up standards of behaviour according to certain rationalities or by producing "governable spaces". Although the various instruments and procedures of development programmes typically refer themselves to the state, they cannot be reduced to it (Brigg, 2002: 428). Transnational forms of governmentality (Ferguson and Gupta, 2002) are also significant here since agencies like the IMF and World Bank have a disciplinary power over development, rendering nation-states visible though processes of surveillance, evaluation and judgement regularly carried out by personnel and consultants during missions and consultancies.

Rather than question what development is "really doing", post-development scholarship tends to focus instead upon how development functions as a discourse: how it was imagined into being, how it has enabled the wishes and worldviews of some groups and societies to become universalised or "imperialised" (Escobar, 1995; Ireland and McKinnon, 2013) and how it became a "thing" that people did, with its own set of rhetoric, practices, literatures and interventions, all taking shape around the problematisation of poverty. Many of its proponents express disenchantment with the term "development" as being simply a "deceitful mirage" (Rahnema, 1997a: x), a "malignant myth" (Esteva, 1985: 78) or "a poisonous gift to the populations it set out to help" (Rahnema, 1997b: 381). For some, post-development is "a set of thinking and doing practices that are guided by a distinctive ethical stance" (Gibson-Graham, 2005: 4); for others, it represents "a field of debate rather than a cohesive body of work with core principles and approaches" (Ireland and McKinnon, 2013: 160). One of the best and most influential works in this corpus of scholarship is *Encountering Development* by Colombian anthropologist Arturo Escobar (1995) which traces the discursive creation in the immediate post-war period of the "third world" as both the needy object of international development intervention and the excuse for the expansion of a new world power's mode of global governmentality. Escobar critiques the idea of modernity as a great singularity in D/development or the "path to be trodden by all trajectories leading to an inevitable steady state" (Escobar, 2004: 225). Escobar's work attempts to imagine a "post-development era" (Escobar, 1992) and seeks to decentre development and displace its "centrality from the discursive imaginary" (Escobar, 2012: xiii), but also to think about the end(s) of development, the emergence of alternatives and the need to transform development's order of expert knowledge and power (Escobar, 1995). Similarly, post-development scholars like Gustavo Esteva and Wolfgang Sachs have noted how "development has evaporated" (Esteva, 1992: 22), how as a field of study/knowledge, it "is a mined, unexplorable land" (22) and how it stands as an "outdated monument to an immodest era" (Sachs, 1999: 1–2), or "a ruin in the intellectual landscape" whose shadows continue to obscure our vision (Sachs, 1992: 1).

Whilst acknowledging there may be common features and lessons for particular development issues, post-development does not aim to "scale up" these examples to a universal model for development practice (Ireland and McKinnon, 2013) and this has been the focus of critique for most of its detractors who argue that the absence of a universalising tendency "can lead only to self-defeating localism, incapable of creating or enacting the kind of global change that development is all about" (164). Such critiques however demonstrate a fundamental misunderstanding of post-development scholarship – it never set out to "replace" development with a new utopia, nor to engage in scripting scenarios for an alternative future but rather to think beyond a paradigm (Escobar, 2000). Geographers have widely criticised post-development, dismissing it as essentialist and reactionary, as relying implicitly on concepts of conspiracy, as insufficiently dialectical (Watts, 1995) and as failing to engage with the institutional practices and processes that (re)produce the discourses of development it sought to critique (Essex, 2013). Wainwright (2008: 10) contends that what is needed is not post-developmentalism but a critique of development and its power through a post-colonial Marxist lens, arguing that writers like Escobar, who begin from a deconstruction of development, end up reifying it as a totalising structure. Citing Spivak's articulation of a concept that "we cannot not desire", Wainwright (2008: 1) argues "we cannot *not* desire development" (emphasis in original).

Of course, there is the trap of setting out to tell yet another story of predatory Northern actors victimising uniquely defenceless Southern actors (Li, 2007) when "predation is worldwide" (Sogge, 2002: 36) and when historically development "seldom pitted North against South or indigenous against foreign; instead, transnational coalitions of expertise, wealth, and political power vied against each other to assert rival 'models' of the future" (Cullather, 2009: 510). This assumption that "development" and the programmes for "improvement" that it promotes always have some hidden agenda or that they are merely a tactic to maintain the dominance of certain classes or to assert control by the global North over the South is, as Li (2007) points out, a problematic interpretation common to dependency theory and its variants. Post-development has also been rather fond of the "colonising metaphor" that appears common "through the employment of inappropriate hyperbolic rhetorical devices linking the operation of power through development to notions of colonisation" (Brigg, 2002: 422). As Foucault's work acknowledges, nothing happens as laid down in the schemes of "programmers" (Li, 2007) yet he insisted they are not simply utopias "in the heads of a few projectors", nor "abortive schemes for the creation of a reality" but are" fragments of a reality" that "induce a whole series of effects in the real", since they "crystallize into institutions, they inform individual behaviour, they act as grids for the perception and evaluation of things" (Foucault, 1991b: 81-82; Li, 2007: 28). The programmers' schemes of development are never just words then, nor are they ever just about a one-way colonisation of life-worlds; they are always subject to contestation.

One of the main problems here has been that post-development writers have often used a rather impoverished version of Foucault's discourse analysis and employed a "somewhat vulgar use of Foucauldian concepts" (Ziai, 2004: 1048) which sometimes miss both the nuances and profundity of a Foucauldian understanding, often using a sovereign, repressive concept of power, for example, rather than recognising the operation of "bio-power" (the management of life and population) or Foucault's relational conceptualisation of it (Brigg, 2002). Foucault understood liberal "government" power as organised around the "conduct of conduct" whereby actions are shaped at a distance by calculated means (in contrast to more direct, disciplinary forms of power that characterised the age of empire). Sovereign power (as in the colonial period) operated by deduction, by taking away and appropriation, by seizing things, time, bodies and life itself whereas biopower fosters, organises, incites and optimises life, redefining and administering life in order to manage it in a more calculated way (Brigg, 2002). Foucault's distinction between *sovereign* and *bio-power* suggests a similar distinction between the colonial and development eras but of course this was never so clear-cut since many colonial regimes pursued forms of development as governmentality. Synthetically bound with bio-power, D/development operates by bringing forth and promoting, rather than repressing, the forces and energies of human subjects (Brigg, 2002). With the progress of decolonisation and the coming end of empire, colonial officials, along with anticolonial nationalist leaders, began to promote the welfare and benefit of the colonies as the operation of a different modality of power in relation to the decolonising countries started to emerge, "one which relied not predominantly on force, but on the mobilisation (including self-mobilisation) of human subjects and nation-states" (Brigg, 2002:

424). It is thus important to recognise that development represented a liberating possibility in the early post-war period for many Third World nationalists (Cooper, 1997: 64; Cooper and Packard, 1997: 9).

Post-development has also perhaps held a problematic notion that power operates through a singular intentional historical force such as "The West". Esteva (1992: 6) states that in the early post-war era, the US "was the master", that "Americans wanted something more", and that they "conceived a political campaign on a global scale that clearly bore their seal". As Brigg (2002: 424) has argued, "Such ascription of agency and intention, regardless of its parsimony, is not adequate to understanding the multidimensionality of social and political relations, including the role of contingency, which led to the formation of the development project." The famous Point Four programme for "underdeveloped" areas launched in President Harry Truman's inaugural speech, rendered by Esteva as a carefully chosen point in the extension of US hegemony, is instructive in this regard. Rist (1997: 70) has shown that Truman's Point Four programme was in fact an afterthought in the scheme of Truman's overall speech suggested by a civil servant and that the idea was taken up as a public relations exercise (and even then, only after some hesitation). Viewing the notion of development solely as a Western imposition or hegemony (e.g. Sachs, 1992: 4–5) elides the fact that many Third World governments and subjects have actively embraced development (Brigg, 2002). If development is, as Gupta (1998: 45) has argued, "Orientalism transformed into a science for action", then it will invariably constitute hybrid formations in its encounter with different peoples and places and with incommensurable meanings.

Many of its critics treat post-development as a coherent school of thought, however, failing to differentiate between the heterogeneous positions subsumed under its heading, and have accordingly not fully grasped their political implications (Ziai, 2004: 1058). It is my contention that the literature on "post-development", though much-maligned, has a great deal to offer any exploration of the nexus between geopolitics and development and that, as Brigg (2002: 433) has noted, "while in some need of rescuing, post-development should not be dismissed because it lacks a programme for development." In many ways, post-development perspectives usefully seek a transfer of power – the power to define the problems and goals of a society – away from the hands of outside "experts" towards the members of the society itself, which, at least for some, adds up to a radical democratic position (Ziai, 2004). For others, post-development thought opens up "plural possibilities of the political beyond the grammar of development" (Nakano, 2007: 65) not only by rejecting the post-World War II development project but also by providing a challenge to (and a critique of) the role of the modern state and in envisioning the possibility of a political community that can be explored beyond the state system (Nakano, 2007). Further, the critical politics of post-development ensures that what Li (2007: 7) calls the "subliminal and routine" anti-politics of technical development is replaced with a more politically acute alternative. It could also be argued that critique remains the prime contribution of critical development studies, rather than a search for alternatives *tout court* (Radcliffe, 2015: 856). Interestingly, programmatic alternatives that have emerged in Latin America's "post-neoliberal" countries, such as *Buen Vivir/Vivir Bien* (living well), have been lauded in many circles as the realisation of post-development agendas although the reality, as Radcliffe (2015) notes, is more complex (see also chapter 5). Some have even argued that an emerging "South–South" dialogue on development in the twenty-first century could potentially be based on, or at least influenced by, post-development thinking (Andreasson, 2007).

Perhaps one of the most significant and valuable critiques of post-development is that informed by psychoanalytic approaches (De Vries, 2007; Kapoor, 2014a, 2014b, 2017), which have suggested that in maintaining that development politics and power are produced discursively in an impersonal way (as an "anti-politics machine" in Ferguson's terms), post-development ignores the fact that such power only takes hold, expands and persists through libidinal attachments:

> it is not enough to critically deconstruct discourses, to point out their gaps, discontinuities and contradictions; it is also vital to identify and come to terms with our libidinal attachments to, and unconscious investments in, these discourses. (Kapoor, 2014a: 1119)

More generally, international development has tended to ignore – or, tellingly, repress – human/social passions (Kapoor, 2014a) despite being:

> replete with disavowed memories (racism, (neo)colonialism, gender discrimination) and traumatic prohibitions (economic recession, poverty), which show up in dreams and fantasies

> (the exoticised Third World, structural adjustment as universal panacea), obsessions (economic growth, "wars" against poverty or terror), or stereotypes (denigration, infantilisation, sexualisation or feminisation of the Third World Other). (Kapoor, 2014a: 1117)

This important question of the unconscious fantasies of development, its gaps, dislocations and blind spots but also the various desires embedded in D/development and its potential to seduce, is very valuable. People's desires for development must be taken seriously and its promises should not be forsaken (De Vries, 2007). Fantasies of and desires for development also motivate many of the volunteers, charity workers and religious actors that get involved in its pursuit and who also become its governable subjects. Development generates the kinds of desires that it necessitates to perpetuate itself and is an autonomous, self-propelling apparatus that produces its own motivational drives (De Vries, 2007). Paradoxically however, the idea of development relies on the production of desires which it cannot fulfil. In other words, "there is a certain 'excess' in the concept of development that is central to its functioning" (De Vries, 2007: 30). Rather than viewing post-development and psychoanalytic approaches as being diametrically opposed, my contention is that the two approaches can very productively complement and offset one another. Both approaches acknowledge the significance of the "development apparatus" (see below) but in some of the earliest post-development interventions (e.g. Ferguson, 1990) the set of institutions, agencies and ideologies that structure development thinking and practice are depicted as a machine-like kind of entity that reproduces itself by virtue of the unintended, unplanned, yet systematic side effects it brings about. De Vries (2007), however, usefully draws attention to the idea of the development apparatus as a "*desiring* machine", a social body constituted by the assembling of heterogeneous desires. Rather than being depicted as a rational, legal-bureaucratic and hierarchical order (and only an apparatus of governmentality that produces the development subject as a contingent side effect), the development apparatus is instead understood here:

> as a crazy, expansive machine, driven by its capacity to incorporate, refigure and re-invent all sorts of desires for development.... The desiring development subject is a response to the lack in the development apparatus. It is the failure to satisfy the desire for development and the impossibility of bringing

> about the promises of development that produce a subject that always already eludes the grasp of power. (De Vries, 2007: 37–38)

Such approaches acknowledge not only that "all sorts of desires" are part of the workings of the development apparatus but also the "excess" that is central to its functioning and the ways in which subjects elude the grasp of power/knowledge. The claim to expertise in optimising the lives of others is a claim to power and collectively the activities of supposedly "enlightened" and "civilised" trustees and experts have played a key role in structuring a field of possible actions around "development" (Li, 2007). The "will to improve" is situated in the field of power that Foucault termed "government" where the concern is the well-being of the population at large and it operates by educating desires and configuring habits, aspirations and beliefs, representing an attempt to shape human conduct by calculated means (Foucault, 1991a). The will to govern for Foucault (1991a: 93) is concerned with "men in their relations, their links, their imbrication with... wealth, resources, means of subsistence, the territory...". Development intervenes in these relations in order to adjust them and the desire to improve populations requires the exercise of what Foucault identified as a distinct governmental rationality – a way of thinking about government as the "right manner of disposing things" – for only then can specific interventions be devised. An explicit, calculated programme of intervention is not the product of a singular intention or will but rather draws upon and is situated within a heterogeneous assemblage or *dispositif* (Li, 2007) that combines "forms of practical knowledge, with modes of perception, practices of calculation, vocabularies, types of authority, forms of judgement, architectural forms, human capacities, non-human objects and devices, inscription techniques and so forth" (Rose, 1999: 52).

Foucault's notion of a *dispositif* (or apparatus) that has overall governing effects and that seeks to normalise, refers to a "thoroughly heterogeneous ensemble" of discursive and material elements (Foucault, 1980a: 194) that may consist of "discourses, institutions, architectural forms, regulatory decisions, laws, administrative measures, scientific statements, philosophical, moral and philanthropic propositions" and so on (Foucault, 1980a: 194). It is particularly useful for capturing both the fluidity and heterogeneity of the development project and for considering relations of knowledge (discourse), power and subjectivity alongside the economic

(Brigg, 2002). Deleuze similarly conceptualises the *dispositif* as a concrete social apparatus and a "tangle, a multilinear ensemble" (Deleuze, 1992: 159). As Foucault (1980a: 194) noted, the *dispositif* is not simply the collection of elements per se but also the "system of relations… established between these elements". In this sense the idea of a *dispositif* is particularly apposite in considering post-war development as it emerged and continues to operate as a complex ensemble of institutions, discourses, resource flows, programmes, projects and practices. Since such a multitude of subjects clearly cannot operate entirely in concert, this conceptualisation avoids the tendency to view development and its effects as monolithic and uniform (Brigg, 2001) and "while a *dispositif* exhibits a certain level of coherence and density, the multiplicity of relations that make up the development ensemble are continually renegotiated and open to contestation, reaffirmation or consolidation" (Brigg, 2004: 60). Such ensembles operate to achieve overall effects, however, thereby serving a dominant strategic function (Foucault, 1980a: 195). In foregrounding the strategic functions and consequences of development, this approach enables us to get closer to a critical appreciation and understanding of the nexus between geopolitics and development. While the pyramidal organisation of relations of power gives a *dispositif* a "head", "it is the apparatus as a whole that produces 'power'" (Foucault, 1979: 177).

It is useful to think of the concepts of assemblage and apparatus, as articulated in the work of Deleuze and Foucault, *relationally* as they "emerge as one and part of each other, but in a continual dialectic" (Legg, 2011: 131). This dialectical approach is also useful in bringing post-development and psychoanalytic approaches together. Alongside the notion of a development *dispositif* it is also useful to think development relationally through an assemblage approach, which can productively be used to capture the more material, embodied nature of geopolitics and to dissolve macro/micro scalar binaries (Dittmer, 2014). In relation to development Li (2007) conceptualises assemblage as an active process which aims to direct social conduct and manage contestation through political techniques of consensus-building, rendering technical, performance and anti-politics "to direct, conduct and intervene in social processes to produce desired outcomes and avert undesirable ones" (Li, 2007: 264). Assemblage here is the continuous work of pulling disparate elements together to "constitute a technical field fit to be governed and improved" (Li, 2007: 286). The assemblage approach suggests a different set of metaphors for the social world (e.g. mosaic, patchwork, heterogeneity, fluidity) and this is ideal for exploring the geopolitics of development since assemblage thinking "foregrounds the ways in which social/political processes are generated through relations between sites, rather than configured through 'internal relations' in sites" (Featherstone, 2011: 140). Assemblages are constantly in process, contingent and fluid processes of assembling, in which development–environment interventions like, for example, a water aid project in India (Mosse, 2005) are analysed as contingent projects held together by political, discursive, embodied and technical practices.

CRITICAL GEOPOLITICS AND DEVELOPMENT

> Critical geopolitics is one of many cultures of resistance to geography as imperial truth, state-capitalized knowledge, and military weapon. It is a small part of a much larger rainbow struggle to decolonize our inherited geographical imagination so that other geo-graphings and other worlds might be possible. (Ó Tuathail, 1996a: 256)

The term geopolitics, which has different meanings in different spatial and temporal contexts and within different kinds of networks of actors, is not only an academic theorising of politics in sub-disciplinary fields like Political Geography, Political Science and IR but also refers to the political action of a wide variety of actors that have sought to write and shape political space. A highly contested term, geopolitics "poses a question to us every time it is knowingly evoked and used" (Ó Tuathail, 1996a: 66). It both informs political practices across the global North/South divide and is shaped by the political ruptures and political rationalities of a given time (Moisio, 2016). The approach taken in this book is informed by the literature on critical geopolitics – a body of scholarship that first emerged in the early 1990s and sought to bridge the disciplines of Geography and International Relations (IR). It was inspired in Geography by the pioneering work of "dissident" scholars including Simon Dalby, John Agnew and Gearoid Ó Tuathail. Its emergence was also coeval with the development of critical theories of IR from the late 1980s onwards, especially in the work of Richard Ashley, James Der Derian, Michael Shapiro and Rob Walker (Campbell, 2009; Power and Campbell, 2010). A key point of departure was a "recognition of plurality, in linguistic terms, that geopolitics is a polysemic sign" (Sidaway, Mamadouh and Power, 2013: 165).

Drawing inspiration from the work of Foucault and Derrida, critical geopolitics did not seek to develop a theory of how space and politics intersect but was concerned with developing a mode of interrogating and exposing the grounds for knowledge production and with seeking to analyse the articulation, objectivisation and subversion of hegemony. It was thus "merely the starting point for a different form of geopolitics" and as such offered "a seductive promise, a putative claim that one can get beyond a baleful geopolitics and recover the real beyond the categorical, the ideological, the dogmatic, the imperialist and the hyperbolic" (Ó Tuathail, 2010a: 316). Through Campbell's (1992) *Writing Security* and Ó Tuathail's landmark text *Critical Geopolitics* common approaches began to develop as an emerging corpus of scholarship sought to radically reconceptualise "geopolitics" as a complex and problematic set of discourses, representations and practices. Since then critical geopolitics has come to encompass a diverse range of academic challenges to the conventional ways in which political space has been written, read and practiced. In reflecting on its emergence Ó Tuathail argued that it is:

> ....no more than a general gathering place for various critiques of the multiple geopolitical discourses and practices that characterize modernity. One important initial vector of critique was the recovery of textuality within practices which are represented as objective or practical, as "beyond the text." Geopolitics is inescapably cultural. A second was the displacement of state-centric readings of world politics and the recovery of the many messy practices that constitute the modern inter-state system. Geopolitics is inescapably plural. A third was the development of critical histories of geopolitical thinkers and discourses. Geopolitics is inescapably traversed by relations of power and gender. (Ó Tuathail, 2010a: 316).

The focus on the displacement of state-centric readings of world politics and the recovery of the complex and prosaic practices that constitute the modern inter-state system has opened up the range of sites/texts/practices where "geopolitics" is seen to take place. The development of critical histories of geopolitical thinkers and discourses has also led to an interrogation of a growing range of formal geopolitical traditions (Atkinson and Dodds, 2000) which include Latin America (Sidaway, Mamadouh and Power, 2013). Rather than assuming critical geopolitics to be a single analytical or methodological endeavour, however, it is important to recognise that this corpus of scholarship encompasses various ways of unpacking the tropes and epistemologies of dominant scriptings of political space. Indeed, for some observers this may be one of its principal weaknesses; that as it has expanded and developed its original concerns have been diluted into the variety of meanings attributed to it. This has meant that critical geopolitics has had something of an "identity problem" (Mamadouh, 2010: 320) where its subject, key theoretical contribution and core methodology become increasingly hard to define as the field diversifies away from the hegemon (the US) and the great powers to examine other states, or moves away from formal and practical geopolitics to explore popular geopolitics or as it shifts away from state-centric approaches to study non-state actors, such as social movements and transnational organisations and various forms of collective action. In this book, I seek to pick up on many of these themes now emerging from a more diversified field but in particular to deepen the engagement between critical geopolitics and development and to "postcolonialise" geopolitics (Robinson, 2003a) in order to take political geography further towards the global margins, the periphery. Interestingly, one of the first explicit attempts to posit the scholarly agenda which subsequently became known as critical geopolitics (Dalby, 2008: 414) was a paper on the "Language and Nature of the New Geopolitics" (Ó Tuathail, 1986), which begins with El Salvador and the culture that supported US "interventions" there. Non-Western political geographies were therefore important to the very foundations of what would later become known as "critical geopolitics" even if they were to become somewhat neglected in its subsequent development and elaboration (see chapter 2).

Like post-development (with which it shares similar poststructural origins and a sense that other geo-graphings and other worlds might be possible), critical geopolitics scholarship has certainly had its critics. Although a diverse and disparate body of work, the focus on the recovery of textuality was particularly popular in the early years and this "mesmerised attention to texts and images" meant that a focus on the practical, mundane and everyday "little things" that actually enable the functioning of a broader sense of geopolitical discourse was often missing (Thrift, 2000). Further, the concept of discourse has often been relatively under-theorised in critical geopolitics (Müller, 2008). Feminist scholars have also noted the tendency towards an elite-centric view of agency as constituted only at the largest scales and argued that the excessive focus on textuality, representation and discourse has meant too little attention to embodied practices and the materiality of geopolitics (Hyndman, 2004; Sharp, 2007). As Dittmer (2014: 394) argues:

> Critical geopolitics has done an excellent job of documenting the role of academics, statespersons, and producers of popular culture in disseminating narratives that produce geopolitical subject positions. What it has not been very good at is tying these subject positions, and the political cognition that they enable, to political affect.

Some scholars have sought to challenge this elite-centric view of agency by developing a concern with "intimacy-geopolitics" (Pain and Staeheli, 2014; Brickell, 2014), which attempts to dissolve the customary boundaries between global–local, familial–state and personal–political as objects of study. This required more sensitive enquiry into the workings of geopower by attending to objects, the human body and matters of percept, affect and emotion, as well as the most ordinary ("precognitive") forms of sociality (Thrift, 2000). One way to address these concerns, as Dittmer (2014) advocates, is to adopt an assemblage approach to geopolitics in order to better understand its embodiment and materiality. Beyond an excessive textualism and (until recently) a limited engagement with questions of affect, critical geopolitics has also often struggled to negotiate the macro/micro scalar divide (Dittmer, 2014), with recent work increasingly emphasising the scale of everyday life in geopolitics. Several geographers have also noted that critical geopolitics, in its quest to destabilise the normative and to decentre the nation-state, "rarely engages in transformative and embodied ways of knowing and seeing" (Hyndman, 2010: 317), or that it has been too often restricted to unpacking discourses and stopping short of a concern for transformation in its quest to develop a transgression of binary oppositions (such as core/periphery, domestic/international etc.). Some critics have also pointed to the presence of "geopolitical remote sensing" (Paasi, 2000, 2006), or an emerging tendency to observe and deconstruct discourses from a distance and out of context but there have been attempts to develop more grounded accounts of geopolitics using ethnographic methods (Megoran, 2006). Indeed, Ó Tuathail (2010b) notes that any serious effort to develop a more geographically responsible geopolitics requires the supplement of regional expertise and fieldwork and argues that critical geopolitics can deepen its critical practice by grounding itself in regional research.

To date critical geopolitics has had very little engagement with issues of global inequality, imperial desires and development, issues it urgently needs to address if it is to become more "radical" in orientation (Mercille, 2008) or if it seeks to be more than "an academic niche chasing America-centric outrages that does not matter much in the arena of global practice" (Ó Tuathail, 2010a: 317). This book seeks to address these lacunae and shortcomings and it does so in part by explicitly considering the entanglement of geopolitics and geoeconomics (Cowen and Smith, 2009; Sparke, 2016) since security relations and political economy have always been inextricably linked. While the formal distinction between the "geopolitical" and "geoeconomic" provides some methodological clarity and analytical purchase, ultimately these logics of power must also be grasped dialectically (Lee, Wainwright and Glassman, 2018). In contemporary Asia, for example, the entwinement between geopolitics and geoeconomics has been complicated by the competition and tension between the US and China which has become more intense in recent years. While China has formulated plans for a "Silk Road Economic Belt" and "Maritime Silk Road" under the title of the "One Belt, One Road" (OBOR) initiative, the US (at least until President Trump took office) has promoted an advanced version of a regional free trade agreement in the region in the form of the "Trans-Pacific Partnership" (TPP) which has served as a discursive, institutional and geopolitical frame to contain and counter China in the Asia-Pacific region (Tsui et al., 2016). Both initiatives represent significant contemporary examples of the imbrication of the geopolitical and the geoeconomic whilst OBOR (see Figure 1.1) in particular reveals a great deal about both the contemporary configuration of development and its complex and shifting spatialities (see chapters 5, 7 and 8).

SITUATING DEVELOPMENT HISTORICALLY

> They talk to me about progress, about "achievements", diseases cured, improved standards of living. I am talking about societies drained of their essence, cultures trampled underfoot, institutions undermined, lands confiscated, religions smashed, magnificent artistic creations destroyed, extraordinary

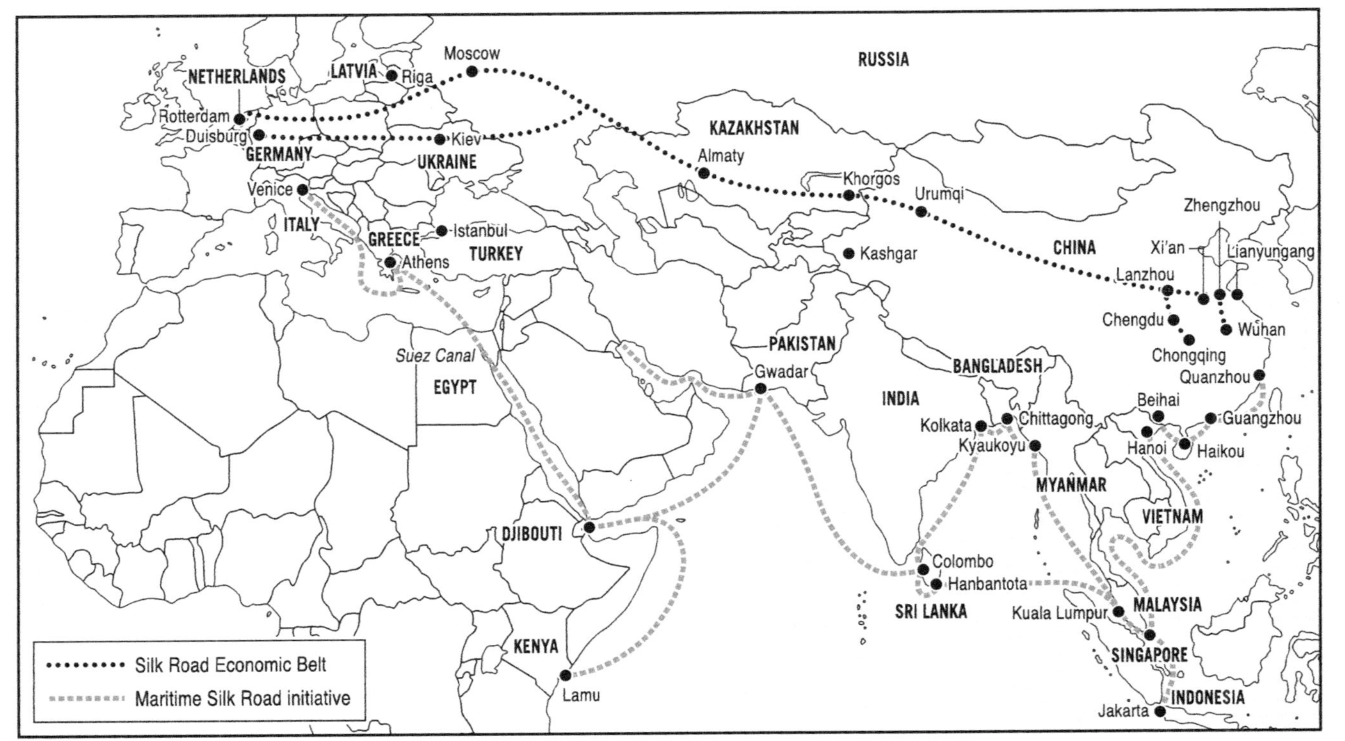

FIGURE 1.1 Map of China's "One Belt, One Road" (OBOR) initiative showing the Silk Road Economic Belt and Maritime Silk Road. Map prepared by Chris Orton, Design and Imaging Unit, Department of Geography, Durham University.

> possibilities wiped out. They throw facts at my head, statistics, mileages of roads, canals and railroad tracks... I am talking about millions of men torn from their gods, their land, their habits, their life... (Césaire, 1972: 21–23).

> An archaeology of development demands a full grasp of location, situating it *historically* (in tracing its complex genealogy and meanings, particularly in the eighteenth century), *geographically* (in relation to sites of productions, routes of movement and patterns of reception), and *culturally* (in relation to the West's self-representation, of reason and the Enlightenment). (Watts, 2003: 8, emphasis in original)

As noted above, many contemporary critiques of development suffer from an inability to properly locate development in historical context (Wainwright, 2008). In many ways development is primarily "forward looking", imagining a better world (Crush, 1995) and as a result there has often been a tendency to dehistoricise development (Power, 2003). As Slater (2004: 224) has noted, power and knowledge "cannot be adequately grasped if abstracted from the gravity of imperial encounters and the geopolitical history of West/non-West relations". This book *insists* upon the need to situate development historically as much as geographically and culturally. In seeking to do this a post-colonial approach to development is invaluable (McEwan, 2009) given its historical focus, its concern with the subaltern and with key themes like knowledge, power and agency. At the time of writing this book in northern England (a country whose own wealth is very much the historical result of the pauperisation of others), such an approach seems particularly important, especially given the way that some British officials have described plans for a post-Brexit trading relationship with the Commonwealth as a kind of "Empire 2.0". Prior to Theresa May's visit to South Africa, Nigeria and Kenya in August 2018, no British Prime Minister had set foot in Africa for over five years since David Cameron attended Nelson Mandela's funeral in 2013, yet Brexiteers have sought to mobilise the idea of formerly colonial nations in Africa (and other parts of the Commonwealth) forming a dynamic, like-minded "Anglosphere" or British global network for trade and prosperity. Such fanciful visions of the future are clearly based on imperial nostalgia but also on a rather distorted misremembering of the past (Kenny and Pearce, 2018). In Geography there have been numerous calls in recent years to decolonise the discipline and geographical knowledges (see chapter 2) and this is occurring precisely at a moment when Britain's foreign relations are increasingly being driven by the resurgence of white supremacy across Anglophone contexts and by post-imperial nostalgia (Radcliffe, 2017; Roy, 2016a).

For many observers, development begins with the post-war creation of ideas and discourses about "underdeveloped areas" (in Truman's "Point Four programme" for example) but its emergence can be traced back much further still, to the seventeenth and eighteenth centuries of European enlightenment rationality (Cowen and Shenton, 1996). During this time, new attitudes to work and capital were formulated (Rist, 1997) in a period of growing commitment to enquiry and criticism (Hampson, 1968) marked by a concern for social reform and the idea of a progression and development of societies built around an increasing secularism (Gay, 1973). The Enlightenment profoundly shaped what development has come to mean – the idea of progress forged during this time remains an article of faith in development thinking today and the history of some of its key actors (like NGOs) can be traced to this period when the societies of Europe and North America were increasingly becoming entwined within global networks of exchange and exploitation (Haskell, 1985a, 1985b). The beginnings of a governmental rationality of development can be traced to the greedy extractivism of the British and Dutch East India Companies (Li, 2007) whose doctors and scientists were among the first to think systematically about the relations between "men and things" as an arena of intervention and mobilised to persuade their employers to do likewise (Grove, 1995; Drayton, 2000; Li, 2007). Many of the voluntary or church-based organisations that represent the birth of contemporary NGOs, were also built through overseas missions during the Enlightenment (Riddell, 2007) as humanitarians began "to formulate new antidotes, new 'cures' for the ills of the world" (Lester, 2002: 278). Indeed, Escobar (1995: 2–4) writes about the post-1945 development "project" as "the last and failed attempt to complete the Enlightenment in Asia, Africa and Latin America". Other key legacies of enlightenment rationality included a tendency to measure progress against the yardstick of technology and a drive to make human society *legible* (Adas, 1989; Scott, 1999). By the 1880s Europeans regarded technical achievement as virtually the sole measure of human worth (Adas, 1989) and this obsession with technology and technical improvement was something that development would never quite shake off.

Cowen and Shenton (1996: 13) remind us that the beginnings of development are located not just in the work of Scottish enlightenment writers like Adam Smith but also in the "rough and tumble of early industrialism" and the emerging need to intervene to confront predations created by notions of "progress" and "improvement". Many enlightenment thinkers viewed the remedy for the disorder brought on by industrialisation as related to the "capacity" to use land, labour and capital in the interests of society as a whole and argued that only certain kinds of individuals could be "entrusted" with such a role (Cowen and Shenton, 1996). The changing social orders brought about by the making of European modernity and the transition from feudal to capitalist modes of social organisation could thus be managed by "trustees" who had the power to harness these capacities for societal good and to manage the "fallout from capitalism's advance" (Li, 2007: 21).

As Westad (2006: 76) puts it, the "new slogans of imperialism at around 1900 were progress and development". With this came a missionary zeal to "civilise" and modernise the colonised and their ways of life. Trusteeship, defined as "the intent which is expressed, by one source of agency, to develop the capacities of another" (Cowen and Shenton, 1996: x), was central to that and has since become one of the most enduring features of development. Its objective was not necessarily to dominate others but to enhance their capacity for action and to direct it (Li, 2007) as development became linked with the imperative to intervene and with the notion of active trusteeship. Trusteeship would be led by "the few who possessed the knowledge to understand why development could be constructive" (Cowen and Shenton, 1996: 117) including a wide variety of actors such as colonial officials and missionaries, politicians and bureaucrats, international aid donors, specialists and "experts" in, for example, agriculture, credit, conservation and hygiene, along with NGOs of various kinds (Li, 2007). Trustees occupied a position defined by the claim "to know how others should live, to know what is best for them, to know what they need" (Li, 2007: 2). In this sense it was a "culturally coded racism" that effectively decided the boundary between the "included and excluded" (Duffield, 2007b: 227). Trusteeship does not however translate into any permanent solution to problems of disorder and decay since, as Cowen and Shenton stress, it cannot, as there cannot be security for everyone caught in capitalism's wake given that capitalism and improvement are "locked in an awkward embrace" (Li, 2007: 21).

Brooks (2017) traces one of the earliest attempts to spur international development to a proposal put forward by Colonel Edward M House (political advisor to US President Woodrow Wilson) in 1913 for a US–European alliance to develop what House termed the "waste places" of the world such as in Africa. During the 1920s and 1930s European colonial powers subsequently sought to maximise the benefits they might accrue from their African possessions, through state intervention and development programmes, regarded by some as amounting to a kind of "second colonial occupation" (cf. Low and Lonsdale, 1976). Cooper (1997: 85) even locates the origins of development in the imperial crisis of the late 1930s and 1940s, brought on by a series of strikes and boycotts in the West Indies and different regions of Africa, arguing that development "did not simply spring from the brow of colonial leaders, but was to a significant extent thrust upon them, by the collective action of workers located within hundreds of local contexts as much as in an imperial economy" (Cooper, 1997: 85). In an attempt to stave off the threat of impending anti-colonial insurrection, late-colonial states increasingly set out to provide native populations with the benefits of improvement and orderly rule through development programmes intended to guarantee economic growth, social welfare and political stability, thus reinvigorating colonialism (Cooper, 1997, 2004). The British government, for example, administered grants for infrastructure projects across poorer countries, passing the Colonial Development Act in 1929 and the Colonial Development and Welfare Act of 1940, the latter releasing funds for spending on sanitation, education, housing, infrastructure and a range of other social projects (see Figure 1.2) designed to both raise living standards and produce a more pacified, more efficient and healthier labour force (Hailey, 1957; Pearce, 1982). In France, the post-war equivalent was the creation of the Investment Fund for Economic and Social Development in 1946 (Cooper, 2002: 36). The interventionist economic management which had begun to take shape in the 1930s proceeded apace after 1945, as Britain, France and Belgium sought to invest in and expand production using new agricultural technologies and practices (and a flood of "experts") in what were depicted as "under-utilised" African territories (in terms of both natural and human resources), involving goods like groundnuts in Tanganyika, cotton in Niger and copper, gold and uranium in the Belgian Congo. Colonial states also increasingly began to envision, plan and implement more comprehensive schemes of public health that included rural African populations (Coghe, 2017).

FIGURE 1.2 Labourers lay shark fillets out to dry in the sun at the Colonial Development Corporation's Atlantic Fisheries Shore Station in The Gambia during the early 1950s. Central Office of Information post-1945 colour transparency collection, courtesy of Imperial War Museum.

The public health, settlement and agricultural development schemes that appeared in Africa during the death throes of empire between the 1930s and the 1970s are good loci from which to observe both the role that science played in the building of the "developmentalist state" (cf. Bonneuil, 2000) but also the emerging role of development in countering disorder and insurgency. It was in these kinds of moments and encounters that we can trace the "imperial DNA of modern development" (Mawdsley, 2017: 109) but also a deepening association between development and the countering of social and political disorder and insurgency, an association that has continued long after the formal end of empire. Key terms in contemporary development parlance like *community* and *citizenship* featured, for example, in the vocabularies of the 1950s colonists in Kenya who sought to "rehabilitate" errant anti-colonial activists through community development programmes that would teach them to become responsible "citizens" (Presley, 1988). Debates about model villages and villagisation in the interwar period were never simply about hygiene and agricultural reform alone but were closely underpinned by anxieties over the potential dangers of colonial "progress" and societal change. Many colonial officials viewed the resettlement of the "native" population, in model-like villages, as a means to better control them and, with them, the way colonial society was changing (Coghe, 2017).

Some of the development projects of the late colonial period were undertaken on a vast scale, as in the cases of, for example, the Cahora Bassa dam in Mozambique, the Suez and Panama canals and the Gezira irrigation scheme in Sudan. Villagisation projects in British, French and Portuguese colonies transformed the conditions of life and modes of production in rural areas (Castelo, 2016) but across them there was a kind of common "quest for legibility", as many of the villages were specifically located in the vicinity of roads in an attempt to make them more "visible" and accessible for tax collection, medical control, labour recruitment and agricultural assistance. For obvious reasons, many Africans did not want to live in the new villages close to roads or administrative posts, where they would be under the constant vigilance of the colonial administration. Portugal undertook a vast villagisation campaign during the wars of decolonisation (1961–75) and in Angola alone Portuguese military and civil administrations forcibly resettled more than a million rural Africans into strategic and rural development villages (Bender, 1978). By doing so, they followed a counterinsurgency strategy that had already been applied by the British and the French in other wars of decolonisation, aimed at preventing contact between guerrilla forces and Angola's rural masses and, simultaneously, at winning the latter's "hearts and minds" through accelerated rural development efforts (Feichtinger, 2017). An embattled and impoverished Portugal fought three insurgencies (in Guinea Bissau, Angola and Mozambique) and by the end of the 1960s had, after Israel, the highest proportion in the world of people in arms (Cann, 1997). Portugal also invested significant public funding, science and technology expertise and state support in a series of rural white settlement schemes or *colonatos* aimed at Europeans and intending to reproduce in Africa Portugal's rural social and material landscape and to construct an imagined Portuguese pastoral way of life that could serve as a model for Africans (Castelo, 2016). The Portuguese brand of the "developmentalist" colonial state envisioned the ideal type of white settler, "modest, rooted in the land, earning only enough to get by" (Castelo, 2016: 267), being emulated by the African peasant, ensuring both social peace and colonial order. Using scientific and technical knowledges and mobilising state resources, "development" (in this case as rural mythology) could be mimicked, transplanted to another location and enrolled in various political projects of bringing order and countering anti-colonial insurgency whilst bolstering an endangered settler colonialism.

Many debates took place in the colonies about which subgroups were more or less "improvable" and about the suitability and readiness of racial others to be governed in a liberal manner, with native difference and deficiency supplying an important rationale for colonial intervention (Li, 2007). Colonial powers were however caught between "the impossibility and necessity of creating the other as the other – the different, the alien – and incorporating the other within a single social and cultural system of domination" (Sider, 1987: 7), thus facing a significant contradiction between difference and improvement. One strategy for dealing with this contradiction adopted by colonial regimes was what Wilder (1999) calls the "structure of permanent deferral" as native society was both "rationalised and racialised", its subjects "destined to become rights-bearing individuals, but always too immature to exercise those rights" (45–47). Li (2007) argues that the structure of "permanent deferral" continues to pervade contemporary development agendas, a situation in which inclusion is accompanied by the continuous production of reasons for exclusion. This has been a central feature not just of the civilising project of colonialism but also of the logic of post-colonial projects of societal improvement (Li, 2007: 15). Planned development is premised on the improvement of the target group but also posits a boundary that clearly separates those who need to be developed from those who will do the developing. Deficient subjects (often racial others) can be identified and improved only from the outside (Li, 2007).

Many countries were just reaching independence in the 1960s (see Figures 1.3 and 1.4), at the "zenith" of international development and during a time of intense Cold War confrontation and as a result both the US and USSR heavily drew upon the idea of development in a variety of African, Asian or Latin American contexts, constituting the "Third World" of newly independent states as an arena of global political and ideological struggle. President Truman's Point Four speech "proposed a complicated merger between development and the Cold War" (Cullather, 2000: 651) and in just a few paragraphs of his inaugural speech of 1949 the principal axes of global opposition, communist/non-communist

FIGURE 1.3 "Redefining the power III", part of a series of images depicting post-colonial performances and futures by Angolan artist Kiluanji Kia Henda. The series shows Angolan fashion designer Shunnoz Fiel dos Santos standing atop various vacated pedestals on colonial monuments across Luanda, Angola. Here Fiel stands in place of Portuguese colonial Governor-General Pedro Alexandrino near Largo Rainha Ginga in Luanda. The image poses questions about Angola's historical memory whilst seeking to introduce into public dialogue more inventive and imaginative notions of the nation's future (Cobb, 2014). Courtesy of Kiluanji Kia Henda/Galleria Fonti, Naples.

and colonial/anti-colonial were enfolded within the overspreading categories of development and underdevelopment (Rist, 1997). In the process, nationalists, communists, expansionists and pan-Africanists all lost the power to define their struggle, their own version of progress, as they "were now forced to travel the 'development path' mapped out for them by others" (Rist, 1997: 79). Following the commencement of worldwide technical assistance programmes by the International Labour Organization (ILO) in 1949, initiated in reaction to communist victories in Asia, development rapidly became an intense arena for debate among member states from the two ideological camps (Maul, 2009). Selecting a development strategy "generally went together with a Cold War alignment, automatically turning dissenting experts and constituencies into either communist sympathisers or counterrevolutionaries" (Cullather, 2009: 511). Despite common assumptions and methods, each side strove to differentiate its own path by nominating surrogates, countries poor enough and "typical" enough to showcase a developmental triumph

FIGURE 1.4 A mural at Largo da Independência in Luanda, Angola, beneath a statue of Agostinho Neto, symbolises the break from the chains of empire. Photo by author.

and "trophy" projects. Although there were various forms of assistance on offer from within the communist world (see chapter 4), engineers from Moscow, Pyongyang and Belgrade largely built the same model farms, trophy stadiums, dams and refineries "with the same extravagant promises and meagre results as their Western counterparts" (Cullather, 2009: 507), although the real stakes "were measured in prestige, state power, and international alignments" (508).

In the process of decolonisation, "development" served as an overarching objective for many nationalist movements and the independent states they tried to form, a "lodestar" (Wallerstein, 1988) invested with the hopes and dreams of many newly emerging states seeking to address inequalities and divisions in their societies (Rahnema, 1997a). Driven on by Cold War imperatives it was also taken up by a range of "developmental states" in East Asia such as South Korea (see chapter 5) that sought modernisation through macroeconomic planning. With the accelerating pace of decolonisation and the creation of independent states in the South, geopolitical questions were thus increasingly addressed from a set of new or "Third World" perspectives, emerging from the perception that "underdeveloped" countries had distinct geopolitical considerations from those of Western societies. Key to this were theories of modernisation (see chapter 3), at the core of which was an assumption of *convergence*, that there is one best form of political economy and that all states are moving towards it. Modernisation was a key part of the Cold War and was an "imperial" struggle between universal rationality and local, contextualised knowledge (*techne* vs. *mêtis*) (Scott, 1998). The way in which the Cold War framed and conditioned the meanings of key terms like modernisation and development is hugely important here, in part because the "storyline of development changes when you put the diplomatic history back in" (Cullather, 2009: 508). This was not a singular or straightforward process since, as Farish (2010: xvii) notes, there were multiple Cold Wars. There were also multiple models of the modern on offer. In the US, modernisation rescued a political consensus for aid that was falling apart in the early 1960s (Cullather, 2009), and revitalised UN agencies struggling to find a mission (Maul, 2009). Moreover, modernisation theory allowed newly independent regimes to signal ideological alignments and identify allies. As the US and newly independent regimes eyed each other with suspicion, development provided room for collaboration (Cullather, 2000).

In the literature on modernisation, technology was theorised "as a sort of moral force" (Escobar, 1995: 36) that would educate and transform but that was at the same time "neutral and inevitably beneficial". Such ideas underlay the Kennedy administration's diagnosis of global poverty: since capital and technology were what the US had to offer. As Galbraith (1979: vi) observed, "poverty was seen to be the result of a shortage of capital, an absence of technical skills. . . Having vaccine, we identified smallpox." Insurgency was almost always the disease for which development would provide the remedy, in both a curative and a preventive sense. Imagining development as a series of technical interventions and a process of incremental, managed change towards a final ideal state, proponents of modernisation were fond of modelling, or the dissection of case studies, with the aim of revealing generalisable principles that could be applied in other circumstances. In the vocabulary of development, a "model" (itself an artefact of the Cold War) was a loose, descriptive analogue pairing a nation with a strategy (e.g. the Taiwan model of export-led growth) and "encapsulated a country's economic or political history as a sequence of strategic moves open to imitation" (Cullather, 2009: 510). From this viewpoint "a particular nation state appears to be a functional unit – something akin to a car, say, or a television set – that can be compared with and used as a model for improving other such units" (Mitchell, 1991a: 29). The penchant for replicable models can be traced to the Enlightenment's search for a legible society that could be known and controlled from the centre (Scott, 1998), which peaked in the mid-twentieth-century period of "high modernism", characterised by a mania for colossal, centrally designed social landscapes – Le Corbusier's planned cities, Stalin's collective farms, the Tennessee Valley Authority – where nearly every aspect of human and natural life could be supervised by experts. In the process, history often got lost as "universal, repeatable features are emphasised while idiosyncrasies, unique circumstances, individuals, or motivations are blotted out like unwanted commissars in a Stalinist photograph" (Cullather, 2000: 645). Politics was also often "blotted out" of modernisers' stories of the stages of growth whilst popular resistance to modernisation was often reduced to a merely "technical" difficulty.

AN AFROCENTRIC FOCUS

In seeking to examine the nexus between geopolitics and development, the book draws upon a wide range of examples, some from Asia and Latin America but the clear majority from Africa, a continent that has long been the focus of intensive D/development and that "continues to be described through a series of lacks and absences, failings and problems, plagues and catastrophes" (Ferguson, 2006: 2). This is partly because of my own research interests, which have largely been centred on Lusophone Africa in particular, but partly also because Africa's complex geopolitics have so often been neglected in IR and Political Geography (see chapter 2). As a continent of "lacks and absences, failings and problems, plagues and catastrophes" it has often had a certain kind of (in)visibility in terms of debates about statehood or globalisation, for example, and in writing this book I wanted to challenge that by placing Africa at the centre of my analysis. IR has typically focused on the way in which marginalised, poor and weak African countries are acted and impacted upon by great powers and international institutions (Beswick and Hammerstad, 2013; Abrahamsen, 2017) and consequently the agency and diversity of African state and non-state actors have often been neglected. The continent also provides an ideal vantage point from which to explore the intersections between geopolitics and development because of the idea that Africa is currently "rising", buoyed by relatively high levels of economic growth combined with the growing regional and even global political power of some of the continent's bigger players (e.g. South Africa). For some observers, this "rise" or "renaissance", given that it is based on an intensification of resource extraction, only serves to further push the continent into underdevelopment and dependency, reifying its peripheral position in the global economy (Taylor, 2014, 2016). In this view, the growing volume of development cooperation between African states and (re)emerging powers like China will do little to change the structural tapestry of Africa's historically entrenched dependency (Taylor, 2016).

The continent has today become an "emerging market", a source of violent threat and a target of various moral crusades and over the past decade or so, the term "frontier", both as a concept and as a metaphor, has been widely used in association with the increasing importance of Africa in international relations and the global economy (Bach, 2013, 2016). There are a number of reasons for this including the discovery of substantial new oil and gas reserves, the proliferation of spectacular infrastructural rehabilitation and development projects, the availability of large stocks of uncultivated arable land and the growing significance of foreign cooperation with "rising powers" like China, India and Brazil but also a wide range of other players like Turkey, South Korea and Malaysia that seek strategic influence in Africa along with access to resources, markets and global alliances. Descriptions of Africa as the "world's last frontier" or as "untapped" or "overlooked" constitute an invitation to "call back the ghosts of the explorers, soldiers, traders and settlers who each in their own way once 'discovered' Africa" (Bach, 2013: 11). The frontier's association with the assertion of control by a core over its periphery has led to talk of a "new scramble for Africa" based on resource extraction, renewed exploitation, accumulation, the marginalisation of African economic actors and the corruption of African elites (Bach, 2013). Historically, the European encounter with Africa has often been quite "bi-polar" and "pendulum-like", swinging between optimism and pessimism towards the continent, its peoples and cultures and engaging in both positive and more explicitly racist and paternalistic registers (Reid, 2014). The idea that massive GDP growth is inherently "good" and something to be worshipped (Hickel, 2017), which prevails in the West's economic engagement with Africa, is a modern manifestation of the early-nineteenth-century perception that all the continent needed was to be "opened up" to free trade in legitimate commodities, whether traded by states or by individuals, and that as a result Africa would "find the peace, stability and prosperity it so badly lacked" (Reid, 2014: 159).

Africa has increasingly become the focus of what Teju Cole (2012) has called the "white saviour industrial complex" in reference to the viral video *Kony 2012* which aimed to stop and apprehend the Ugandan leader of the Lord's Resistance Army (LRA), Joseph Kony. In particular, "saving" the continent has become a favourite hobby for celebrity humanitarians (Mathers, 2012), through self-righteous Africa-related campaigns and events such as those concerning "conflict minerals" or debt (e.g. "debt, AIDS, trade, Africa" or DATA), the Save Darfur Coalition (SDC), Live 8 and Make Poverty History, although arguably these often do more harm than good (Moyo, 2009) in approaching international problems in a way that satisfies the sentimentalities of (white) audiences (Cole, 2012), but diverts attention away from the structural causes and (geopolitical) roots of conflict or impoverishment. *Kony 2012* promoted "the belief that western involvement in weak states in order to protect individual and group rights arises from unquestionably altruistic motives and is the answer to addressing human suffering worldwide" (Belloni, 2007: 454). Similarly, the SDC campaign presented Darfur as a moral rather than a political issue and ignored its complex history, whilst simplifying the issues to racial and religious binaries (of Arab Muslim perpetrators and black Christian victims) and reporting fictional data (Mamdani, 2009). Celebrities mete out a "politics of pity" (Boltanski, 1999) where structural and social problems are forcefully refracted through the neoliberal lens of the "heroic individual", thereby taking responsibility off international financial institutions (IFIs), states and the economic structures of inequality more generally (Goodman, 2010). As Shivji (2007: 43) asks, "how can you make poverty history without understanding the 'history of poverty?'" U2's lead singer Bono, who has been very vocal about African poverty, has attempted to gain credibility by conjoining his past with that of Africa by invoking the history of Irish colonial dispossession under the British and by transferring his Irish underprivileged background and post-colonial citizenship to Africans and to the African present (Magubane, 2008). As someone born in Ireland myself (and a big fan of U2 growing up), I have always found this fascinating but also how celebrity humanitarians like Bono or Bob Geldof are themselves part of a long tradition that continually objectifies "Africa" as a place of "ungovernability" where "horrendous things happened to benighted people, and where the West could display its full panoply of moral and material powers to positive ends" (Reid, 2014: 144).

In several celebrity humanitarian accounts "Africa" appears as a feminised object, something beautiful to be admired, gazed at and tamed, rather than a speaking subject of world politics in charge of its own future or representation (Repo and Yrjölä, 2011). Africa has been consistently infantilised and homogenised, repetitively reduced to what is "seen" to be deficient (Andreasson, 2005). This is in part the result of a dominant "scopic regime" that plays a major role in enacting a place in the world called "Africa", largely through the repetition and reiteration of colonial tropes (Campbell and Power, 2010). The changing nature of these artificial "Africas" is apparent in critical analyses of museums and colonial exhibitions, photographs, advertising, Hollywood movies and media images (Coombes, 1997; Ryan, 1997; Landau and Kaspin, 2002; Mayer, 2002; Ramamurthy, 2003; Chouliaraki, 2006; Repo and Yrjölä, 2011). The writing of Africa through the metaphor of darkness and light has been a central and recurring theme (Brantlinger, 1988), as has the visualisation of Africa as a space of backwardness and underdevelopment that requires external intervention. Similar representations of Africa as the "dark continent" continue to play out in contemporary media representations (e.g. concerning China–Africa relations; Mawdsley, 2008; Power, Mohan and Tan-Mullins, 2012). Africa has also been a focus for the production of "poverty porn" by NGOs, charities and aid agencies that have sought to exploit visual images of African poverty in order to generate the necessary sympathy or support for a given cause. The subjects are overwhelmingly children (usually depicted in a pitiable state with a swollen belly, staring blankly into the camera, waiting for salvation). Drawing upon post-colonial and post-development literatures, this book seeks to problematise and challenge these persistent colonial representations which depict "Africa" as an object rather than a subject of world politics, or as a feminised and infantilised space of deficiency, poverty, backwardness and underdevelopment. Rather than seeing Africa as acted and impacted upon from outside by great powers and international institutions or as in need of saving by benevolent white saviours, it highlights the agency of African political actors and insists on the complexity, heterogeneity and rich diversity of African cultures and polities.

Finally, as a continent constructed as a source of violent threat, Africa reveals a great deal about contemporary configurations of the geopolitics/development nexus, in part because of the changing (and increasingly covert) military relationships between the continent and a range of external actors. Since 9/11 Africa's perceived fragility and marginality have been increasingly securitised, with growing concerns in the US in particular about Islamic extremism and the dangers of state failure and underdevelopment on the continent. The (re)emerging economies have also made security and defence a key part of their cooperation with African states (alongside other key actors like the EU). African political leaders now regularly link their own domestic struggles (e.g. with a particular group of insurgents) to larger global security agendas and have proved especially adept at persuading Washington that they are the best guarantors of stability in their particular region or can be relied upon to sign up to larger anti-terrorist projects (Reid, 2014). Association with the US may however render countries more vulnerable than previously to outside attack and, in the eyes of many of their citizens, erode rather than increase their legitimacy, particularly following US President Trump's reference in January 2018 to African nations as "shithole countries" (*The Guardian*, January 12th, 2018). Again, this is not just about the US, however. Saudi Arabia, the United Arab Emirates (UAE), Qatar and Iran have also been seeking support from countries in Africa in an attempt to advance their own domestic and international political and security agendas. Saudi Arabia and its allies have been concerned about jihadist groups that have taken root in East Africa and the Horn, including Islamic State and al-Qaeda affiliates, gaining strength in the Arabian Peninsula. Saudi Arabia and the UAE's break with Qatar in 2017 (claiming that the Gulf state supports terrorism) upended traditional alliances and both countries have since urged African states to break relations with Iran, enhancing their development cooperation with Africa and establishing ports and military bases at sites in Somalia, Djibouti and Eritrea whilst Qatar and Turkey, which support a different model of political Islam and are closer to Saudi Arabia's arch-rival Iran, have also been increasing their presence in Somalia and Sudan (Stevis-Gridneff, 2018).

THE STRUCTURE OF THE BOOK

The book comprises a total of eight chapters. Chapter 2 explores the ways in which questions of development and (geo)politics in the non-Western world (particularly Africa) have historically been conceptualised in Political Geography and International Relations. It examines the emergence of "tropical geographies" (and the orientalist representations that shaped them) but also considers the emergence of the Area Studies complex and in making the case for a post-colonial geopolitics of development calls for greater attention to subaltern spatialities and imaginaries. Chapter 3 focuses on one of the most significant and enduring metageographies of development, the "Third World", and examines the theorisation, discursive practices and institutional architecture of modernisation along with the way in which they were closely shaped by Cold War geopolitics. Chapter 4 then considers the Cold War history of foreign aid practices with specific reference to the US, China and the USSR and examines their use of development as a means of either countering communist insurgency or fomenting it. Chapter 5 then turns to a consideration of the role of state and non-state actors in development, exploring the post-colonial crisis of state-led developmentalisms in the South and the consequent proliferation of social movements that have sought to resist and contest state-led narratives of development and to construct alternatives. Chapter 6 examines the emergence of the security–development nexus and the increasing securitisation and politicisation of overseas development and official aid practices since 9/11 and explores the rediscovery of development-based counter-insurgency in Afghanistan, Iraq and Palestine. It also seeks to map the contemporary configuration of the expanding US military assemblage in Africa and the contemporary scripting of the continent as a "swamp of terror" infested with Islamic insurgencies. Chapter 7 then examines the importance of economic risers or (re)emerging economies like China, India, Brazil and South Korea who have increasingly sought to develop an internationalist profile and to assert themselves as humanitarian, peacekeeping and peacebuilding actors in (and champions of) the South, along with the growing importance of SSDC and asks what this means for the existing landscape of development cooperation and for Africa. Finally, chapter 8 distils the key conclusions from the book and maps out a research agenda for furthering the study of geopolitics and development.

Chapter 2

Post-colonialism, geopolitics and the periphery

INTRODUCTION: THE CHANGING METAGEOGRAPHIES OF DEVELOPMENT

> The true power of the West lies not in its political and technological might but in its power to define. (Nandy, 1987: ix)

> The geographer's "South" is not exactly the same as the "South" in UN trade debates, or the "third world," or the "less developed countries," or the economists' "periphery" or the cultural theorists' "post-colonial" world, or the biologists' "southern world," or the geologists' former Gondwana – though there is some overlapping along this spectrum... there is enormous social diversity within each; recognizing the polarity is only the beginning of analysis, not the end. (Connell, 2009: 3)

THROUGHOUT its history, the complex spatialities of development have often been collapsed down into shorthand geographical imaginations and designations like "the tropics", the "Third World" and the "Global South" and whilst these practices vary within academic and policy circles (albeit with "some overlapping") (Connell, 2007, 2009), "disciplinary languages have a history of upholding national boundaries, regional differences, and geopolitical hierarchies" (Levander and Mignolo, 2011: 2). As such, "[i]ntellectual inquiry can all too easily naturalize territoriality" (2). This chapter seeks to critically examine the scripting and mapping of some of the major regionalisations and spatialisations of the non-Western world that have historically guided development theory and practice, such as "the tropics", the "Third World" and the "Global South", and explores their (in)significance to theorisation and scholarship in IR and Political Geography.

During the emergence of modern Geography as an academic discipline over a century ago a number of schemes were developed that attempted to regionalise the world, at macro as well as sub-national scales (Sidaway et al., 2016). Indeed, the urge to classify and compare has been a founding feature of Geography, remaining a point of departure for the discipline and one that requires further reflection on its origins and fortunes (Sidaway, Woon and Jacobs, 2014: 6). Much of what has travelled under the banner of "comparison" has tended to be "deeply retrograde" (Hart, 2018: 372) but comparison can operate as a means of critical engagement as well as a tool of oppression (Hart, 2018). As Connell (2009: 3) notes, recognising the polarity expressed by terms like the "Third World" or "the South" is "only the beginning of the analysis, not the end". In seeking to further understand the nexus between geopolitics and development it is first necessary to interrogate the *metageographies* that have historically shaped understandings of international relations and political geography – the spatial structures and "unconscious frameworks" (Lewis and Wigen, 1997) that have organised studies of development past and present:

> Every global consideration of human affairs employs a metageography, whether acknowledged or not. By *metageography* we mean the set of spatial structures through which people gain their knowledge of the world: the often unconscious frameworks that organize studies of history, sociology, anthropology, economics, political science, or even natural history. (Lewis and Wigen, 1997: ix)

The concept of metageographies refers to the relatively unexamined and often taken-for-granted spatial frameworks through which knowledge is organised within all fields of the social sciences and humanities. The distinction between the merely geographical and the metageographical is not always clear-cut. In one sense "Africa", for example, is merely a geographical label that is defined in different ways by different writers, but if one essentialises the category and begins writing about "African politics", one clearly moves into metageographical terrain. Thus, as Lewis (2015: 1) explains, "geographical concepts become 'metageographical' concepts to the extent that they lose their specific spatial coordinates and become imbued with extraneous conceptual baggage". It is not rare to associate an entire continent fully with a specific label (such as "underdeveloped Africa"), thus ascribing a high degree of "homogeneity" to a continent of 54 states, a practice that is not uncommon inside the "black box" of development or underdevelopment (Grant and Agnew, 1996). Mignolo (2005) calls for a shift away from the assumption that history takes place in continents, towards an understanding of how such geographies came about – as labels, designations, and identities that are themselves historical constructs – reflecting wider geopolitics. In part, this involves a focus on the "locus of enunciation", or the "epistemic location from where the world was classified and ranked" (Mignolo, 2005: 42). The idea of Latin America, for example, can be traced from its emergence in Europe under France's leadership, through its appropriation by the Creole élite of South America and the Spanish Caribbean in the second half of the nineteenth century and supposes that there is an America that is Latin, which can be defined in opposition to one that is not, creating the image of a homogeneous region with defined borders in ways which exclude indigenous peoples and overlook complex geographies of migration (Mignolo, 2005).

As important concepts and partitions of the world, the wide circulation of these metageographies plays a fundamental role in the building of our personal geographical imageries (Vanolo, 2010) and imaginations; using the language of Baudrillard (1983), hyper-realities (representations) are often more determining than "hard facts" in influencing our actions. In this way it is important to consider the ideas, methods and categories that foreign scholars bring as part of a process of intellectual pacification and ordering of the world (Cooper, 2005). This also involves understanding the variety of metaphors used in development as "culturally learned ways of looking at places, not as 'mirrors' of the territory" (Vanolo, 2010: 27). As noted in chapter 1, an important example of this is the metaphor of "Africa" as the "dark continent" which is in part the result of a dominant "scopic regime" that plays a major role in enacting a place in the world called "Africa", largely through the repetition and reiteration of colonial tropes (Campbell and Power, 2010). We might also trace the historical emergence of other "geopolitically determined domains" such as Southeast Asia, the Southern Cone of Latin America and Southern Africa (Sidaway, 2012). "Southeast Asia", as Keyes (1992) has argued, largely amounts to an artificial grouping of diverse lands and peoples projected as a distinct region by colonial administrators and by allied leaders as a theatre of operations during World War II. Tyner (2007: 1) has also shown how "the construction of Southeast Asia as a geographic entity has been a crucial component in the creation of the American empire". Arguably the idea of the Middle East has also had comparable significance (Culcasi, 2010).

According to post-development scholar Wolfgang Sachs (1992: 3), "the scrapyard of history now awaits the category Third World to be dumped". Before the category is completely discarded, however, it is important to fully appreciate its significance as a geopolitical and epistemological category in relation to development and its spatialities (see also chapter 3) as well as its historical "decomposition" as a category (Sidaway, 2013) and the alternatives that have been put forward to replace it. The "Third World" was not a place but rather a project (Prashad, 2007: xv) and one that came to embody something of the radical spirit associated with national struggles, revolution and liberation (Sidaway, 2012). For Vanolo (2010) this means that the "Third World" may be considered as a "creative metaphor" while "North" and "South" are dismissed as "just denominative marks", designating a global divide that is conceptually similar to "rich–poor". Just as the Third World was not a place, however, neither is the "Global South" simply "a directional designation or a point due south from a fixed north" (Grovogui, 2011: 176). Like the term "Third World", the "Global South" is also "a symbolic designation that captures the possibilities of cohesion that can emerge when former colonial entities engage in political projects of decolonisation" and one that "continues to invite re-examinations of the intellectual, political, and moral foundations of the international system" (Grovogui, 2011: 176). The South is thus "an idea and a set of practices, attitudes, and relations" but also represents a "disavowal of institutional and cultural practices associated with colonialism and imperialism... [both] a call and a label signifying the coming into form of a different world based on responsibility toward self and others" (Grovogui, 2011: 178). More than an existing entity to be described by different disciplines, the "global South" has been "invented in the struggle and conflicts between imperial global domination and emancipatory and decolonial forces that do not acquiesce with global designs" (Levander and Mignolo, 2011: 3). In examining the geopolitics–development nexus it is important to attend to the historical construction of such invented metageographies along with the way in which they became both the focus of "global designs" but also key sites of struggle between imperial domination and anti-colonial emancipation.

This chapter begins with a discussion of one of the earliest metageographies to have shaped geographical research on development: "the tropics". This became a powerful and enduring imaginative geography and a particular kind of Orientalism (Power and Sidaway, 2004) but, particularly through the work of scholars like French geographer Yves Lacoste, it also arguably became an important site for the early development of critical geopolitics. The chapter then traces the historical emergence of Area Studies out of imperial projects of classification, ordering and power in the context of post-war concerns to contain the spread of communism before exploring the ways in which questions of development and (geo)politics in the non-Western world have historically been conceptualised in Political Geography and IR. In particular, the chapter attempts to set out the situated basis of their claims and vantage points and directly engages with a number of critiques of "Western IR" and Political Geography. The principal focus here is on the historical silencing of a Southern or "Third World" Other through constructions of "the West" as the only subject with a right to speak and through the mapping of the global South as a space of exception. The chapter then examines the ways in which Africa in particular has been discussed and debated in IR and Political Geography and seeks to draw out the ways in which these (sub)disciplines have struggled to capture the complexity of politics and political spaces on the continent, along with the role they have played in the normalisation of particular kinds of states as the benchmark for analysis through the creation of certain assumptions and teleological arguments in which many African states have been depicted as "deviant", "weak" or "failing", with no "real" sovereignty. Finally, the chapter concludes by making a case for a post-colonial geopolitics of development focused on subaltern geopolitics, which, it is argued, offers a more nuanced sensibility towards the varied range of post-colonial trajectories and forms of politics in the South.

TROPICALITY AND ORIENTALISM

> whether viewed as the exotic site of a noble innocence and simplicity that the West has lost, or as a fertile yet primitive estate awaiting the civilising and modernising intervention of the West, the tropics have been affixed to Western frameworks of meaning, desire and knowledgeable manipulation – a framework in which tropical peoples have been deemed to be unable to represent themselves. (Clayton and Bowd, 2006: 209)

The concept of "the tropics" has been seductive, powerful and enduring and, like the Orient, needs to be understood "as a conceptual, and not just physical, space" (Arnold, 1996: 141–142). Tropicality, as a discourse or complex of Western ideas, attitudes, knowledges and experiences, can be traced back to the fifteenth century (Clayton and Bowd, 2006) and one of the "Western frameworks of meaning, desire and knowledge manipulation" (211) it gradually became affixed to was that of development. Tropicality, like Orientalism, operates as what Said (1978: 55, 327) described as a system of citation, with a select number of tropes and motifs ("typical encapsulations" as Said called them) about the Orient, or tropics, taking on the mantle of truth through repeated use (Bowd and Clayton, 2013). The language of development was key to these systems of citation and the metageographies and truths they gave rise to. As Said (1978: 55, 327) argued, partial and value-laden ideas and images of regions such as the Orient and tropics become taken-for-granted texts and "create not only knowledge but also the very reality they appear to describe". Orientalism worked through an "imaginative geography" that rendered the Orient as "an enclosed space" and "a stage affixed to Europe" (Said, 1978: 63–68). The tropics also provided a similar kind of stage in the early nineteenth century, as the British, for example, "affixed" India to the tropics through the deployment of alien European categories of nature and landscape (Arnold, 2005: 225). In this sense, the ideas and methods that Western observers and scholars brought to foreign and colonial regions were "less a natural means of analysis of bounded societies located elsewhere than part of a process of intellectual pacification and ordering of the world" (Cooper, 2005: 15).

Naming a part of the world "the tropics" became a way of identifying a space that was separate from the West (Arnold, 2000, 2005) and of judging this space against the northern temperate zone (Clayton and Bowd, 2006) constructed as the normal, against which the tropical world is perceived and evaluated. A false singularity was ascribed to the tropics as "other" where climate was used to make sense of cultural difference and in the process to project moral categories onto global space. The tropics here were depicted as altogether "other" – climatically, geographically and morally (Driver and Yeoh, 2000) – as Western identities (and discourses of development) were elaborated in opposition to foreign lands and peoples. From the earliest photographic attempts to represent tropical hybrid races to depictions of disease in new tropical medicines, there were multiple practices through which "the tropics" were known, practical and bodily as well as intellectual and discursive (Power, 2009). By the beginning of the nineteenth century it was commonplace to proclaim that the luxuriance and heat of the tropical zone made its indigenous peoples wild and indolent and therefore ill equipped to harness the natural riches of the tropical world or to become "civilized". As landscapes of desire, the tropics were understood as a domain of largely untamed nature that served, by contrast, to demonstrate the moral and material "superiority" of northern climates, civilisations and "races". Whether considered a sublime landscape, malignant wilderness, or the endangered site of environmental conflicts, the tropics are thus largely a construct of the Euro-American imagination (Stepan, 2001) although geographical formations of "tropicality" came in many varieties and were refracted through various national intellectual traditions.

As an orientalist discourse that articulated with wider imperial visions of non-Western places and subjects, the notion of a "tropical geography" has a somewhat complex history through the nineteenth and twentieth centuries but became more formalised and widely recognised after World War II (Power and Sidaway, 2004). Tropical visions and representations extended far beyond particular national research schools and traditions whilst debates about "tropical geography" drew in contributions from geographers based in a wide variety of locations. In France, geographers' concern with the tropics and tropicality first emerged in the context of French colonial engagement with monsoon Asia, specifically Indochina (now the states of Cambodia, Laos and Vietnam), and the construction of scientific and geographical knowledge that ensued was intrinsically bound up with French power there (Bruneau, 2005). In Britain, World War II was a key turning point, increasing the exposure of geographers to the non-Western world. Some British geographers like Charles Fisher and Bertram Hughes Farmer found themselves in the service of the military and some even became prisoners of war. Others worked for the Naval Intelligence Division (NID), which produced a series of geographical handbooks intended to provide Commanding Officers in the Navy with information on countries they might be called upon to serve in. These military handbooks went on to become important points of reference, regarded by Farmer (1983: 73) as "very useful indeed to the first generation of post-war British geographers struggling to write lectures in their demob suits and to prepare themselves for fieldwork overseas". Military institutions, practices and personnel (along with experiences of war) thus played an important role in the emergence of teaching and research on tropical geographies. Commenting on geography's age-old military penchant "for targeting places and people", Wisner (1986: 212–215) noted over 30 years ago how geographers had showed "remarkably little awareness of how central their knowledge and methods are to military adventures". Geographical investigation in the tropics was often depicted as a kind of militarised adventure dedicated to progress and modernity, the realisation of "tropical potential" and the enlightenment of tropical spaces and usually relied on an implicit geographical tradition of exploration, which has historically played a key role in legitimising whiteness given Geography's imperial disciplinary roots (Abbott, 2006).

Geographical scholarship on the tropics gained momentum after World War II and following the creation of the *Malayan Journal of Tropical Geography* in 1953 and the publication in English of French geographer Pierre Gourou's *The Tropical World*. Gourou became a key architect of twentieth-century tropicality and played an important role in the complex post-war reconfiguration of the colonial world as a "backward" and "developing" space (Arnold, 2000). In Gourou's scripting "the image of the tropics as a world set apart by nature, a world characterised by poverty, disease, and backwardness... acquired a new scientific authority and specificity" (Arnold, 2000: 16). Gourou marginalised issues of colonialism and revolution in much of his work (although it was full of ambivalences) (Clayton and Bowd, 2006) but instead stressed the importance of scientific knowledges for tropical development and focused on issues like tropical diseases, soils, plantations, population densities and the potential for white settlement. The Martinican intellectual and politician Aimé Césaire (1972) attacked Gourou's "tropicality", arguing that its zonal imaginary preserved the temperate/tropical binary that underpinned European and American colonial and neo-colonial exploitation of the tropical belt (Clayton and Bowd, 2006) and that it implied that significant progress could only come from the science and expertise of the temperate Western world, making Gourou a "watchdog of colonialism" (Césaire, 1972: 32). Echoing Enlightenment rationalities and unable to escape a Eurocentric commitment to the superiority of "western civilization" (Bruneau, 2005), Gourou saw the West as the epitome of "civilisation", with India and China in secondary roles and the rest more or less outside history – the subjects of "tropical geography" (Clayton and Bowd, 2006).

A distinct field of geographical enquiry quickly began to emerge, supported by new conferences, journals and funding possibilities. In a way, the notion of "the tropics" was crucial to the very creation and professionalisation of the discipline of Geography (Power and Sidaway, 2004), imbuing it with a sense of itself as a "sternly practical science" (Livingston, 1993). Geographical practices and knowledges provided a set of lenses through which "the tropics" were known, understood and represented and what began to emerge was a new spatialised domain of intellectual enquiry underpinned by (colonial) geographical imaginations and heavily racialised constructions of otherness around which crystallised a distinctly "modern" set of truths, assumptions and hierarchies. People of "the tropics" were routinely positioned as "backward"; their "clumsy" practices were seen as requiring the kind of modern economic development that was anticipated to come with the emergence of modern states.

A focus on issues of land use, agrarian change, population growth, mobility and environmental conservation (amongst others) ran through much of the early thematic approaches of these kinds of geographies and many geographers were often uncritical of the impact of colonial rule on indigenous societies, some even dedicating large sections of their texts to "colonial achievement" and (in general) viewing colonialism as something of a material success. However, in a discussion of French geographical scholarship (and its anti-colonial and anarchist influences), Ferretti (2017) challenges the notion that nineteenth-century European writing about tropicality was always straightforwardly in the service of empire. Nonetheless, many geographers worked for the late colonial state, undertaking surveys, administering forestry and agriculture, overseeing new development projects and censuses, although many also worked for newly independent post-colonial states where they carried out similar roles (Craggs and Neate, 2018). Indeed, colonial and post-colonial state boundaries delineated many of the geographies that were written, as the white male geographers of yesteryear became "local specialists" in the colonies of their nations of origin, focusing their attention on the management challenges of emergent statehood (Gibson-Graham, 2016: 800–801).

Craggs and Neate (2018) trace the passage of a cohort of careering geographers through the Geography Department at University College Ibadan (UCI), which had strong links to the Nigerian Colonial Service, showing how geographical knowledge, through survey and agriculture, contributed to wider projects of colonial development. Focusing not just on the publications written by these geographers or their contributions to theory but on their contributions to the everyday work of academic geography (e.g. hiring, teaching, textbooks, course outlines, departmental administration, university regulations, subject associations), Craggs and Neate (2018) show how geographers' work at colonial universities and in Britain was both influenced by and itself part of the process of decolonisation. As employees of universities that started as late-colonial development projects and became much-prized post-colonial institutions, academics, as well as the universities they worked in, were visible embodiments of the decolonising and developmental state (Livsey, 2017). Only recently has Geography begun to explore the discipline's mid-twentieth-century political entanglements, and decolonisation, arguably one of the most important geopolitical transformations of that century, has remained largely absent from disciplinary histories (Craggs and Neate, 2018).

In the 1960s the study of "developing areas" gradually became more formalised in the UK, reflecting broader geopolitical and educational shifts, along with a desire on behalf of the UK government not to lose its competitive advantage in expertise about developing areas as it withdrew from the colonies, and a concomitant rise in interest in area studies in the US, reflecting wider Cold War concerns (Craggs and Neate, 2018). Yet the founding of such area studies centres and associations, as well as the building up of African or Southeast Asian expertise in particular departments, also reflected the interests of a large group of academics, geographers amongst them, who had worked in the colonial universities. Many of the contributions to what was becoming known as "development geography" still lacked any major theoretical undercurrents, however. Much of the geographic scholarship of the 1950s and 1960s was framed by some variant of modernisation theory (see chapter 3), or the presumption that processes of modernity were shaping indigenous institutions and practices. Many geographers interested in Africa sought to model modernisation surfaces and attempted to map patterns of modernity by charting the diffusion of indices of modernity (from roads and schools to mailboxes) through the settlement pattern. This work at best raised only limited questions about the legacies of colonial transport systems or the character of African urbanism. One of the things that changed this however was accelerating revolutionary pressures in the South (epitomised by the insurgencies in Vietnam and the Portuguese colonies and the lurch into Mao's "cultural revolution" in China) which some radical geographers began to embrace (Power and Sidaway, 2004; Bunnell, Ong and Sidaway, 2013).

The philosophical and theoretical foundations for the growing radicalism in development geography were found in the work of Marx, with which radical geographers began to increasingly engage in the 1960s and 1970s in response to wider shifts in development thinking across several disciplines. Dependency theory had begun to offer popular structuralist explanations of the causes of "underdevelopment", a kind of "anti-geopolitics" (Routledge, 2003), and to locate the production of dependent relations within the structure of the capitalist world economy. The establishment of the radical journal *Antipode* in 1969, which published a number of significant contributions concerned with geography and uneven development, also provided a further stimulus. The flows and networks of knowledge creation in emerging debates around the political geographies of development were complex, however. Consider, for example, the circulation of ideas developed by the Brazilian critical geographers exiled by the military dictatorship (1964–78), such as Milton Santos, who played a leading role both in critical urban studies (e.g. Santos, 1979) and in development studies as demonstrated by the abundant material he published and supervised in French (Santos, 1969, 1970, 1971, 1972), or Josué de Castro who inspired the reutilisation of the word "geopolitics" in France in a critical sense (de Castro, 1952, 1965, 1966) (see also Lévy, 2007; Ferretti, 2018). Publishing in journals like *Hérodote* in France and *Antipode* in the English-speaking countries, scholars such as Santos worked with and influenced radical and critical geographers in France and in North America, helping to draw attention to the "Third World" in working on themes of social geography, decolonisation, poverty and underdevelopment (Davies, 2018; Ferretti, 2018; Ferretti and Pedrosa, 2018). These authors also contributed to important South–South flows and networks of knowledge creation on themes like decolonisation, by engaging with French anti-colonialist geographers like Jean Dresch and with African scholars like Akin Mabogunje (Figure 2.1) (see Dresch et al., 1967; Dresch, 1979; Mabogunje, 1980, 1981, 1984).

In the post-war context of decolonisation and Cold War, a potent image of the tropics as militant, combative, belligerent and revolutionary began to emerge as the idea of the tropics was increasingly resisted in various ways by the "tropicalised" (Clayton, 2013). Cuban revolutionaries, in particular, propagandised their Marxist position through deft manipulation of tropical imagery and narrative (Sacks, 2012). Tropicality was also combated in "altogether more visceral ways" in a series of jungle and guerrilla wars in Malaya, Indochina, the Congo, Kenya and Vietnam (Clayton and Bowd, 2006: 218). Clayton (2013) deploys the term "militant tropicality" to identify this

FIGURE 2.1 Akin Mabogunje presents a paper on the theme "Geography and the Challenges of Development in Africa: A Personal Odyssey" at the Festival international de géographie (FIG), at Saint-Dié-des-Vosges, in France on September 30th, 2017. Source: Wikimedia commons.

suite of counterhegemonic knowledges, practices and experiences emanating from the tropical world that challenged the way the West judged "the tropics" against the presumed normality of the temperate north. This represented:

> an eclectic body of counter-hegemonic thought and practice which, from World War II, brought "the tropics" into the air of revolutionary discourse and anti-imperialist struggle, and challenged the West's construction of the tropical zone as an exotic and bountiful space at its behest. (Clayton, 2013: 188).

French geographer Yves Lacoste's 1972 exposé on the American bombing of the Red River Delta of North Vietnam (Lacoste, 1972; see Figure 2.2) and its impact on opposition to the Vietnam

FIGURE 2.2 Yves Lacoste at a congress on Vietnam held in Nijmegen in October 1972. Photo by Hans Peters. Source: National Archive of the Netherlands.

War is particularly significant here (Bowd and Clayton, 2013). The International Commission of Inquiry into US War Crimes in Indochina, which Lacoste came with, sought to investigate allegations that the US Air Force (USAF) was deliberately bombing the region's dike system, threatening catastrophic flooding. In a series of reports and newspaper articles using various tools of classical geography (first-hand observation, mapping and the integrated analysis of physical and human factors), Lacoste disclosed the troubling political connections between law, war and environment (or what he termed "geographical warfare") (Bowd and Clayton, 2013; Lacoste, 1973). Lacoste showed how the exotic imagery of the tropics that had operated as a mode of othering and Western dominance could also serve as means of opposition and critique. Lacoste's exposition of geographical warfare depended heavily on Gourou and aspired to be a mode of analysis that could make geography a means of stopping rather than waging war, as tropicality became a critical instrument (Bowd and Clayton, 2013).

Gourou (1961) had been heavily critical of the physical determinism that pervaded some accounts of the tropics [such as Wittfogel's (1957) study of Asian hydraulic systems entitled *Oriental Despotism*] and Noam Chomsky subsequently identified important radical extensions to Gourou's "tropicalist" critique of Orientalism, with North Vietnamese resistance to American imperialism "now paired with peasants' ingenious, yet fraught, mastery over the Red River" (Bowd and Clayton, 2013: 637). This resilience was, Chomsky (1970, 1971) noticed, at odds with what Porter (2009) has since termed "military orientalism": US stereotypes about "war against Asian hordes". As Bowd and Clayton (2013: 637) have argued, Lacoste's tropicality should be seen as "ambivalent or strategic", which (again to borrow Said's terms) was "filiated" to the discourse of tropicality but "affiliated" to a global anti-war movement (Said, 1983: 157). Lacoste saw *Hérodote* as a vehicle for bringing French geographical research to a wider audience and whilst the journal's inaugural issue is best known for its interview with Foucault on geography, the Vietnam War also looms large. Its front cover gives an aerial view of an American B-52 bomber traversing a pockmarked Vietnamese landscape (see Figure 2.3), and Lacoste used his experience in Vietnam to reflect on "the links between certain geographical representations and certain forms of ideological behaviour" (Lacoste, 1973 cited in Bowd and Clayton, 2013: 639). Lacoste then was acutely aware how geographical knowledges and methods had been central to military adventures (Figure 2.4). His exposé (Lacoste, 1973) has been linked with the development of "critical geopolitics", yet strangely, as Bowd and Clayton (2013: 635) point out, there has been "a lack of dialogue between the literatures on tropicality and critical geopolitics". The engagement between critical geopolitics and the non-Western world is

FIGURE 2.3 The front cover of the inaugural issue of the *Hérodote* journal gives an aerial view of an American B-52 bomber traversing a pockmarked Vietnamese landscape. Copyright of La Découverte, Paris.

considered later in the chapter but it is the use of area-based knowledges in the post-war struggle to contain the spread of communism to which we now turn.

THE RISE OF THE AREA STUDIES COMPLEX

The birth of area studies can be traced back to Enlightenment efforts to support theories of human progress by comparing Europe to other

FIGURE 2.4 Front cover of the *Hérodote* journal for a themed issue on Geopolitics in Africa. Copyright of La Découverte, Paris.

regions of the world and elaborating the contrast between Europe and other areas (China, India, Africa, America), a tradition of universal comparison and ranking which has been carried into the twenty-first-century theories of development (Ludden, 2003). European imperial expansion also played an important role (Ludden, 2000) as Area Studies emerged out of, and continued to reflect, imperial projects of classification, ordering and power. The emerging Area Studies map included the former Soviet Union, China (or East Asia), Latin America, the Middle East, Africa, South Asia, Southeast Asia, eastern and central Europe, and, much later, western Europe but strangely it "did not include the United States, despite the fact – or because of the fact – that it was the principal site of area studies scholarship" (Mirsepassi, Basu and Weaver, 2003: 2).

A range of different branches of Area Studies were developed in the Cold War concerned with communist areas, coinciding with various policies of containment that were adopted by the US. As an academic discipline Geography withered to some extent in the US in the first half of the twentieth century (with Harvard Geography disbanding in 1948), creating space for the emergence of Area Studies which quickly became "the dominant academic institution in the US for research and teaching on America's overseas 'others'" (Goss and Wesley-Smith 2010, ix). An international division of labour emerged during the early stages of the Cold War within the social sciences based upon the idea of three separate worlds, which excluded other kinds of participation and narration (Pletsch, 1981). In the USSR Soviet knowledge of the "Third World" had also begun to develop (see chapter 4), particularly under Nikita Khrushchev, as the Institute of Oriental Studies at the Russian Academy of Sciences expanded, and new institutes were set up for the study of Africa and Latin America in 1960 and 1961, respectively (Westad, 2006: 68). The Soviet intelligence services also expanded during this time and were given briefs relating to information gathering about "third world" countries (Westad, 2006).

Following the Cold War division of the globe into three worlds and as Area Studies expanded, an emerging alliance was forged between modernisation theory and classical orientalism (e.g. through language training for social sciences) (Ludden, 2000). Said (1978: 17) noted how Orientalism was characterised by "a distribution of geopolitical awareness into aesthetic, scholarly, economic, sociological, historical, and philological texts" which had "less to do with the Orient than it does with 'our' world". Area Studies constituted not a simple copy of Orientalism "but another original, an afterlife and an afterimage" (Harootunian, 2002: 153). At the discursive level, the approach of Orientalism towards the non-West is almost directly inherited by Area Studies with knowledge perceived in a highly instrumental manner, serving the purpose of monitoring and controlling the non-West (Kolluoglu-Kirli, 2003). Southeast Asia, for example, which soon became the scene of the biggest US effort to contain Third World revolution (and perceived communist influence), "was more real, in the 1950s and 1960s, to people in American universities than to anyone else" (Anderson, 1998: 10) as Area Studies thrived during US hegemony (Evans, 2002).

In parallel with emerging work on "the tropics", most of those who were later to become leading figures in Area Studies had served during the war as area experts for the Army (Naft, 1993). World War II had brought large numbers of social scientists into government work, with many deciding not to return to an isolated Ivory Tower after they had seen Washington, whilst the Japanese attack on the US naval base at Pearl Harbor also ushered in an era of total mobilisation, sending social scientists to war in large numbers and in a variety of capacities (Engerman, 2010). The rise of Area Studies as a Cold War innovation thus had its roots in wartime intelligence as well as Army and Navy training programmes and in the "pervasive and persistent militarization that characterised the United States during the 1940s and 1950s" (Farish, 2010: xii). In this sense "area studies as a mode of knowledge production is, strictly speaking, military in origins" (Chow, 2006: 39). One interesting example of this was Project Camelot, a US military-sponsored study of the revolutionary process planned in 1964 to be executed by the Special Operations Research Office (SORO). With a projected cost of US$6 million (Solovey, 2001) it sought to assemble an eclectic team of social scientists to enhance the US Army's ability to predict and influence developments in target countries (mostly in Latin America and the Middle East) (Wallerstein, 1997), although it did not move beyond the planning stage due to a series of controversies. SORO itself was created at American University in 1956 by the Army's Psychological Warfare office and initially focused on creating handbooks for US personnel overseas, before expanding into studies of the social context for counterinsurgency. As a result, studies of counterinsurgency proliferated significantly, and Vietnam, Laos and Cambodia became obvious targets for the new techniques of social and psychological warfare that had begun to emerge.

In 1941 President Roosevelt created a civilian intelligence and propaganda agency attached to the White House, the Coordinator of Intelligence (COI), directed to centralise intelligence on matters of national security. It was re-named the Office of Strategic Services (OSS) in 1942, creating a model for collaboration between intelligence and academe (Cummings, 1997) whilst its Research and Analysis Branch became a crucial source of wartime service for many scholars. The OSS, dissolved in 1945, had many prominent post-war development scholars of the 1950s and 1960s amongst its alumni, including the economists Walt Rostow and Paul Sweezy who became key contributors to modernisation and dependency theories, respectively (Barnes and Crampton, 2011). MIT's influential

Center for International Studies (CIS), a key source of Cold War modernisation thinking which had co-evolved with Area Studies (see chapter three), was underwritten in its early years in the 1950s by the CIA, almost as a subsidiary enterprise (Cummings, 1997).

For Rafael (1994), what is significant about Area Studies is not so much the unsurprising point that they are tied to Orientalist legacies but rather how, following the end of World War II, Area Studies were increasingly integrated into larger institutional networks, ranging from universities to foundations, making possible "the reproduction of a North American style of knowing, one that is ordered toward the proliferation and containment of Orientalisms and their critiques" (Rafael, 1994: 91). In the US, the "area studies complex" was spear-headed first by the Rockefeller Foundation and later the Carnegie Corporation and the Ford Foundation as universities, state agencies and foundations became increasingly interwoven in the pursuit of knowledge about communist areas. The 1958 Defense Information Act had made substantial funding available for the study of languages, histories and geographies of remote places (Mirsepassi, Basu and Weaver, 2003) whilst the Ford Foundation invested a total of US$270 million in 34 universities for area and language studies from 1953 to 1966 (Cummings, 1997). The CIA (the OSS's successor) also collaborated with Ford and some of the other major foundations (such as the Social Science Research Council or SSRC) and hired prominent Area Studies scholars from the leading research universities to engage in consultation and recruitment. Large numbers of geographers helped gather and interpret intelligence for the US during World War II (Barnes and Crampton, 2011), along with all the social sciences which were represented in the Research and Analysis branch of the OSS. In this sense, "no nation had ever made such systematic use of the social sciences in the gathering and interpretation of military and strategic intelligence" (Barnes and Crampton, 2011: 232).

While the Area Studies enterprise was explicitly designed to serve national needs, "those needs were not necessarily 'Cold War' in origin; the programs started during World War II, reflecting a sense that Americans knew little about a world with which they would be more deeply engaged" (Engerman, 2010: 397). The Association for Asian Studies (AAS) was the first "area" organisation in the US, founded in 1943 as the Far Eastern Association and reorganised as the AAS in 1956. In both direct and indirect ways, the US government and the major foundations traced the boundaries of Area Studies by directing scholarly attention to distinct places and to distinct ways of understanding them (for example, communist studies for North Korea and China and modernisation studies for Japan and South Korea). In the US, area studies of Russia and the predominantly Slavic societies, or of East and Southeast Asia, were deemed to be of strategic importance, receiving significant federal funding and circulating widely within military communities. Countries inside the containment system, like Japan or South Korea (often constructed as "success stories" of development), and those outside it, like China or North Korea, were clearly placed as friend or enemy, ally or adversary (Cummings, 1997), and the key processes were things like modernisation, or what was called "political development" towards the explicit or implicit goal of liberal democracy (Cummings, 1997: 8).

Viewing Area Studies across the board and in the *longue durée*, we can arguably identify three "waves" of interest, punctuated by three "crises" (Sidaway, 2017). Imperial Area Studies lost much of its original rationale when the European empires yielded to post-colonies and its descriptive style was eventually replaced by the social scientific American-led Area Studies of the Cold War. The end of that conflict, in tandem with globalisation and the critique of Area Studies' enduring orientalism, have yielded a third wave of Area Studies, more conscious of the politics of representation, questioning of putative boundaries around areas and attendant to transnationalism (Sidaway, 2017). In recent years, critical appraisals have reconsidered the role and status of Area Studies within Geography and cognate disciplines (Gibson-Graham, 2004, 2016; Roy, 2009; Sidaway, 2012, Sidaway et al., 2016) and there has been a rethinking of Area Studies such that the emphasis is no longer on "trait geographies" but on "process geographies" (Appadurai, 2000) or on the forms of movement, encounter and exchange that confound the idea of bounded world-regions of immutable traits. As will be argued in the next section, however, the engagement with questions of geopolitics in the non-Western world has been rather "backward" and "underdeveloped" in the discipline of IR and the sub-discipline of Political Geography.

IR, POLITICAL GEOGRAPHY AND DEVELOPMENT

Writing a few years after the establishment of the journal *Political Geography*, Perry (1987: 6) noted that, "Anglo-American political geography poses and pursues a limited and impoverished version of the discipline, largely ignoring the political concerns of four fifths of humankind." Eleanore Kofman reiterated this in the mid-1990s, noting "the heavily Anglocentric, let alone Eurocentric, bias of political geography writing" (Kofman, 1994: 437). These limitations are not unique to Political Geography, however; "Anglo-American" human geography more widely has periodically been subject to similar critique (Berg, 2004; Jazeel, 2014, 2016; Kitchin, 2005; Minca, 2003; Radcliffe, 2017; Robinson, 2003a; Slater, 1989). The concern articulated in some of these interventions is that there are dominant "parochial forms of theorising" (Robinson, 2003a) in the discipline as a whole, centred upon particular intellectual traditions and contexts, leading to "a geography whose intellectual vision is limited to the concerns and perspectives of the richest countries in the world" (Robinson, 2003a: 273). This "view from the West" has clearly shaped a range of theorisations in Political Geography such that "parochial knowledge" has continued to be "created in universal form" (Robinson, 2003b: 648). This parochiality has been seen as based upon "a US-UK configuration" (Mamadouh, 2003: 667) or "Euro-American axis" that has come to prominence in a way that potentially narrows the base of political geographical thought and obscures "the situated basis of its claims and vantage-point" (Sidaway, 2008a: 51).

Critiques of IR have also suggested the presence of a similar "Euro-American axis" and even that Eurocentrism is constitutive of International Relations (Gruffydd Jones, 2006). Here, the narratives of IR are seen to establish Europe as the central referent and main actor of history, and events beyond its borders become derivative of events that have already happened in Europe (Barkawi and Laffey, 2006). Common theories of IR (e.g. liberalism, constructivism, realism) all rest on Western conceptions of statehood, civil society, political processes and rationalities, and have been developed with reference to Western historical processes of state formation (Harman and Brown, 2013). Tickner (2003a: 300) writes of the developing world as an *agent* of IR knowledge rather than an object of IR study and even goes as far as to say that IR is "autistic", in that it "ignores problems and perspectives that fail to resonate with its own worldview". IR has shared something of a Eurocentrism and reductionism with the discipline of Development Studies, where people and places in the South have often become the objects of history and modernity (Pieterse, 1995; Hart, 2001; Escobar, 1995), foreclosing a wide range of different forms of political agency along the way. Both have also (at times) had a tendency towards the silencing of a Southern or "Third World" other through their constructions of the West as the only subject with a right to speak and through their mapping of the global South as a space of exception. Further, both have been characterised by an implicit and Eurocentric statism that places the state at the centre of explanations (Dunn, 2001).

For all its concern with the realm of inter-state relations, IR as a discipline is fundamentally shaped by a particular vision of what a state *is* (Brown, 2006) and has therefore played a role in the normalisation of particular kinds of states as the benchmark for analysis, creating certain assumptions and teleological arguments in which many states in the South can be depicted as "deviant", "weak" or "failing", with no "real" sovereignty (Mercer, Mohan and Power, 2003). As Doty (1996) once noted, texts about Third World sovereignty and statehood in IR scholarship have often shared a fear or sense of danger regarding the entry of the "Third World" into the international society of states and in trying to make sense of "Third World" sovereignty scholars drew on "a whole array of hierarchical oppositions" (149) with "weak states" in the South needing to live up to the Western ideal model. Similar representations of states in the South as bedevilled by corruption, chaos and disorder and in need of modernisation have also been articulated by key global development agencies like the World Bank and IMF (thus legitimising intervention in the affairs of those states). The benevolent, democratic "international community" thereby replaces the superior "West/white man" of earlier imperial encounters (Doty, 1996). As Robinson (2003b: 651) suggests, "what if these kinds of states were allowed to coexist, to be exemplars of state-ness everywhere, to speak to what states elsewhere might also become?" Traditions of political thought from the ancient Greeks to modern Europe and America are important foundations in the development of Western IR theories and have often been taken for granted as the foundations of IR knowledge, but they displace other worldviews that are equally salient to understanding IR (Cheung, 2014).

Inayatullah and Blaney (2004: 2) have noted that the "current shape of IR" is "itself partly a legacy of colonialism" whilst Saurin (2006) even conceptualises IR as "imperial relations" and questions whether the strategy to overcome its Eurocentrism can be achieved through post-colonialism (see also Biccum, 2009). In taking aim at the core concepts in IR theory, Darby (2004: 6) notes that the "decolonisation of the international has barely begun". Similarly, in Geography, critical geopolitics initially began with a sense of needing to "decolonise our inherited geographical imagination so that other geo-graphings and other worlds might be possible" (Ó Tuathail, 1996a: 256). There have been several attempts in recent years at thinking past "Western" IR which has increasingly been seen as "ethnocentric, masculinised, northern and top-down" (Booth, 1995: 125), with many critics arguing that it has consistently ignored or misrepresented regions of the "global South" and Africa in particular (Neuman, 1998; Nkiwane, 2001; Chowdhry and Nair, 2003; Thomas and Wilkin, 2004; Lemke, 2011). IR (and to an extent Political Geography) has been structured by a global division of labour that is deeply entangled with racial/colonial hierarchies and so thinking about the value of Southern knowledges in this context necessitates a clear sense of the geopolitics of knowledge production: the intersection of the epistemic and social location of knowledge (Grosfoguel, 2010). Given the parochiality of this scholarship it may also be necessary to radically reimagine our modes of engaging Southern knowledges outside the logics of modern academia (Santos, 2014: 190).

Major Western IR (and Political Geography) journals continue to be dominated by scholars based in North America and Europe and the Western self thus "remains the author and authority of IR" (Dunn, 2001: 3), but it is important to remember that IR is not just Western – it is also liberal – whilst the "liberal underpinnings of IR theory are used to interpret and support liberal programmes of reform in Africa promoted by western states" (Harman and Brown, 2013: 72). Efforts to insert the periphery into IR are often based on an attempted reversal of "Western" theorising yet such attempts should not limit their task to looking beyond the spatial confines of the "West" in search for insight understood as "difference", but also ask awkward questions about the "Western-ness" of ostensibly "Western" approaches to world politics and the "non-Westernness" of others (Bilgin, 2008).

In an article criticising the American study of IR, Biersteker (1999) argues that one way to overcome its provincialism is to engage with scholarship from other parts of the globe and from other disciplines. Similarly, Inayatullah and Blaney (2004: 2) call for "an IR based on the creation of conversations among cultures". The idea that scholars in the core of the field (mainly the US and UK) are the innovators of theory, while scholars in the periphery (e.g. Africa, Asia and Latin America) are mere consumers of theory, has however been widespread. Unfortunately, as Mallavarapu (2005: 1) points out, this view is not only held in the core: scholars from the South "have been complicit in viewing themselves as mere recipients of a discourse shaped elsewhere". As Tickner (2007: 5) has also noted, just "as members of an academic community accept its respective rules and power arrangements as a precondition for admission, [some] academic elites in the south internalize and reproduce this hegemonic arrangement by favouring core knowledge as more authoritative and scientific in comparison to local variants".

Rather than focusing on the passive reception of theories developed in the West it is also important to recognise the ways in which these bodies of scholarship have been adapted and reworked as they have travelled to new locations. Acharya and Stubbs (2006: 128) note how scholars of Southeast Asian IR, for example, have adapted IR theories to make them more appropriate for understanding the particularities of the region. Similarly, in Latin America, dominant US discourses have been appropriated and moulded to the Latin American context, suggesting that the flow of knowledge from the US has been adjusted to fit conditions in the region (Tickner, 2003b: 326). The literature on autonomy produced in Latin America during the 1980s, for example, succeeded in establishing a "conceptual bridge" between dependency theory and mainstream IR theory (Tickner, 2003b: 330). Autonomy came to be regarded as a *sine qua non* for economic development (as per dependency theory) but was also linked to Latin American foreign policy and became viewed both from the inside in "as a mechanism for guarding against the noxious effects of dependency on a local level, and from the inside out as an instrument for asserting regional interests in the international system" (Tickner, 2003b: 330). Similarly, Ayoob's (2002) response to the narrow focus of conventional IR theory was to develop a "subaltern realism" (which focuses on state-making in the Third World) to better account for the experiences of "subalterns in the international system" (40).

In Korea, the call to develop IR theory based on the Korean perspective has been driven, since the 1980s, by a desire to counter the external influence of foreign perspectives (Chaesung, 2010: 69). There have also been efforts by Chinese IR scholars to contemplate the possibility of developing IR theories from a national perspective (Cheung, 2014). The development of IR as an academic discipline in China has been closely tied to the country's changing domestic and international political circumstances and first began when IR departments were established in the Universities of Peking, Renmin and Fudan in the 1960s where the revolutionary experience in Asia, Africa and Latin America was often the focus (Cheung, 2014). To lend support to the contemporary foreign policy rhetoric of China's "peaceful rise", Chinese IR scholars have drawn upon various intellectual traditions in ancient China to demonstrate an alternative understanding of power and international order, and to argue specifically that a powerful China is non-threatening. Ancient Chinese political philosophies in the pre-Qin Warring States period, the concepts of *tianxia* ("all under heaven") and Confucianism have been used by Chinese scholars to construct a positive and pacifist image of a powerful China as a benevolent force in a hierarchical international order (Callahan, 2010). China is not just concerned with geopolitical ambitions overseas, however, and also faces substantial challenges posed by fissures within its sovereign space (notably Tibet, Xinjiang and Taiwan's status) (Callahan, 2011). In this context, it is worth remembering the call from Chinese historian Emma Teng (2004: 7) for a corrective to the assumption that imperialism was "essentially a Western phenomenon".

In trying to meet the broader challenge of reimagining IR as a global discipline (Acharya, 2014) and in developing a post-colonial geopolitics of development, it is important to attend to the subaltern geographies of IR (and Political Geography) and to engage with other ways of thinking politically about space, other geographies "that may be considered lower ranking in the context of disciplinary geography's Eurocentric hegemony" (Jazeel, 2014: 96). According to scholars like Arturo Escobar and Walter Mignolo who have been concerned with conceptualising "decoloniality" (see chapter 5), it is important not to hierarchise knowledges, but to bring them together, in a more horizontal relation, from different settings in juxtaposition with each other (Connell, 2007). As Jazeel (2014: 88) has noted, there is:

> much to gain from treating subalternity not just as a reference to subordinate subjects and/or groups, but also as a more figurative reference to geographies occluded by the hegemonic conceptualizations of space that pervade our discipline.

Critical geopolitics in some ways also remains a particularly Western way of knowing which has been much less attentive to other traditions of thinking through international politics and the role of the nation and citizen within these narratives (Sharp, 2013: 20). As Tyner has argued:

> Our geographies, and especially our political geographies, remain largely distant from non-European theorists and theories. Our texts on nationalism and identities, in particular, are woefully ignorant of Pan-African nationalism and other African diasporic movements. (Tyner, 2004: 343)

Africa still does not figure prominently in Political Geography literatures and debates and relatively little African political geography has been published in the flagship journal of the sub-field, *Political Geography*, or the other major English-language Political Geography journal, *Geopolitics* (Myers, 2014). Geopolitical representations originating in Africa also rarely make much of an impact on political theory (Sharp, 2011) whilst "few western theorists assume that Africans, their modes of thought, ideas and actions have been integral to the dramas of modernity" (Grovogui, 2006: 8) or that African discussions of the inadequacy of modernity "provide useful grounds for thinking differently about international relations, modern political forms, their modalities, and implications" (11). Julius Nyerere's continental thinking, for example, is a form of geopolitical imagination that challenges dominant neorealist projections and is significant as a contribution to geopolitical thought (Sharp, 2013). Kwame Nkrumah's pan-African ideology and work on neo-colonialism are another example (White, 2003) and there is also the Non-Aligned Movement's alternative geopolitical vision for development (see chapter 3), which consciously rejected the totality of both Soviet and US projections of modern futures (Sharp, 2013). More generally, it is also necessary to acknowledge that certain concepts and traditions attributed to Western scholarship (such as postmodernism or liberal democracy) in fact have non-Western roots, or in the case of democracy "multiple places of birth" (Bilgin, 2008: 7–8). Even realism does not have its roots exclusively in Western history or thought (Clark, 2001: 88) and includes, for example, the work of China's Shang Tzu and Han Fei-tzu or India's Kautilya.

Writing just over 20 years after Perry's original intervention, the editors of *Political Geography* noted in 2008 that "[m]ost political geographers in their discipline's North American and European core still know fairly little about the evolution of political geographies in relative peripheries" (O'Loughlin, Raento and Sidaway, 2008: 2). There does however seem to be a growing number of scholars keen to challenge the hegemony of the English language in Political Geography and, more generally, to learn from other regions (O'Loughlin, Raento and Sidaway, 2008). This is about more than just extending the geographical scope of the kinds of empirical studies which dominate the discipline and the kinds of places which are paid attention to in the course of theoretical innovation and scholarly discussion. As Sidaway contends, there is more at stake here than simply "supplementing the range of case studies and terminologies that characterise Anglophone political geography" (Sidaway, 2008a: 41), especially if such non-Western political geographies are offered as "supplements that remain as examples, footnotes or exceptions to the Anglo-American mainstream" in the absence of the mainstream becoming "more attendant to its own situatedness" (Sidaway, 2008a: 48–49).

In many ways, the work of Edward Said provides a model for IR and Political Geography scholars to "be critical" (Duvall and Varadarajan, 2007). In Said's work, there is "an uncommon articulation of the postcolonial and the global, a suturing together of a global moment of humanism and a postcolonial moment of listening to and hearing – contrapuntally reading – the voices of/from alternative loci of enunciation" (Duvall and Varadarajan, 2007: 83). These could be constitutive principles for critical approaches to geopolitics and IR. Typically work on post-colonial geographies has tended to favour studies of long-ago high imperialism as well as present-day colonialisms but Craggs (2014: 40) urges geographers to turn their attention to the era of the mid-twentieth century, "during which people, institutions and states negotiated, performed and experienced becoming postcolonial". Geographers have however increasingly brought a post-colonial sensibility to researching geopolitics through a focus on the subaltern and on the "politics of representation from the margins" in order to highlight those voices that are too often rendered silent in political accounts (Sharp, 2011, 2013; Dittmer, 2010; Harker, 2011). Spivak (1985) defines the subaltern as "a group of people whose voices cannot be heard or are wilfully ignored in dominant modes of narrative production" (in McEwan, 2009: 61). Academic engagement with the subaltern began with the Subaltern Studies collective in the 1980s, following their criticism of Marxist and elitist narratives within India that disregarded the historical role and agency of the Indian masses and subsequent identification of non-elites, specifically peasants in India, as "agents of political and social change" (in McEwan, 2009: 60). Their methodology of "history from below" engages with the masses rather than state elites and with specific events and incidents rather than grand state narratives (McEwan, 2009).

This focus on subaltern geopolitics is a valuable framing that enables us to explore how political actors outside the main power centres do not simply bear witness to dominant geopolitical discourses, but also produce their own narratives (Harker, 2011; Woon, 2011). It recognises that "subaltern imaginaries offer creative alternatives to dominant (critical) geopolitical scripts" (Sharp, 2011: 271) and thus challenges critical geopolitics to "find space for this subaltern agency, to reverse the gaze, to look back" (Sharp, 2011: 271–272). Sharp's (2013: 22) focus on the original military meaning of the subaltern as a subject position that is "neither the commander, nor outside of the ranks" presents it as an ambiguous subject position of marginality "that refuses to be seen purely as the 'Other'" (Woon 2011: 286). Rather than being seen solely within the realm of the dominated and the resistant, the term "subaltern" thus provides space for a more complex rendering of geopolitical identities, processes and institutions, "suggesting a position that is not completely other to dominant geopolitics, but an ambiguous position of marginality" (Sharp, 2011: 271).

PLACING AFRICA IN IR AND POLITICAL GEOGRAPHY

> At best, Africa remains a case-study in which to explore international relations; at worst it is still, depressingly, wheeled on to the stage as representative of whatever delinquency, from state failure to the drugs trade, is exercising the analyst. (Harman and Brown, 2013: 70)

The very emergence of African Studies can be closely linked to questions of geopolitics, as the processes of decolonisation gave birth to a spate of new countries precisely at the time of simmering Cold War tensions. At its inception the dominant scholarship and research agendas of African Studies mapped readily onto the geopolitical concerns of the West (Zezela, 2006, 2007). According to the German-born American political scientist and historian Hans Morgenthau (1973: 369), Africa did not have a history before World War II and prior to the interventions of great powers as it was "a politically empty space". In IR Africa has often been pushed to the margins of some mainstream approaches by a focus on great powers (Harman and Brown, 2013) or "the states that make the most difference", as Kenneth Waltz (1979: 73) once put it. As Taylor (2010) has argued however, it is not the case that sub-Saharan Africa has been *irrelevant* in international politics, as the slave trade, the "scramble for Africa", the colonial period, the proxy wars of the Cold War and the continent's natural resources have all contributed to the region's geopolitical and geoeconomic importance. For Ayoob (2002), if theories of IR are focused solely on the great powers the likelihood is that this will result in political theories which reproduce the visions of those same powers. The study of geopolitics is also arguably problematic in this sense as it has historically been framed as the study of the spatialisation of international politics by core powers and hegemonic states (Ó Tuathail and Agnew, 1992: 192). Not only does such a fixation with the "great powers" ignore geopolitical narratives emerging from the majority world, it also means that critical geopolitics has remained a largely Western way of seeing and knowing (Sharp, 2013).

Despite the undoubted relevance of Africa as a space where IR is played out, from colonial rule to resource competition and post-colonial aid dependency (Harman and Brown, 2013), there have been numerous critiques of the way in which Africa has historically been marginalised and neglected in IR. African Studies has arguably always suffered from an inferiority complex (Abrahamsen, 2017) with the continent viewed by some as IR's permanent "other", serving to reproduce and confirm the superiority and hegemony of Western knowledge, epistemologies and methodologies. For Dunn (2001: 3) Africa provides the "ever present and necessary counterpart that makes the dominant [IR] theories complete", serving as a kind of mirror, the other to a mythical Western "self". Here Africa is the voiceless space upon/into which the West can write and act and "exists only to the extent that it is acted upon" (Dunn, 2001: 2). To an extent Marxism, Dependency and World Systems approaches refocused IR's gaze on Africa but they have also often replicated Western biases by viewing an "agency-less" Africa as the victim of manipulation by the great powers in the core and by seeing it solely as part of the periphery. There has often been an assumption then that Africa "does not have meaningful politics, only humanitarian disasters" (Dunn, 2001: 1). As an object of study, the continent becomes a place apart, a place for the application of theories, or a source of raw data, but "not a site for the generation of ideas and theoretical insights that have widespread and general relevance for the world"

(Abrahamsen, 2017: 129). Depictions of Africa in international politics as "agent, bystander or victim" (Van Wyck, 2016) have often informed both policy-making representations of the continent in the media and academic literature (Bach, 2016).

Africa is so much more than a case study with which to explore international relations or the geopolitics of development, and this book seeks to get away from the tendency to wheel the continent on stage as "representative of whatever delinquency" (Harman and Brown, 2013: 70) only for it to be dismissed as an undifferentiated exemplar of the more disorderly areas of the international system. Africa is not just *acted upon* by global systems of governance (made up of international institutions, global policies, foreign aid flows etc.) but has its own actors in global geopolitics which have more fully acquired *subject* status in the international relations system (Cornelissen, 2009: 24; Abrahamsen, 2017). As Bayart (2009: 269) argues, as a mirror, Africa (however distorting it may be) "reflects our own political image" and as such has "a lot to teach us about the springs of our western modernity". The application in Africa of basic concepts key to traditional IR such as sovereignty, the state, the market, anarchy and the international/domestic dichotomy can be very revealing in this regard, with many critics arguing that these concepts become problematic if not highly dubious when used in the African context (Dunn, 2001). Due to its neorealist insistence on placing the state at the centre of explanations and on using conceptions of states as coherent and clearly delimited entities, Dunn (2001) argues that IR is incapable of comprehending the "real" political dynamics of the continent (since defining where statehood begins and ends in Africa is too empirically uncertain). Others point to "a lack of 'fit' between the discipline's theoretical constructs and African realities" which is particularly problematic since "the state, sovereignty and statehood are not fixed categories of analysis when understanding Africa and IR" (Harman and Brown, 2013: 71). Similarly, although critical geopolitics has sought from the outset to unsettle the central position occupied by the state in traditional Political Geography and IR writing and has directly challenged "conventional demarcations of foreign and domestic, political and non-political, state and non-state" (Dodds, Kuus and Sharp, 2013: 7), it has often remained quite state-centric. This is particularly problematic in studies of African politics where more inclusive conceptualisations, that explore non-state and particularly sub-state actors as important agents of state dissolution and state formation, are needed (Malaquias, 2001).

Although IR concerns itself with other actors and non-state processes, Hirschman (1989) once cautioned that the complexity of the Third World may threaten the discipline:

> in relationship to the Third World, one has a sense of a discipline looking for a coin in a floodlit street in a country where 90% of the people have no electricity. The very great possibility of this field of study seeming to be irrelevant to the poorest people of this world appears to be inherent. Even when it goes beyond relations between states... it remains a discipline which is distant in the extreme from the concerns of the vast majority of the Third World, and, therefore, from some of the most fundamental problems facing the world community... A field of study that is so concerned with the behaviour of those who have power has precious little to say about the powerless... International relations, then, beyond appearing irrelevant, also looks like an unconcerned discipline. (Hirschman, 1989: 53)

Hirschman (1989) noted that at the time IR was not participating in debates surrounding development, neo-colonialism, the nature of the state etc., that other disciplines such as Sociology, History, Development Studies and even Political Science have engaged in at least since the 1960s. Similarly, Murphy (2007) praises the progress made by development and feminist scholars and laments the failures of IR scholars to engage in scholarship which places human suffering at the centre of its theoretical project, pointing to the "disconnection" between IR scholars and issues and experiences of inequality:

> given the vastness of the inequalities that exist at a global level, the social worlds of critical IR scholars and those we wish to serve are so disconnected... there is no social group of the world's least advantaged with which we have any particularly close connection; it is very unlikely that we understand much at all about their life-worlds, self-understanding or struggles. (Murphy, 2007: 131–132)

Noting that IR has long ignored a number of paradigms and discourses that Africa has been at the centre of, such as development, Dunn (2001: 2) has argued that the continent "has long been absent in theorising about world politics". Placing the interpretive grid of development at the centre of our analyses of African politics is not without its own problems, however, and the effect of arguing that Africa underlines the limits to theory and is so different that it requires an as yet unspecified "new" theory only serves to further exoticise Africa and to marginalise the continent from the core debates of IR (Brown, 2006). Further, many of the issues which critics cite as problems of "IR theory" in Africa, are not unique to Africa and are in fact problems in IR theory *wherever* it is applied (Brown, 2006). The weakness and absences of statehood in Africa have often been overstated then and what is required is a more relational understanding of states (and of the connections between states and societies) so that the broad issues that emerge are ones that can and should be applied to thinking about state agency in general (Brown, 2006). There is no one-way process of *imposition* of the Western ideal state onto Africa as if Africans themselves had little to do with it and not only was the course of colonisation shaped by the *interaction* between Africans and Europeans, but decolonisation and the foundation of independent states was a process in which Africans were actors, not simply "acted upon" (Brown, 2006). Critics of the problems of IR theory in Africa thus often inadvertently exoticise and essentialise both Africa and Europe and their histories and differential participation in the international system.

There is consequently a need to recognise the diversity and character of post-colonial sovereignties (Sidaway, 2003) and not to interpret this as a hierarchy with the putatively "strong" Western states at the apex and post-colonial states (especially those fractured by insurgencies and secessionist movements) at the bottom as somehow "abnormal" or lacking the features of the Western state. This is part of a wider need to unsettle the ideology and imagined history of a beneficent West, intent on "spreading democracy and prosperity" since "established conceptions of the political underwrite western dominance" (Darby, 2004: 3) and given that as Adebajo (2008b: 236) has poignantly observed, the West's self-representation is "repugnant in its hypocrisy and historical inaccuracy". Rather, the supposed "weaknesses" of some post-colonial states might be interpreted not as arising from a lack or absence of authority and connection but rather as an *excess* of certain forms of them (Sidaway, 2003). Further, if we continually see Africa as a space of exception, as *apart from the world*, we miss its wider embeddedness and relationality. As Mbembe and Nuttall (2004: 348) put it: "the obstinacy with which scholars in particular (including African scholars) continue to describe Africa as an object *apart from the world*, or as a failed and incomplete example of something else, perpetually underplays the embeddedness in multiple elsewheres of which the continent actually speaks" (original emphasis).

As outlined in the previous chapter, this book adopts a Foucauldian approach to the study of development and it is worth noting here that the applicability of Foucault's thought to global politics, particularly the concept of governmentality, has come in for sustained criticism with African politics often invoked as a "limit" to liberal forms of government, beyond which analyses predicated on advanced-liberal or neoliberal formations of power cannot go. As the argument goes, "areas like sub-Saharan Africa are relatively bare spots on the map. The networks of capital and information associated with post-industrial progress are sparse and stretched in these zones" (Joseph, 2010: 236). Africa is used here to demonstrate the inadequacy of governmentality-inspired engagements with international politics, with Joseph concluding that "[i]f we are concerned with how techniques of governmentality build lasting social cohesion, then clearly areas like sub-Saharan Africa are currently non-starters" (2010: 238–239). As Death (2011) notes however, this usage of Africa echoes the stale, tired dismissals of life on the "dark continent" as nasty, brutish and short. A governmentality approach to understanding the forms of rule and the practices, technologies and mentalities of government in Africa can be very useful in resisting the simplistic dichotomy between dominators and dominated and in showing that dependency, subjectivity and autonomy can be related and co-constitutive categories, rather than analytical opposites, as can resistance and complicity (Bayart, 2009: xxiii, 208, 250–253). A governmentality approach can provide illuminating insights into the operation of politics in societies outside Western liberal democracies, as well as into the operation of contemporary global politics, since it can help to map the fragmented, uneven, heterogeneous, overlapping, fractured spaces of global politics, not just in Africa:

> it is precisely in terms of what might be called spaces of contragovernmentality, ungovernability, anarchical governance, or the borderlands of global politics, that a governmentality approach to African politics can contribute most to our understandings of world politics. (Death, 2011: 23–24)

Foucault has frequently been caricatured as a Eurocentric, inward-looking theorist obsessed with textuality, discourse and representations, and having little of value to say to those outside metropolitan café culture (Williams, 1997). Scepticism towards the applicability of poststructuralist or postmodernist approaches has also come from within African studies (see discussions in Abrahamsen, 2003; Ahluwalia, 2001; Shani, 2010). Foucault's Eurocentrism has perhaps been overemphasised, however (Escobar, 1984–85: 378; Jabri, 2007). Foucault lived and worked in Tunisia from 1966–68 and was closely involved with student anti-government protests in Tunis against the Bourguiba regime. Foucault's methodological transition from archaeology to genealogy can be attributed to his period in Tunisia, and "it was the student revolts of Tunisia that had the effect of politicising his work" (Ahluwalia, 2010: 605). Foucault also appears to have considered a move to Zaire, and was attracted to Africa's "tropicality", being drawn to "the sun, the sea, the great warmth of Africa" which he believed allowed him a sense of perspective and a better vantage point to reflect upon European social and political institutions (Ahluwalia, 2010: 599).

CONCLUSIONS: TOWARDS A SUBALTERN GEOPOLITICS OF DEVELOPMENT

> The Global South is everywhere, but it is also always somewhere, and that somewhere, located at the intersection of entangled political geographies of dispossession and repossession, has to be mapped with persistent geographical responsibility. (Sparke, 2007: 117)

> The question of Africa's place within IR is not then simply a question of "add Africa and stir". (Abrahamsen, 2017: 127)

> political geography is richest when reworked, resituated, redeployed and re-imagined. (Sidaway, 2008a: 51)

An important part of the project of extending Political Geography's engagement with the non-Western world and the periphery is, I want to argue, an intensification of the dialogue between critical geopolitics and development theory which began with a series of interesting exchanges in *Transactions* between Gerard Ó Tuathail and David Slater in 1993–94 (Slater, 1993, 1994; Ó Tuathail, 1994). Slater's important intervention challenged the circumscription and disciplining of the political by Western development agencies by exposing the meta-politics and geopolitical imaginations that enframe their orthodoxies. Slater's contention that all conceptualisations of development contain and express a geopolitical imagination which condition and enframe its meanings and relations is a critical one, suggesting that it is impossible to understand the contemporary making of development theory and practice *without* reference to geopolitics and the geopolitical imagination of non-Western societies (Slater, 1993). In response Ó Tuathail suggested that to develop this engagement further it was also necessary to document how Cold War discourses were a condition of possibility for post-war development discourses and to document the disciplinary, practical and popular geopolitics that the pursuit of development has involved.

In this book I want to show that there are many different ways that we can begin to do this. Typically, the study of "development" in Geography has been kept apart from other sub-disciplines like Political Geography or Economic Geography by a well-established division of labour which casts an engagement with the geographies of the non-Western world as "area studies" or constructs development as a technical or managerialist domain, shorn of all politics. Political geography is richest when "reworked, resituated, redeployed and re-imagined" (Sidaway, 2008a: 51), particularly when resituated in relation to the global South and when brought into conversation with the critical study of development and its multiple and contested spatialities. Similarly, the study of development in Geography can also productively be brought further into dialogue with economic geography, particularly since geopolitical and geoeconomic logics of power must be grasped dialectically, which can help bring to the fore "specific – and sometimes unduly neglected – aspects of development dynamics" (Glassman, 2018: 412). The scholarly division of labour has also kept apart Development Studies and IR which, as Gruffydd Jones (2005: 75) notes, are in many ways "both natural and uneasy bedfellows". Despite the important and insightful exchanges between Slater and Ó Tuathail in the 1990s, to a significant extent the domain of the (geo)political is *still* quite often regarded as discrete and separable from the (geo)economic and the technical domain of "development". The dialogue between critical geopolitics and development theory that Slater and Ó Tuathail's interventions began remains somewhat underdeveloped but in the chapters that follow I seek to reinitiate, deepen and intensify it.

In this chapter I have argued that we might begin with the ways in which sociocultural and political formations in the South have been categorised, classified, compared, mapped and represented but also with how people from the South have generated counter-representations of their own realities (Slater, 2004), contested representations of North–South relations or produced their own "geographies of repossession" (Sparke, 2007). This means attending to the metageographies, or the spatial structures and "unconscious frameworks", that create the worlds of development, past and present, "acknowledging the power of the dominant imaginative geographies while also disclosing the critical possibilities of the other geographies that are covered-up" (Sparke, 2007: 122). We might also trace the *decomposition* of important metageographies over time. The complex geographies of (post-)development raise important questions about the composition and decomposition of the "Third World" as a meaningful geopolitical and epistemological category and underline the need for more sustained attention to the interactions of enclosure, boundaries and subjectivities (Sidaway, 2008b). This focus on the new metageographies of development including enclaves (Sidaway, 2007) and other "spaces of enclosure" (Vasudevan, McFarlane and Jeffrey, 2008) is very valuable here (see chapter 7).

It is also necessary to engage with recent critical reworkings and reimaginations of area studies and with the critical rethinking of comparison, which in part requires being deeply suspicious of efforts that assert overarching processes and reduce spatio-historical difference to empirical variation, along with a commitment to an approach that is closely attentive to constitutive processes arising out of multiple arenas of practice (Hart, 2018). Geographers have used a range of approaches to do this. Robinson (2016a, 2016b) deploys a Deleuzian assemblage approach in contrast to the more deconstructive approach used by Roy (2015, 2016b) and the Marxist post-colonial geographies developed by Hart (2018). This work is grounded in conceptions of different regions of the world as "always already interconnected" (Hart, 2018: 389) and starts with important processes and practices rather than with any bounded units of analysis (nation, city, village etc.). Hart (2018) suggests a relational approach to comparison where "relational" refers to an open, non-teleological conception of dialectics at the core of Marx's method, one that draws on both critical ethnography and spatio-historical analysis of conjunctures and interconnections.

The global South "is everywhere" as Sparke (2007) rightly observes and cannot easily be identified with or anchored to simple geographical locations but instead it can and should be located at the "intersection of entangled political geographies". All too often, as

Comaroff and Comaroff (2012: 1) note, the "non-West… now the global south" is presented:

> primarily as a place of parochial wisdom… of unprocessed data… as reservoirs of raw fact: of the historical, natural, and ethnographic minutiae from which Euromodernity might fashion its testable theories and transcendent truths.

The South is not a stable ontological category symbolising subalterneity (Roy, 2014a) and is best understood as a temporal category, as "an emergence that marks a specific historical conjuncture of economic hegemony and political alliances" (15). In this book my intention, following Sparke (2007: 117), is to utilise the category "global South" as a "concept metaphor" since such an approach interrupts the "flat world" conceits of globalisation (Roy, 2014a), enabling us to think about the locatedness of all theory and to map the geographies of theory responsibly. As Comaroff and Comaroff (2012: 47) suggest, "the south cannot be defined, *a priori*, in substantive terms. The label bespeaks a *relation*, not a thing in or for itself. It is a historical artefact, a labile signifier in a grammar of signs….it always points to an ex-centric location, an outside to Euro-America" (emphasis in original). It is this sense of relationality in thinking about geopolitics and development in the South that I want to take forward in what follows since it is in this "ex-centric" space, this "outside", that "radically new assemblages of capital and labour are taking shape" that "prefigure the future of the global North" (Comaroff and Comaroff, 2012: 12). Theory from the South is not about narrating modernity from its "undersides" but rather "revealing the 'history of the present' from the 'distinctive vantage point' that are these frontiers of accumulation" (Roy, 2014a: 15).

The move towards a more "post-colonial" (or "post-tropical") geography and geopolitics of development requires a much greater engagement with the complex and rich experiences and scholarship of different places. This requires being attentive to other geopolitical traditions and other modes of thinking about international politics as well as tracing the complex South–South and South–North flows and networks of knowledge creation around the political geographies of development. The course of Area Studies, for example, looks different when viewed from a range of sites and vantage points (even within the Anglophone academy) or through different contests (see also Berg, 2004; Kratoska, Raben and Nordholt, 2005; Sidaway, 2012; Smith, 2010). Tropicality was also plural and contested (Driver, 2004) and there were multiple practices through which "the tropics" and later the "Third World" came to be known. IR theorising in particular has strong Western and liberal foundations and needs to deepen its engagement with non-Western spaces and the post-colonial world as the Western self "remains the author and authority of IR" (Dunn, 2001: 3) with Western conceptions (e.g. of statehood, civil society) and experiences (e.g. of histories of state formation) taking centre stage. Some scholars have consequently argued that it is necessary to "bring Africa in from the margins" of how we think about international relations (Brown, 2006) since there is a growing recognition that Africa is a source of "theoretical and conceptual innovation for International Relations as a discipline" (Bach, 2016: 144). IR is however becoming more self-reflexive and aware of its parochialism and shortcomings and similarly, African Studies is showing an increasing engagement and rapprochement with IR. In both IR and Political Geography there has been talk of decolonising knowledges, but decolonisation is not a metaphor (Tuck and Yang, 2012) and, as Esson et al. (2017) argue, the emphasis on decolonising knowledges rather than structures, institutions and praxis risks reproducing coloniality, as it re-centres non-Indigenous, white and otherwise privileged groups in the global architecture of knowledge production. In this sense, we might take up Rivera Cusicanqui's (2012: 100) call for an actively "decolonizing practice" in academic work that goes deeper than vocabulary. This also requires confronting white supremacy and privilege both past and present and a commitment to anti-racism (Esson et al., 2017).

Neither is it sufficient simply to "bring Africa in" or to demonstrate the inadequacy or failure of IR theory to capture African realities. It is not a question of adding interesting, anomalous, different and esoteric empirical cases from the continent, or a case of "add Africa and stir" (Abrahamsen, 2017: 127), but rather of recalibrating theory itself since Africa is no longer some "distant or deviant locale whose relevance needs to be demonstrated within or to the disciplines but instead a window on our contemporary world" (139) and constitutive of it. Such interventions echo that of Comaroff and Comaroff (2012: 2) who have argued that the post-colonies might offer "privileged insights into the workings of the world at large" (see also Mbembe and Nutall, 2004). An assemblage approach usefully offers one way of meeting the challenge of foregrounding Africa in IR in a manner that appreciates both its specificity and its globality but thinking politically with assemblages also enables a better sense of how the political orders of contemporary Africa come into being and of what forms of agency and power different actors, actants, norms and values have (Abrahamsen, 2017). By studying Africa "from the ground up", as it is being constantly assembled by a multiplicity of local and global forces, the continent's politics and societies "can be captured as both unique and global, as a window on the contemporary world and its articulation in particular settings" (Abrahamsen, 2017: 127).

Just as scholars have increasingly come to look at the colonial construction of "development" and its key concepts so too might geographers give further consideration to the colonial construction of some of the key categories used in Political Geography (such as sovereignty, territory, states and so on) (cf. Sidaway, 2000, 2003). Similar work has begun in IR (Inayatullah and Blaney, 2004; Muppidi, 2012) where scholars have traced how colonial forms of knowledge have impacted modern understandings of international relations. In this sense, a post-colonial critique in Geography must "take aim at the theoretical heart of the discipline" (Robinson, 2003b: 649). Also important here is the formulation of a more nuanced sensibility towards the varied range of post-colonial trajectories and forms of politics which in itself ought to introduce caution into some Western narratives about their universality and value across diverse contexts (Robinson, 2003b). This requires greater attention to the subaltern geopolitics of development, listening to and engaging with voices from alternative loci of enunciation, exploring geopolitical representations and imaginations that emanate from the non-Western world and, in addition to the post-colonial methodology of "history from below", which focuses on ordinary people (and not just elites) as agents of social and political change, what is needed here is "a geopolitics from below" (Routledge, 1998) that captures how development is rejected, reworked, resisted and reimagined in the everyday, from the ground up. The ways in which tropicality was contested through counterhegemonic knowledges, practices and experiences emanating from the tropical world, such as "militant tropicality" (Clayton, 2013), represent important examples of the subaltern geopolitics of development. Unlike "anti-geopolitics" (Routledge, 2003), a focus on subaltern geopolitics does not position its subjects outside of the state and associated institutions but instead recognises the possibility that political identities can be established through geographical representations that are neither fully "inside" nor "outside" and enables a more complex rendering of geopolitical identities, processes and institutions. The subject of subaltern spatialities is taken much further in chapter 5 in relation to the role of the state, insurgency and social movements in development but it is to the mid-twentieth-century subaltern spaces of mobilisation around the three worlds schema that we now turn in chapter 3.

Chapter 3

Modernising the "Third World"

INTRODUCTION: A GLOBAL HISTORY OF MODERNISATION

THE emergence of the Third World, together with the bloody, conflict-ridden process of decolonisation that it brought forth, not only coincided temporally with the Cold War but was inextricably shaped by it (McMahon, 2001). It was the Cold War struggle between the US, the Soviet Union and their respective allies for global power, influence and ideological supremacy, itself premised on an "endless performance of multiple boundaries" (Farish, 2010: xix), that gave birth to the term. Consequently, for some, without the Cold War the Third World ceases to exist since it was so heavily defined by its geopolitical structure and discourses (Roy, 1999). Backed with enormous power and spurred on by an intense mutual rivalry, after 1945 the superpowers held a strong sense of destiny and formulated agendas that they thought might appeal to political leaders in the Third World, with the US and Soviet dominions constructed as very different in aim to those of the retreating European empires. Both the US and USSR identified vital national interests in Third World territories (Leffler, 1992) and for both Washington and Moscow, "developing areas appeared critical to the achievement of basic strategic, economic, political and ideological goals" (McMahon, 2001: 2). The "most important aspects of the Cold War", as Westad (2006: 396) notes, were thus "neither military nor strategic, nor Europe-centred, but connected to political and social development in the Third World". IR scholarship has often viewed the early Cold War largely through a Eurocentric lens, as essentially a struggle over the fate of Europe, but revisionist scholarship in the 1960s and 1970s changed this by highlighting the centrality of the Third World in the US drive for global hegemony (McMahon, 2001). The "Third World" was not incidental or peripheral here – most of the East–West crises of the Cold War erupted in the Third World, as did most of the armed conflicts that broke out after World War II.

This chapter examines the emergence of the three worlds schema and the theories, practices, ideologies and discourses that led to the construction of a Third World as a space in need of modernisation and "development". There was no singular modernisation "theory" led by a single proponent, only a more amorphous modernisation school of thought which contained within it various development strategies, ideologies and theorisations (not all of which originated from the US). In some ways, modernisation thinking is almost inseparable from the Cold War and was not "merely some adventitious appendage to the idea of three worlds, it is constituent to the structural relationship among the underlying semantic terms" (Pletsch, 1981: 576). Critically examining the theorisation of modernisation in the "Third World" in the context of its Cold War political and historical geographies can tell us a great deal about the geopolitical enframing and imagination of development. The "Third World" was much more than an arena for Cold War superpower rivalry – it witnessed popular mobilisation on a vast scale and the subjugation and oppression of its peoples across Asia, Africa and Latin America was often invoked in calls for revolution and tricontinental solidarity. It was articulated as an idea, as a set of commonalities and collective demands and as a project of mobilisation around race and identity, in important diplomatic spaces and sites such as the "Afro-Asian solidarity" meetings in Bandung and Cairo (in 1955 and 1961, respectively), the creation of the Afro-Asian People's Solidarity Organisation (AAPSO) (founded 1957), the Non-Aligned Movement (founded 1961) and the Tricontinental Conference in Havana in 1966 (Prashad, 2007). During the Cold War, the international institutional architecture of Development was also structured around the idea of three worlds (along with many parts of the "area studies complex") and the United Nations, from its inception in 1948, was increasingly regarded as the main institution for the expression of "Third World" demands.

The Third World became the object of theories and ideologies of modernisation that sought to develop its peoples, industries and infrastructures in the name of various socialist and capitalist visions of modernity. In the context of widespread decolonisation, modernisation became either a way of exporting socialist revolution or of containing and countering the spread of the communist "contagion". This chapter traces the emergence of the Third World as a series of ideological projects and seeks to understand plans for its modernisation as they unfolded through a variety of nodes, sites and practices and in the specific encounters between "modernisers" and their subjects (Engerman and Unger, 2009). Western countries like the US certainly did much to manage, control and in many ways, *create* the Third World politically, economically, sociologically and culturally (Escobar, 1984–85; DuBois, 1991) but the "thoroughly teleological" idea of three worlds (Pletsch, 1981: 576) was not solely a Western construction and its history and geopolitics need to be viewed and understood from a range of different sites and vantage points. For all its power in the Cold War, the US "was not everywhere and at every moment the most important part of the 'battle for the hearts and minds' of the Third World" (Engerman and Unger, 2009: 378).

Modernisation programmes in particular have often been seen as "at best plays for geopolitical loyalty, ploys to help the American economy, or playing fields for academics eager to try out their theoretical models in practice" (Engerman and Unger, 2009: 376). Modernisation was very much a transnational enterprise with complex spatialities, however, and was closely shaped by a wide variety of localities, projects and individuals. Modernisation theories and practices transcended national and regional borders, transferred knowledge across continents and sought to establish significant political and economic connections between societies and individuals. As Connelly (2001) has shown in his work on the Algerian struggle for independence, throughout the Cold War local actors could be authors of and full participants in their own history. Citino (2008) similarly emphasises the need to consider the role of non-US actors, arguing that Cold War modernisation relied in part on an "Ottoman legacy". In the pursuit of modernisation Middle Eastern post-war underdevelopment was (re)imagined in terms of an Ottoman imperial geography, with Turkey, for various geopolitical reasons, depicted as further along in the development process than the Arab successor states (Citino, 2017: 89). The Middle East was also the source of important challenges to modernisation thinking, such as in the work of Ali Shariati, an Iranian scholar, sociologist and intellectual of the 1970s. Highly influenced by the Third Worldism that he encountered as a student in Paris (Matin, 2010), Shariati argued that modernisation had replaced the idea of civilisation in the Third World and that colonialism had convinced colonised peoples that the two were one and the same, separating them from their own traditions and cultures and replacing them only with an empty, rootless modernisation focused on creating and then catering to desires that would benefit the expansion of Western capitalist production (Shariati, 1979). In order to acknowledge the significance of such contributions, what is needed is a more "global history" that explores modernisation:

> not as an American export, but as a global phenomenon that was hotly contested, between blocs but also within them; it examines the intersections between modernisation and geopolitics, considering them analytically distinct but often overlapping; and it starts from the assumption that modernisation was a global project in character and scope. (Engerman and Unger, 2009: 376–377)

The chapter is divided into four parts. The first section examines the origins of the "Third World" as an ideological project refracted through various national contexts, with a particular focus on China and some of the diplomatic sites where "Third Worldism" was performed and enacted. The next two sections then explore the role of the Soviet Union and the US, respectively, in constructing a "Third World" as a space of Cold War intervention. Attention then turns to the Cold War political geographies of modernisation in the US and the USSR and the historical experiences and models of modernity they were based upon before the final section explores the geopolitical enframing and imagination of modernisation as a form of inoculation against the "contagion" of communism.

THE THIRD WORLD AS IDEOLOGICAL PROJECT

> this ignored, exploited, scorned Third World, like the Third Estate, wants to become something, too. (Sauvy, 1952: 14)

> The Third World was not a place. It was a project. During the seemingly interminable battles against colonialism, the peoples of Africa, Asia and Latin America dreamed of a new world. . . . They assembled their grievances and aspirations into various kinds of organizations, where their leadership then formulated a platform of demands. . . . The "Third World" comprised these hopes and the institutions produced to carry them forward. (Prashad, 2007: xv)

The concept of the three distinct worlds has a long historical pedigree and does not come to us "as a mere descriptive category"

(Ahmad, 1992: 308) but carries within it multiple and sometimes contradictory layers of meaning and political purpose. Not surprisingly (given the imprecise nature of the term itself) there are several competing explanations of its origins but typically it is most often used in reference to "the former colonial or semi-colonial countries in Africa, Asia and Latin America that were subject to European. . . economic or political domination" (Westad, 2006: 3). The majority view is that the term was coined by French demographer and anthropologist Alfred Sauvy in 1952 by analogy to the revolutionary "third estate" of France – that is, the commoners of France before and during the French revolution – as opposed to priests and nobles, comprising the first and second estates, respectively. Like the third estate, Sauvy suggested, the Third World "is nothing" and "wants to become something", implying that both had a revolutionary destiny. Without a world government, Sauvy believed, the very notion of a single "world population" was meaningless and instead he proposed that there were in fact three worlds: a capitalist "first world" of Europe, the US, Australia and Japan and a communist "second world" (both of which were headed towards convergence in modernity) with the remainder belonging to a "Third World", caught in a "cycle of misery" (Sauvy, 1952: 14).

Although Sauvy called for assistance to the Third World, his vision was driven by the fear that accepting the idea of a "world population" would legitimise demands for unfettered migration and the global redistribution of land and resources (Connelly, 2008). The concept of the Third World, then, was forged to fence non-European populations within their own regions. The Third World was however, as Prashad (2007: xv) has argued, not a place. It was a project. Other than political equality it included a variety of demands for the redistribution of the world's resources and the benefits that accrue from them, a more dignified rate of return for the labour power of its people and a shared acknowledgment of the heritage of science, technology and culture (Prashad, 2007: xvii). The Third World also referred to a position and a desire, its purpose to decolonise international practices on questions of foreign policy and development but also to chart a path towards political, social, and cultural emancipation away from those prescribed by the superpowers. The (geo)politics that Third Worldism gave rise to took "distinctly different routes in various national contexts" (Nash, 2003: 101) and unfolded through a variety of different nodes and sites from Beijing to Belgrade, Havana to Hanoi.

The origins of the concept of three worlds can also be traced to Chairman Mao's division of China's social and political forces into three categories, first applied to the international situation in his conversation with American correspondent Anna Louise Strong in 1946 (Mao, 1961) and based on what he considered to be patterns of exploitation rather than of diplomacy or formal ideology. Mao stated that US "reactionaries" were striving to dominate the world but that their global march was halted by the Soviet Union, the "progressive" socialist state and a "defender of world peace". Following the Sino-Soviet split (1956–66) the Soviet Union was being labelled by Beijing as a "social-imperialist" state which had replaced the US as the main threat to world peace and in 1964 Mao modified his view of the world order, identifying two "middle zones" in a forerunner of what would later be his theory of three worlds, one comprising the "developing" countries of Asia, Africa and Latin America, and a second made up of the "developed" countries from Western Europe plus Japan, Canada, Australia and New Zealand (Mao, 1969: 514). For Mao the key battlefield lay in the zone separating the two superpower rivals and including many capitalist, colonial and semi-colonial countries across Europe, Asia and Africa (see Figure 3.1). Mao's emphasis on the existence and importance of a third force enabled China to

FIGURE 3.1 Chinese propaganda poster (1964) entitled "Resolutely support the anti-imperialist struggle of the peoples of Asia, Africa and Latin America". Source: Stefan R Landsberger collection, International Institute of Social History, Amsterdam.

develop its own identity and expand its influence in international relations (Lin, 1989), constructing itself as part of the Third World and at the head of a united international proletariat battling against imperialism (see Figure 3.2). In further developing a

FIGURE 3.2 Chinese propaganda poster (April 1968) entitled "Chairman Mao is the great liberator of the world's revolutionary people". Source: Stefan R. Landsberger collection, International Institute of Social History, Amsterdam.

sense of contrast with the USSR, China repeatedly pointed out that revolution could not be exported and that all communist parties were independent and should make their own decisions (Dee, 1983: 242). Mao advocated a Chinese brand of socialism that would serve in similarly placed societies (colonial or semi-colonial) as a model not for emulation but for inspiration so that people might find their own "third ways" (Mao, 1965). In a speech to the UN General Assembly held in April 1974, Deng Xiaoping expounded Mao's strategic thinking on the question of three worlds and this came to operate as a major "geopolitical compass" for China and its leaders (Kim, 1980).

China was also able to outline its aspirations for the Third World and its vision of "Afro-Asian solidarity" at the Bandung conference in West Java (Indonesia) in April 1955. A meeting of representatives from 29 African and Asian nations, Bandung aimed to promote economic and political cooperation and to oppose colonialism whilst its symbolic meaning can be explained as a "collective crowning" or "inauguration ceremony of post-colonial Asia and Africa" (Shimazu, 2014: 232). For Samir Amin (2014: 131) Bandung began a kind of "counter-geopolitics, defined by Southern states, to push the geopolitics of the triad [the US, Western Europe, Japan] back". Bandung was thus in many ways the "launching pad for Third World demands" where countries distanced themselves from the "big powers seeking to lay down the law" (Rist, 1997: 86). Initially led by several key statesmen of "Third Worldism" including Sukarno (Indonesia; see Figure 3.3), Nehru (India) and Nasser (Egypt), the event was sponsored by Burma, India, Indonesia, Ceylon (Sri Lanka) and Pakistan and sought to cut through the layers of social, cultural, political and economic difference that separated nations of the "Third World" in order to think about the possibility of common agendas and actions. For all the talk of solidarity and cooperation, however, there were several notable absentees. The USSR, South Africa, North Korea, Israel, Taiwan and South Korea were all not invited, conspicuous in their absence.

One of the key axes of mobilisation around Bandung was race. The "racialised assemblages" (Weheliye, 2014 of empire that disciplined humanity into full humans, not-quite-humans (such as black and native populations within the colonies) and non-humans created commonalities around which a post-colonial basis for Afro-Asian solidarity could be formed. In his opening speech to the conference on April 18th, 1955, Sukarno

FIGURE 3.3 President Sukarno of Indonesia addresses the Bandung Afro-Asian conference in 1955. World History Archive/Alamy.

urged participants to remember that they were all united by a common "detestation" of colonialism and racism (Sukarno, 1955: 1) and pointed out that colonialism was not dead or in the past but also had its "modern" (neo-colonial) forms. In the US, Bandung raised alarm partly because of recurring fears of the "vast" populations of Asia being mobilised in a racial alliance against the West, which remained a feature of the American cultural and political imagination for much of the early Cold War period (Jones, 2005). More generally, by the time of Bandung race had become a key factor in the way the US' role in world affairs was perceived by both domestic and global audiences (Jones, 2005). As Connelly (2000) has shown, alongside the anti-communist fixations of the Cold War, the Eisenhower administration (1953–61) was often fearful of the emergence of North–South and more general racial tensions when it surveyed the international scene. US authorities were also concerned that the pan-Asian "contagion" might spread domestically to the African-American population. Richard Wright, an African-American writer that attended the conference, wrote in an account entitled *The Colour Curtain* that it was not hard for countries with histories of colonial exploitation to find something in common since the "agenda and subject matter had been written for centuries in the blood and bones of participants" (Wright, 1956: 14). For Wright, there was something almost "extra-political" about Bandung but the event itself aroused considerable concern and suspicion on the part of the superpowers and was dominated by discussion of Cold War issues. US Secretary of State John Foster Dulles even toyed with the idea of a US-sponsored "reverse Bandung" conference, in which the Western-oriented Third World countries would have the upper hand (Gilman, 2003). Dulles, concerned about the rise of a pan-Asian movement, feared that Bandung "might establish firmly in Asia a tendency to follow an anti-Western and 'antiwhite' course, the consequences of which for the future could be incalculably dangerous" (Dulles, 1955: 83). Department of State officials sought to undercut the impact of Bandung with a set of carefully timed announcements (Engerman, 2004) and there were fears in both the US and USSR that too much independent international organising among Third World nations would limit their own strategic influence.

Following Bandung, the Non-Aligned Movement (NAM) was founded by its leading advocates from India, Indonesia and Yugoslavia on a principle of non-alignment with either the US or the USSR in the Cold War. Its members did however include countries that had in fact aligned themselves (or were members of security pacts) with one superpower or the other such as Indonesia, Ethiopia and Egypt, despite the movement's stated aims of asserting independence from East and West. By 1970 all of NAM's founding members had either died or been overthrown and it had become fractured. A year after the formation of NAM, the "Group of 77" was also established (mostly made up of the attendees of previous conferences) to press for changes from UN agencies, hastening the arrival of new international institutions explicitly dealing with "development" (Rist, 1997). The principles of the Bandung conference were also taken forward by early "Bandung regimes" (Nehru's India, Nasser's Egypt, Sukarno's Indonesia, Nkrumah's Ghana) and led to what Samir Amin (1994) has called the "Bandung Era" (1955–75) in recognition of the extent to which, following the conference, the "Third World" became the site of intense debates regarding options for "development" and non-capitalist paths to socialism (Scott, 1999). From the late 1960s through the 1970s radicalisations of the Bandung project emerged in Chile, Tanzania, Jamaica, Grenada and Sri Lanka. In each case "socialism" was the name of a "variously configured oppositional idea of political community defined largely in terms of anti-imperialism, national self-determination, and anti-capitalism" (Scott, 1999: 144). Marxism, in particular, appeared to many in the Third World to offer a code of legitimation against imperialism, a justification of power, the promotion of new structures of rule and the possibility of "catching up with history" (Laïdi, 1988: 11).

As a space of diplomacy Bandung was also an important site for the performance of elite, "non-aligned" and anti-colonial political identities in the era of decolonisation. In recent years Diplomatic History and IR have begun to take the sociocultural more seriously, and to engage more carefully with a variety of different performances and spatialities of geopolitical relations at diplomatic events like conferences and summits (Shimazu, 2012; Craggs and Mahony, 2014; Mamadouh et al., 2015). As part of a wider re-engagement with the historical geographies of the international as a concept, a scale, and a political and cultural affiliation, work has also usefully focused attention on the international conferences of the interwar period as important sites where the international was created and challenged (Goswami, 2012; Hodder, Legg and Heffernan, 2015) and where significant "imaginary futures" of development were elaborated. It is important to recognise that such event spaces become key sites of knowledge creation, public performance, legitimation and protest (Craggs and Mahony, 2014). In some of this work the "diplomatic stage" is shifted from the abstract sphere of high politics to the concrete sphere of the local milieu (Shimazu, 2012: 335) and a variety of diplomatic spaces and performances are taken in: not only the conference centre, but also the street, the hotel, the evening dinner dance and the angry press conference. The focus here is on overt *performances* and their representation in the popular press but also on the smaller, less public practices through which atmospheres (of agreement and discord, trust and alienation) were made. Sukarno, Nehru, Zhou Enlai and Nasser all understood the value of performance "in their role as new international statesmen, representing the *esprit de corps* of the newly emergent post-colonial world" (Shimazu, 2014: 225).

Beyond its international significance Bandung also had an important domestic subtext and served as a "political theatre" where Sukarno's Republic mobilised many signs of Indonesian nationalism in order to muster popular support and acceptance of the new regime, enacting them for the benefit of regions where rebellion had been rising (Shimazu, 2014). These "performative" acts of independence and nationhood helped to reinforce the sense of unity in the Republic by rekindling recent popular experiences of the Indonesian revolution, whilst at the same time acting as symbols of the unity of the Afro-Asian world by reinforcing the sense of solidarity of the 29 nations in attendance. In this way, the Republic's leaders, seen together with a variety of world statesmen at the event, made a considerable impression on the Indonesian electorate (Shimazu, 2014), enabling Sukarno to appear as the great unifier and leader of a united nation, which was, itself, emerging as a leader of the Third World. For some of its key figures then like Sukarno, Nasser and Nehru, Bandung was as much electoral *realpolitik* as an ideological call to arms (Ahmad, 1992).

In addition to Bandung, another important event in the making of the "Third World" was the Tricontinental: the first Conference of the Organization of Solidarity of the Peoples of Africa, Asia and Latin America (OPSAAAL), held in Havana in January 1966 (see Figure 3.4), which brought together the anti-colonial struggles of Africa and Asia with the radical movements of Latin America and marked the initiation of a global alliance of the three continents against imperialism (Young,

FIGURE 3.4 A roadside billboard advertises the Tricontinental conference in Havana, Cuba in 1966. The message reads "This great mass of humanity has said enough! and is already on the march". Photo 12/Alamy.

2005). The extension, into the Americas, of the Afro-Asian solidarity movement begun at Bandung was considered by Washington to be a direct threat to the security of the US. The Tricontinental marked both a high point in the emergence of a non-aligned movement and the construction of a Third World anti-imperialist project but also a break with those earlier efforts. Whereas Bandung was a relatively modest affair, in which the various political currents in the Third World came together to articulate a minimum programme, the Tricontinental was avowedly more radical, explicitly attempting to align anti-imperialism with a wider challenge to capitalism. This conjunction was mediated at that time by the worldwide fight against imperialism represented by the American intervention in Vietnam, where resistance to US intervention reminded many in the Third World of their own anti-colonial revolutions. Although by the end of the 1960s the high point of Afro-Asian solidarity had passed, and much post-colonial optimism had faded, protracted and violent decolonisation struggles in Southern Africa and apartheid in South Africa provided a continuing focus for action into the 1970s (Lee, 2010).

THE SOVIET UNION AND THE "ROMANCE" OF ECONOMIC DEVELOPMENT

The Cold War existed outside of military conflict, diplomatic standoffs and superpower summits; it also encompassed crucial agreements and disagreements about economics (Engerman, 2004). The principal Cold War adversaries shared a desire for higher levels of economic production and agreed that "economic *performance* was a defining element of modern life and an important measure of national success" with industry seen as the "prerequisite for higher levels of economic performance" (Engerman, 2004: 24). The conflict between East and West in the Third World was thus an expression of two hegemonic models of modernisation, democratic and socialist, free market versus central planning (Westad, 2006). Both sought widespread appeal to the new states being created from the ashes of colonialism. One, "symbolised" by the US:

> promised intensive urban-based growth in both the private and public sectors, the import of advanced consumer products and the latest technology through joining a global capitalist

> market, and an alliance with the world's most powerful state. The other, that of the Soviet world, offered politically induced growth through a centralised plan and mass mobilisation, with an emphasis on heavy industry, massive infrastructural projects, and the collectivization of agriculture, independent of international markets.... Both, however, offered a road to high modernity through education, science, and technological process. (Westad, 2006: 92)

The Cold War "mapped certain traditional Orientalist stereotypes onto the Russians" (Pietz, 1988: 69) but for many Third World leaders, the Soviet model "was more in line with the state-centred and justice-oriented ideals they themselves had for the development of their new countries" (Westad, 2006: 92–93). Stalin's "great break", which sought to make a radical change from an agrarian past by advancing "full steam ahead" along the path of collectivisation and industrialisation to socialism and to leave behind an age-old Russian "backwardness", along with ambitious Five-Year plans implemented by a centralised planning apparatus, resonated with many in the "Third World" in the 1950s and 1960s. The Soviets' journey from a "backward" exporter of agricultural products to an industrial society and "a nation of metal" through the construction of huge new steel plants, gigantic dam projects and vast industrial enterprises with newly mechanised collective farms feeding the workers had considerable appeal (Engerman, 2004). Leftists and economic planners from overseas flocked to observe the Soviets' plans in action with many of these "pilgrims of planning" (Engerman, 2004) succumbing to what American diplomat and historian George Kennan once termed "the romance of economic development" (Kennan, 1932) in Russia which, he argued, had led young people to "ignore all other questions in favour of economic progress". Even Kennan, however, balanced his criticisms of Soviet policies with an appreciation of the Soviets' goal of modernisation (Engerman, 2004). Experiences with national development planning in the first three decades of the twentieth century in post-Revolutionary Russia preceded the beginning of debates in America about the possibility of "modernisation" (Engerman, 2000) whilst many of the pioneers of development economics were from East Central Europe (Berger, 2001). North American economists and policymakers became increasingly interested in the Russian "version" of the national development project and from the onset of the Cold War there were fears that the USSR was providing a better example of development than anything the West had to offer (Gilman, 2003).

In the context of decolonisation and in places like Addis Ababa, Luanda and Havana, the Soviet Union and its developmental methods appeared to occupy the rhetorical high ground during the early post-war period as the Soviets came without the same colonial baggage that accompanied Western powers. For those in the midst of an insurgency, Russian communists had a model for how to overthrow the former regime and a pattern for a new state that was "just" and "modern" whilst in little more than a generation Lenin, Stalin and their compatriots had "transformed a backward, underdeveloped country into a military-industrial powerhouse" (McMahon, 2001: 7). Economic plans in India, Indonesia and elsewhere in the Third World sought to emulate Soviet economic priorities, emphasising industry over agriculture. India's Jawaharlal Nehru, for instance, wrote enthusiastically about Russian industrial progress, drawing explicit comparisons between Russia's circumstances and India's and claiming that Russia's determination to overcome its economic backwardness would make India's struggle all the easier. Nehru sought to learn the strategies of rapid industrialisation (Nehru, 1949 [1929]) and keenly watched the "Soviet experiment", arguing that its economic ambitions provided a sharp contrast with the bleak outlook of the 1930s, its Five-Year Plans "a bright and heartening phenomenon in a dark and dismal world" (Nehru, 1941: 230–231).

For the intelligence cadres of the liberation movement known as the Zimbabwe African People's Union (ZAPU) that were trained in Moscow by the Soviets in the 1960s, Soviet egalitarianism, anti-racism and state provision for basic needs held a powerful appeal due to the dramatic contrast to settler-ruled Rhodesia (Alexander, 2017). The foundation for the communists' claim to have modernised Russia (and the basis of its claim to provide a model for "backward" countries of the Third World) was the collectivisation of agriculture and the process of making the peasants create the state-controlled surplus that was necessary to jump-start an industrial economy. There was a belief that the success the Soviet Union had in technology and production in the 1950s would win support for communism abroad. In particular, the Virgin Lands campaign (Nikita Khrushchev's plan to dramatically boost the Soviet Union's agricultural production in order to alleviate the food shortages plaguing the Soviet populace) and the space programme were two key projects that inspired this as Soviet know-how in agriculture and industry was held up as something that would help the transition to socialism move faster.

The USSR also played on its ambiguous position as both "inside" the international system and "outside" it, presenting itself as the natural "midwife" for completing the independence of newborn states (Laïdi, 1988). The Soviets chose to support liberation movements not only because of the Cold War but also because it regarded these struggles as part of the global "anti-imperialist struggle" (Shubin, 2008: 3) which was not understood as a battle between the two "superpowers" assisted by their "satellites" and "proxies", but rather as "a united fight of the world's progressive forces against imperialism" (Shubin, 2008: 3). Under Khrushchev (1955–64), there was a rediscovery of Lenin's argument that the peoples of the colonial world represented de facto allies of the proletariat and of the first proletarian state, the Soviet Union (Kanet, 2006), along with a belief that their struggle for independence from the imperialist West would contribute to the weakening of the major opponents of the Soviet Union, including ultimately the US (Kanet, 1974).

While Americans celebrated the market, the Soviets denied it (Westad, 2006). For Lenin's followers, modernity came in two stages: a capitalist form and a communal form, reflecting two revolutions, that of capital and productivity and that of democratisation and the advancement of the underprivileged. Communism was the higher stage of modernity and it was the destiny of Russian workers to lead the way towards it. The Communist International (Comintern) set up in 1919 would be the vehicle through which communists would set off rebellions against colonialism, although its influence in the Third World declined between 1928 and 1943 and there was some stifling of dissident voices within it (Westad, 2006). Lenin began bringing "Third World" socialists to Moscow in the immediate aftermath of the 1917 Communist coup and Mongolia, as the first People's Republic, became a testing ground for much of communist policy where many of the same methods of cultural work, education collectivisation and propaganda that would appear later in other countries, were first introduced by Soviet advisers. Many future leaders of anti-Western resistance passed through the Comintern (including Nehru and Ho Chi Minh) and their encounters with communism and the Soviet Union provided succinct ideas about how to construct their movements and subsequent states (Westad, 2006: 55). The Soviet Union also sought to expand its influence abroad by means other than diplomacy. The Peoples' Friendship University of Moscow, for example, founded in 1960 and renamed after Congolese leader Patrice Lumumba, following his assassination in 1961, granted scholarships and degrees to thousands of students from Africa, Asia and the Middle East while indoctrinating them with communist ideology in an effort to influence the future elites of the Third World (Staar, 1991). Many foreign students, however, primarily those from Africa, experienced racism by members of Soviet state institutions in the 1960s and 1970s and African students and African-American residents in the Soviet Union found that opportunities for economic and spatial mobility were extremely racialised and that racial slurs were often used outside official antiracism discourse (Fikes and Lemon, 2002: 503).

Khrushchev's first major visit abroad in 1954 was to Beijing (followed by India, Burma and Afghanistan) and stressed Soviet willingness to cooperate in the national development of Third World countries with colonialism and imperialism as the enemy. Building an alliance with China was seen as a priority through the assistance programme carried out under the Sino-Soviet Friendship Treaty, a kind of Soviet Marshall Plan, agreed in 1950. The relationship with China had been lauded as the ultimate proof of socialism's suitability for the Third World but increasing rivalry between the two led to more intense competition for strategic influence there. In an ambitious attempt to stamp Soviet socialism on China, Moscow agreed in 1953 to increase aid sevenfold over two years and "in every department of every ministry, in every large factory, in every city, army or university there were Soviet advisors or experts who worked with the Chinese to modernise their country" (Westad, 2006: 69; see Figure 3.5). In 1954, the USSR proposed 156 key projects to assist China, including plans for the development of the Yellow River basin and the huge Russian-designed Sanmenxia dam project which began construction in 1957 (Li, 2003). Mao had wanted "more, faster, better and cheaper" socialism, however, and by designing the Great Leap Forward in 1958 "broke with all Soviet advice about caution and stages" (Westad, 2006: 69).

In July 1953 the Soviet representative to the UN Economic and Social Council (ECOSOC) announced that the USSR would contribute four million rubles to the UN programme of technical assistance to underdeveloped nations (having had a long history of opposition to it) (ECOSOC, 1953: 142). Over the next three years the USSR worked intensively towards the restoration of Lenin's vision of "forging a united front between the nationalist aspirations of the developing world and the revolutionary, anti-Western objectives of the Soviet regime" (Porter, 1984: 16). A five-year trade agreement was signed with India in September 1953 and in the following year Afghanistan became the first Third World nation to receive credits from Russia since World War II (see Figure 3.6). In

FIGURE 3.5 Chinese propaganda poster (1953) entitled "Study the Soviet Union's advanced economy to build up our nation". Source: Stefan R Landsberger collection, International Institute of Social History, Amsterdam.

February 1955, one of the first and largest Soviet economic aid projects in the Third World was announced: the Bhilai steel mill in India costing over US$100m (Westad, 2006: 17).

The Soviet cause was boosted significantly when Fidel Castro declared his commitment to building a Marxist–Leninist political system in Cuba and turned to Moscow for military and financial support. Following the Cuban Revolution in 1959, Fidel Castro's regime began a programme of actively supporting anti-imperial revolutions across the so-called "Third World". This brand of Cuban Internationalism was heavily influenced by the call for international solidarity among the proletariat; a key tenet of Marxism–Leninism, carried out according to the axiom of guerrilla warfare developed by Ernesto "Che" Guevara. It was under this ideological umbrella that communist Cuba became not only an exporter of revolutionary rhetoric but a material supporter and participant in the armed struggle against imperialism. In December 1961 Cuba sent its first international aid – a cache of rifles, machine guns and mortars – to the Algerian FLN (*Front de Libération National*). No African liberation movement was denied Cuban solidarity, whether expressed in material

FIGURE 3.6 Soviet pilot and cosmonaut Gherman Titov (second right) at the construction site of the Naghlu hydropower station in Kabul province, Afghanistan. The project was financed and supervised by the Soviet Union between 1960 and 1968. Source: Sputnik/Alamy.

and arms or in the training of military and civil technicians and specialists (Deutschmann, 1989: 45). Mozambique, Guinea-Bissau, Cameroon and Sierra Leone all at one time or another requested and received forms of solidarity aid (Babbit, 2014) and Cuba's intervention in Angola's liberation struggle (although not always in synchrony with the Soviet Union) proved decisive (Marcum, 1978). In March 1977, Fidel Castro set out on a triumphal tour of Africa, both celebrating the Cuban victory in Angola (see chapter 4) and confirming his role as chief spokesman for the Third World (George, 2005).

In the 1960s, Cuba and Vietnam challenged not only Washington in defence of their revolutions, "they also challenged the course set by the USSR for the development of socialism..." (Westad, 2006: 158) and were certainly not, as various US administrations liked to depict them, Soviet "puppets". After 1949 US officials viewed Beijing's expansive inclinations with "nearly as wary an eye as Moscow's" (McMahon, 2001: 5) and worried that they posed just as dangerous a test. Maoist China's criticism of Moscow for lacking revolutionary fervour had paved the way for independent brands of communism and while countries like Cuba and Vietnam were recipients of Soviet aid, they were also ready to embarrass the Kremlin in the lengths to which they would go to challenge the West (Young, 2006).

THE US AND THE THIRD WORLD

The US itself became a Third World power (though the term had not yet been coined) when it seized possession of several Pacific and Caribbean territories following the Spanish–American war of 1898 (McMahon, 2001), although its extensive trade links with the non-Western world long pre-dated the imperial surge of the 1890s. In particular, during the administrations of Theodore Roosevelt (1901–09) and Woodrow Wilson (1913–21) the US substantially deepened its diplomatic, military and commercial involvement with non-Western areas. World War II also further deepened US awareness of (and interests in) the raw materials and markets available in "developing areas". In this sense, the US "fixation" with the Third World that developed during and after World War II "had strong historical antecedents" (McMahon, 2001: 2). After 1945, a big part of what furthered this obsession was a growing concern that the "nationalist aspirations of dependent peoples be accommodated so as to defuse more revolutionary tendencies" (McMahon, 2001: 3). US beliefs in individual liberty, free market economics and "progress" led to policies of intervention in East Asia, the Americas and Africa that long pre-dated the Cold War, but which became much easier to justify once communism could be portrayed as a threat (Westad, 2006). Initially under Roosevelt and later the Truman administration (1945–53), what increasingly began to take hold was the idea that the US should assume a more activist role in the Third World. US strategists of the 1950s, often referring more to the "free world" than to the Third World (Sewell, 2010), set about integrating "developing" regions more fully into the global economy and spurring free trade but also aimed to establish a worldwide network of US military bases. The Cold War alone was not responsible for creating US interests in and priorities for the Third World then, it merely served to reinforce them.

President Truman's Point Four programme, announced in his inaugural address on January 20th, 1949, was based on a strong belief in sharing American know-how, technology and capital in order to share the "American experiment/experience" with nations around the world (McVety, 2008). On that important day of Truman's second inauguration some "two billion people became under-developed... transmogrified into an inverted mirror of others' reality" (Esteva, 1992: 7). Under President Truman, "underdevelopment" became the incomplete and "embryonic" form of development and the gap was seen as bridgeable only through an acceleration of growth (Rist, 1997). Truman vowed to make a central part of the US government's national security agenda the development of the underdeveloped world, where there were "untold resources" needing only "somebody who knows the technical approach" (Truman, 1949). Technical assistance was seen as contributing to economic development, which in turn contributed "to a host of politically good things – democracy, peace, non-Communist governments, good will, international understanding" (Packenham, 1973: 47–48), although no one had proven that economic development would in fact lead to political development. The Truman administration made technical assistance a crucial part of Point Four (McVety, 2012) although it requested only US$45 million for the programme and in practice the State Department struggled to define its scope (Latham, 2011; see Figure 3.7). Rather than offering a grand plan for economic development, Point Four technicians taught classes on public health and irrigation, distributed chickens and vaccines, and helped build schools and water treatment facilities in 34 countries around the world (McVety, 2012). Green revolution technologies, for example, sought to dramatically increase yields of wheat and rice and although gains in productivity were typically attributed to plant breeding science, security concerns and management of foreign exchange were prime motivators of the new technologies (Perkins, 1997).

After 1945 the US' role in the Third World was deeply contradictory, on the one hand claiming its anti-colonial credentials and pushing for decolonisation and the liberty of colonised peoples whilst on the other acting as an imperial power itself, through the creation of a trade and investment empire, through a series of controversial wars, invasions and occupations and through projections of its power and authority. In practice the US also often struggled to "balance its desire for friendly, cooperative relations with the emerging postcolonial states with the need to maintain harmony within the Western alliance" (McMahon, 2001: 6), seeking to avoid alienating European NATO allies that held colonies. US analysts attempted to

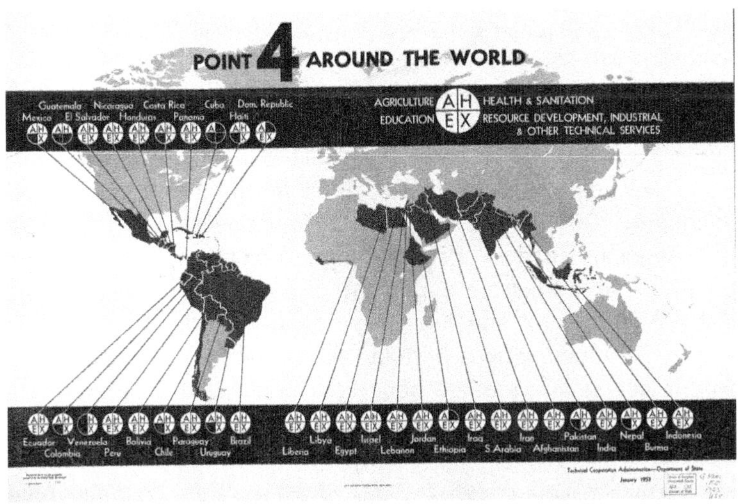

FIGURE 3.7 A projection of 'Point IV around the world' prepared by the Technical Cooperation Administration (TCA) in 1953 maps out the global extent of the programme's interventions. Courtesy of US Library of Congress.

demonstrate the efficacy of the capitalist pathway to development and prove its superiority over the communist route, combating the alternative form of modernity the Soviets had been developing since 1917 (Westad, 2006: 25). Containing the Soviet Union and constructing a "healthy international environment" were, as National Security Council Paper 68 (NSC-68) pointed out in April 1950, the two most basic policy goals of the US, distinct but overlapping (National Security Council, 1993: 40–41). The Third World was crucial to each (McMahon, 2001). Until the end of the 1950s, US policies of communist containment mostly focused on Asia, with Africa regarded as "secured" by the maintenance of repressive colonial regimes such as Portugal and South Africa (Young, 2005).

Condescending and paternalistic, US development theory in the post-war period recast older notions of racial hierarchies and "drew its inspiration from the old American vision of appropriate or legitimate processes of social change and an abiding sense of superiority over the dark-skinned peoples of the Third World" (Hunt, 1987: 160), rating the amenability of certain cultures to modernisation. Development in this sense needs to be understood not only as a geopolitical tool of the Cold War, but also as a biopolitical technique of governance that took shape within the realm of the domestic and through a racialised gaze

(Domosh, 2015, 2018). Several recent studies have traced US development practices to the first decades of the twentieth century, if not earlier, and to events that occurred as much within the US as outside of it (Ekbladh, 2002, 2010, 2011; Sneddon and Fox, 2011; Domosh, 2015, 2018; Nally and Taylor, 2015). High-profile US domestic infrastructure projects provided inspiration and served as models of what US aid could potentially achieve through development of the Third World (Ekbladh, 2010), such as the Tennessee Valley Authority (TVA) established in 1933 as a grand scheme to develop the natural and human resources of the Tennessee Valley region (see Figure 3.8) and the New Deal's Rural Electrification Administration which became prototypes of how the state could act as a rational, benevolent enforcer. Further, in the 1960s, as popular mobilisation proliferated across the Third World, America's civil rights movement was mobilising domestically and its cities were beset with widespread racial violence which also provided an important, if often neglected, context for the making of US development discourses and strategies.

Key elements of US international development practices in the post-war era can be traced back to the US South in particular, a region considered "undeveloped" in the first decades of the twentieth century. The domestic agricultural extension service run

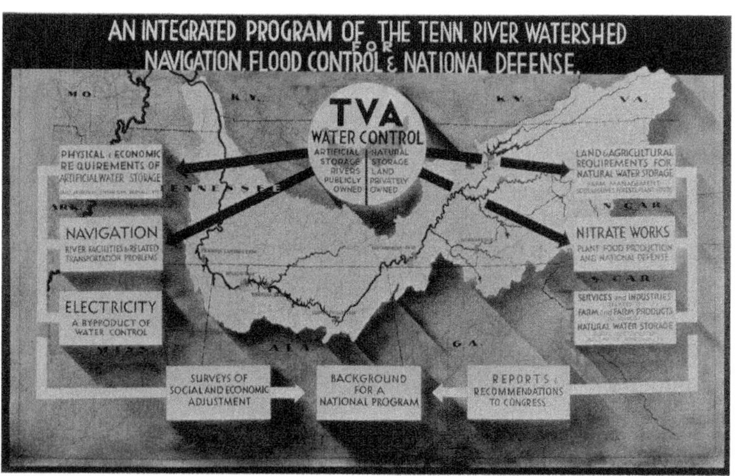

FIGURE 3.8 A comprehensive planning chart from the Tennessee Valley Authority in 1940 maps out the construction benefits from several hydroelectric dams. Everett Historical/Shutterstock.

by the US Department of Agriculture (USDA) targeted the rural farm home and farm women in a series of heavily gendered and racially segregated interventions that then informed US interventions in other countries under the guise of agricultural modernisation (Domosh, 2015). The USDA's African-American extension service in the early decades of the twentieth century also served as a model for early articulations of liberal ways of development (where it was also taken up by experts in the British Colonial Office concerned with the ecological and economic crises looming in their African colonies) since this biopolitical approach to governance seemed to hold the promise of social engineering that would create self-reliant individuals and self-sustaining communities (Domosh, 2018). Similarly, the Rockefeller Foundation's extension service, which attempted to spark agrarian change in the US South through the inculcation of modern habits and aspirations among farmers, provided a model for its international rural development projects which sought to spread "green revolution" technologies of self-help as part of an anti-revolutionary geopolitics, using the idea of "revolution within" to contain the threat of "revolution without" (Nally and Taylor, 2015).

The US was buoyed by the perceived success of the first major US foreign aid programme – the Economic Recovery Program proposed by Secretary of State George Marshall in 1947, which seemed to demonstrate that the US could kick-start growth in impoverished economies. The Marshall Plan, which had proposed US$13.3 billion of spending over four years in quite "recipient-friendly terms" (Sogge, 2002: 1), had sketched out the principles of "helping" poor countries to "recover" and restructure their economies and societies. US policies were also subsequently shaped by (and pursued through) an emerging post-war international institutional architecture of Development. The role of agencies like the World Bank, the UN Educational, Scientific and Cultural Organisation (UNESCO), the Food and Agriculture Organization (FAO) and the World Health Organization (WHO) in negotiating the Cold War has usefully been traced in a number of critical reassessments (Engerman, 2003; Jolly, 2004; Kraske, 1996; Marshall, 2008; Mason and Asher, 1973; Shapley, 1993; Connelly, 2008). Geopolitical agendas can be glimpsed behind both the problems these agencies identified and the solutions they adopted as US officials began to recognise that they could perhaps accomplish more by masking their actions behind multilateral institutions and technical interventions. These institutions played a subtle, nuanced role in waging the Cold War in that multilateral development offered an understated, "constructive" alternative to the clumsy, overtly self-interested and "contentious" manoeuvres of national diplomacy (Staples, 2006: 62–63).

Until 1957, more than 50% of World Bank financing was to "developed" or "Part I" countries, although this fell to zero by 1968 (Kapur, Lewis and Webb, 1997). World Bank loans to "developing countries" were often subordinated to Cold War objectives promoted by the US, with the agency financing programmes of support for dictatorships (such as in Brazil, Chile, Nicaragua, Guatemala, the Philippines and Zaire) whilst turning a blind eye to mismanagement and corruption (Bello, Kinley and Elinson, 1982). Cold War geopolitics deeply influenced the World Bank's efforts to promote economic development in the "Third World" with the Bank championing the cause of private property and free markets in the face of growing Soviet influence (Escobar, 1995). Many US appointees to the World Bank presidency, such as John McCloy, Eugene Black, Robert McNamara and Paul Wolfowitz, were also seasoned career cold warriors. Together with the IMF, the Bank played an important role in the scripting of spatial entities (through the discourses of US foreign policy) as democratic or totalitarian, developed or underdeveloped, East or West, civilised or barbaric, and ultimately, based on these dichotomies, as friendly or hostile, creating the possibility and justifications for certain types of intervention (Popke, 1994).

The FAO similarly acted as a US surrogate until 1965 (McLin, 1979) and every executive director of the UN Children's Fund (UNICEF) since 1947 has been American whilst some of UNICEF's policy actions are taken in consultation with the US State Department (US Department of State, 2003). International efforts to control population through organisations like UNICEF, the WHO and the UN Population Fund (UNFPA) also quickly became embroiled in the campaign to contain the spread of communism (Connelly, 2008), with the CIA investigating UNESCO activities as early as 1947 and claims circulating in 1953 that UNESCO was under communist control (Dorn and Ghodsee, 2012). As such, the idea of teaching poor people to read "became hopelessly entangled with fears of socialist revolution" (Dorn and Ghodsee, 2012: 398). Cuba's success with literacy campaigns (and its "Third World" appeal) led the US to take an interest in developing its own literacy campaign to "promote capitalist economic development by providing trained manpower for nascent industries and increasing peasant productivity by teaching 'modern' farming methods" (Dorn and Ghodsee, 2012: 387). Fears of the spread of communism were thus increasingly at the heart of the Development apparatus, particularly as it became clear in the early 1970s that the US was losing the war in Vietnam.

ARRESTING THE COMMUNIST "CONTAGION": THEORISING MODERNISATION IN THE US

> If national security was the sum of our fears, a nightmare vision of an American garrison state cowering before a hostile Eurasia, development spoke our dreams: a transparent, modernising world mastering man and the environment with American technology. (Cullather, 2000: 652)

Key modernisation theorists often described communism as a "pathological" or "deviant" form of modernity, hoping that the Soviets would "converge" with the liberal version extant in the West (Gilman, 2003). Whilst containment focused on the problems of keeping the Soviets and their leftist allies at bay, development was intended to provide "long-term immunity against the contagion of communism" (Hunt, 1987: 159–60). It was believed that what would stop the disease in its tracks was an alternative, capitalist form of modernity. In this way development theory became the "younger sibling of containment" (Hunt, 1987: 157). The modernisation school turned anti-communism "from the hysterical red-baiting populism of McCarthy into a social-scientifically respectable political position" (Gilman, 2003: 13). The social sciences had come under particularly heavy attack from McCarthyite inquisitions, as did their institutional sponsors, such as the Ford Foundation which was accused by a 1954 congressional investigation of financing "the promotion of Socialism and collective ideas" in the US (Deconde et al., 2001). In a way then the quest for the improvement of "developing areas" thus provided a safe outlet for social engineers whose New Dealism had become suspect (Cullather, 2000). The modernisation school continued pre-war progressive themes that emphasised the meliorist impact of government social programmes and when some of these, such as Truman's Fair Deal, began to fall apart, liberals found themselves unable to continue their social engineering projects domestically (Gilman, 2003).

The "developmental state" with its paternalistic vision of government assuming responsibility for economic growth and social and economic equity and well-being for all its citizens thus became "the third world analog to the welfare state" (Gilman, 2003: 17). The US, in effect, exported its liberalism in the form of the energy and ideas of its best social thinkers and the funds of its largest philanthropies (such as Ford, Rockefeller and Carnegie), which were increasingly focused on issues affecting "backward areas" (Packenham, 1973). Together with government agencies and universities they jointly created the institutional arrangements for generating theorisations of modernisation. Third World elites, underlined the Social Science Research Council in 1957 (SSRC, 1957), were looking for a new concrete form for their states and societies, and "it was the duty of American social scientists to produce one" (Westad, 2006: 32). In this sense, modernisation continued the long-standing idea of the "manifest destiny" of the US in embodying the conviction that the country "could fundamentally direct and accelerate the historical course of the postcolonial world" (Latham, 2011: 2). As such, it represented "the most explicit and systematic blueprint ever created by Americans for reshaping foreign societies" (Latham, 2011: 5). Modernisation was in part a "will to spatial power" (Slater, 1993: 421) and certainly helped to construct an American world order (Ekbladh, 2011), using social science to rationalise the post-World War II drive to achieve global free trade and American geopolitical hegemony (Gilman, 2003). Through a growing number of area studies framed around modernisation, orientalist discourse came to play a key role in the attempt to subordinate, contain and assimilate the Third World "other" (Slater, 1993).

The need to "do something" for these post-colonial regions was very much a political imperative but one that demanded a "scientific" justification. Key to the development of US modernisation thinking was the creation in the early 1950s of the Center for International Studies (CIS), the result of a top-secret anti-communist propaganda project conducted at the Massachusetts Institute of Technology (MIT) called Project Troy. CIS scholars made several interventions around US aid policy and saw it as an integral part of the anti-communist containment policy that had begun with the Truman doctrine in 1947, regularly arguing for increases in the aid budget. Founded by economist Max Millikan, its first study on "Soviet vulnerability" (Rostow, 1953) was conducted by US economist and political theorist Walt Whitman Rostow, one of the founding fathers of modernisation thinking. During World War II Rostow served in the Office of Strategic Services (OSS), described by Gardner (2009: 60) as "a nursery of

American cold war intellectuals", and he participated in (among other tasks) the selection of targets for US bombardment. In 1947 Rostow served as assistant to Gunnar Myrdal, the then Executive Secretary of the Economic Commission for Europe, giving him a key role in the development of the Marshall Plan. Rostow's seminal text entitled *The Stages of Growth* (Rostow, 1960) offered up a paternalistic and prescriptive five-stage typology that envisaged a "take-off" from "traditional society", through "transitional society" and the "drive to technical maturity" into an age of high mass consumption (Rostow, 1960). Rostow worked closely with the JFK administration to establish new aid programmes that would help nations of the Third World proceed from stage to stage in an orderly fashion.

The CIS was an important source of regional and area studies expertise emerging during the early Cold War and its officials lobbied for and publicised the need for American development aid overseas. Focusing on India in the late 1950s and early 1960s (Engerman, 2004), CIS scholars typically explained the urgency of their work in technical and economic terms but also in geopolitical ones, emphasising the necessity of keeping India out of the Soviet bloc. A key focus of CIS scholarship was why some countries were especially susceptible to communism and at what point they might fall prey to what Rostow called the "communist disease". For Rostow communism was "an opportunistic virus that took out infant nations not yet blessed with a constitutional 'maturity'" (Gilman, 2003: 195) and was not the agent of modernisation but a side effect of it, or a "disease of the transitional process" likely to spread in any nation during the early, difficult stages of development. Instead of accelerating growth, he argued, communism disfigured it, producing an unbalanced and dysfunctional modernity. The political scientist Lucian Pye joined the CIS in 1956 on the strength of his newly published book *Guerrilla Communism in Malaya* and his work similarly regarded post-colonial nations "as essentially rebellious adolescents, potentially susceptible to Communist delinquency, which in turn might lead to a life of international crime" (Gilman, 2003: 170). For Pye, communism was an ideology that appealed to "disturbed" individuals, adrift amidst the transition of modernisation (Gilman, 2003: 196). These representations of post-colonial nations as "young" or "immature" and as somehow "endangered" by the transitional stages of growth appear throughout the literature on modernisation. Development, Rostow and Millikan assured the CIA in 1954, could create "an environment in which societies which directly or indirectly menace ours will not evolve" (Millikan and

Rostow, 1954: 41) as foreign aid quickly became a key weapon for waging the Cold War (see chapter 4).

JFK, THE "DECADE OF DEVELOPMENT" AND THE RISE OF "DEVELOPMENTESE"

The inclination under the Truman and Eisenhower administrations to defer, at least in part, to the European colonial powers in areas like Southeast Asia and the Middle East (Suez excepted) gradually gave way to a growing desire and active effort under the Kennedy administration (1961–63) to "woo" and secure the states of the Third World. In JFK's White House the Third World "was still a collection of trouble spots where freedom was a mixed good and sometimes even downright dangerous" (Hunt, 2007: 208), his administration clutching at modernisation theory as a means to manage these same trouble spots (Hunt, 2007: 208) whilst "groping for an explanation of the United States' place and responsibilities amid the uncertainties of the post-war world" (Gilman, 2003: 5). For Rostow the most politically dangerous period in a nation's development was the "preconditions" stage and he argued that it was on the weakest nations, facing their most difficult transitional moments, that Communists concentrated their attention as "the scavengers of the modernisation process" (Rostow, 1961a: 234). The Democratic administrations of the 1960s took Rostow's theory to mean that if the US could quickly help shepherd underdeveloped countries safely through the transition to modernity and into the take-off stage, by whatever means necessary, then the Communist "contagion" could safely be arrested.

The main mode of transmission for the Communist "virus" for Rostow was guerrilla warfare, since, as Schlesinger (2002: 341) has noted, "Guerrillas were an old preoccupation of Walt Rostow's." Indeed, Rostow advised JFK to turn Fort Bragg in North Carolina into a centre for teaching a counter-guerrilla operations course, arguably making him the "primary theorist" and "grand designer" of counterinsurgency phenomena in the United States (Gray, 2015: 114). Rostow recommended destroying what he called the "external supports" to guerrilla insurgents, as "the most hawkish civilian member of the Kennedy and Johnson administrations" (Milne, 2008: 6). Averell Harriman, one of America's most celebrated diplomats, once described Rostow as "America's Rasputin", comparing him to the Russian mystic who advised Russian czar Nicholas II, for the powerful and sometimes unsavoury influence Rostow exerted on the decision making of Presidents JFK and Lyndon Johnson (he also served as a speech writer for President Eisenhower) (Milne, 2008). Rostow was the first to advise Kennedy to send US combat troops to South Vietnam, and the first to recommend the bombing of North Vietnam and an invasion of Laos. A State Department white paper on Laos in 1959 even insisted that the landlocked country of three million people constituted a "front line to the free world" (United States Department of State, 1959: i).

Vietnam in particular was an important test case for the modernisation thesis laid out in Rostow's *Stages* (Gardner, 2009). In 1966 Rostow was named Special Assistant for National Security Affairs (the post now known as National Security Advisor) where he was a main figure in developing the government's policy in the Vietnam War, a position he approached with a "missionary zeal" (Gardner, 2009: 60). Rostow was a bundle of contradictions, however (Milne, 2008). In his earlier work outside of government Rostow argued that the US needed to assist developing nations economically, not militarily, and that this was the best path to eventual democracy and alliance with the US; but in his later work, he repeatedly argued that military regimes could supply the stability and administrative competence needed for development and that eliminating the enemy's capability to wage war by destroying factories, power plants and logistical networks was the best path to victory (Milne, 2008). Rostow's supposedly "visionary" rhetoric would thus soon "end up justifying the bombing of a number of countries back through several 'stages of growth'" (Gilman, 2003: 199). Drawing on his experience as an army captain in World War II and his work in the Economic Warfare Division of the American Embassy in London, where he had the principal task of identifying suitable enemy targets for Allied bombing missions, Rostow championed massive escalatory strategic bombing of North Vietnam, much like the bombing of the Axis powers in World War II (see Figure 3.9). Rostow had very little understanding however of Southeast Asian political or cultural history and was analytically deficient in perceiving the conflict as a nationalist civil war first, and a war between communism and a fledgling democracy second (Milne, 2008). Along with Max Milikan, Rostow was asked by JFK to draft the initial proposal for the Peace Corps (see below) and played a key role in the development of the US Agency for International Development (USAID) and the Alliance for Progress (AfP).

FIGURE 3.9 In the situation room at the White House in Washington, DC, George Christian, President Lyndon B Johnson, General Robert Ginsburgh and Walt Rostow look at a relief map of the Khe Sanh area in Vietnam on February 15th, 1968. Photo by Yoichi Okamoto. Courtesy of LBJ Presidential Library.

The AfP was set up in 1961, along rigidly Rostovian lines, providing economic, technical and educational assistance to Latin American countries and suggesting that it was only by becoming more like the US that Latin America could really begin to develop (see Figure 3.10). In anticipation of imminent take-off, the AfP called for a big push to raise investment rates, foster social capital and induce thorough reforms in institutions, land tenure and income distribution. Rostow predicted a transition to self-sustaining growth within the decade, allowing the administration to keep the problem of underdeveloped areas "off our necks as we try to clean up the spots of bad trouble" (Rostow, 1961b). The anticipated follow-on effects of the alliance failed to materialise, however, and despite US$20 billion in aid support, growth rates after the first year were far below the ambitious target figures. As Latham (2011) points out, Rostovian modernisation thinking was never held to account for the alliance's failures and instead setbacks only appeared to some observers to reinforce the validity of concepts that provided explanations and remedies for failure. The AfP's problems were blamed on stubbornly anti-modern Latin

FIGURE 3.10 President John F Kennedy introduces the First Lady at La Morita, Venezuela, accompanied by President Romulo Betancourt of Venezuela and others on December 16th, 1961. US State Department. Jacqueline Kennedy Onassis Personal Papers. John F Kennedy Presidential Library and Museum.

leaders who had refused to carry forward the reforms or were simply chalked up to bureaucratic ineptitude and personnel problems (Latham, 2011). Since the theory was rooted in the historical experience of the industrial core, there was a belief that it could not be invalidated by contradictory experience in periphery nations (Deconde et al., 2001). In this way, modernisation funnelled dissent into its own framework, where it could challenge the execution but not the conception of the plan.

Theorisations of modernisation often "echoed and amplified unfolding American sentiments about the condition of modernity at home" (Gilman, 2003: 12) and were as much about defining

America as attacking communism. It resonated with Americans' cultural understandings of themselves and their country's destiny (Latham, 2000). As cultural historians have shown, Americans interpreted and imagined the developmental encounter through multiple literary and cinematic forms (Klein, 2003; Slotkin, 1992; Nashel, 2000; Reynolds, 2008). Across fiction, reportage and political commentary, an American version of the underdeveloped world began to emerge between the 1940s and 1960s as US writers like W E B Dubois reflected on foreign cultures and on their own complex positions as Americans in a global context. In the works of these authors, the ideals of the US as "apostle of modernity" and sponsor of "development" feature as central in the decades after World War II (Reynolds, 2008). The Rogers and Hammerstein musical *The King and I* and bestsellers such as Eugene Burdick and William Lederer's *The Ugly American* and Tom Dooley's *Deliver Us From Evil* (1956) validated modernisation thinking by associating it with mythic conventions in which "a 'hostile' [Asian] is converted into a 'friendly' one by the White American's display of honour and competence" (Slotkin, 1992: 449).

Writers and artists framed a tolerant, inclusive and sentimental role for the US and its citizens (Latham, 2011: 61), celebrating their nation's altruistic efforts to reach across the lines of race and culture and putting forward a liberal, internationalist ideal of integration and education. *The Ugly American,* co-written by William Lederer, a publicist working for the US Navy and the CIA, remained on the *New York Times* best-seller list for 78 weeks, sold an astonishing four million copies and was made into a blockbuster movie starring Marlon Brando. The ensuing media frenzy almost put development on a par with the space race in terms of its popularity with the US public (Deconde et al., 2001). It also created a new strand of populist internationalism that JFK seized upon to boost his presidential bid but also the atmosphere in which JFK declared America's willingness to "bear any burden", building up US special forces and emphasising new tactics of counter-insurgency to combat communist "people's war" in South Vietnam (Hellman, 1986). Under JFK, "...the non-western world [was] to be the site of Western adventures, the battlefield on which Westerners tired of domestic routines [could] find urgency, adventure and glory" (McLure, 1994: 45).

Under the influence of men like Rostow and Pye then, "developmentese became the Kennedy administration's court vernacular" (Cullather, 2000: 641). JFK told campaign aide Harris Wofford that he had wanted to run for President to initiate a "new relationship" between the US and the developing world (Kennedy quoted in Rice, 1985: 23). Concerned about the existence of an "economic gap" in Asia that was being filled by Soviet aid, in the first year of his presidency, Kennedy launched the AfP, the Peace Corps, Food for Peace (which took established agricultural surplus disposal programmes and organised them around a developmental mission) and USAID, declaring the 1960s to be the "decade of development" in a challenge taken up by the UN in 1961. JFK's administration saw decolonising countries as central to a changing world, even claiming that the USA was an older anti-colonial sister to the twentieth-century anti-colonial movements. There was also his Irish Catholic minority status which led Norman Mailer in 1957 to label him a "white negro" (Mailer, 1957). Dubbed "Secretary of State for the third world" (Schlesinger, 2002: 509), JFK's engagement of non-aligned countries and courting of Third World leaders was unprecedented (see Figure 3.11). As Rostow (1985) noted, Kennedy had been deeply moved and influenced by exposure to poverty in visiting some parts of the Third World. JFK also badly wanted to deploy American "expertise" and "liberalism" abroad and the modernisation school gave him a justification for doing so (McVety, 2015; see Figure 3.12).

The Peace Corps idea was conceived during the Presidential campaign and JFK saw them as helping to overcome the efforts of "Mr Khrushchev's missionaries" (as he often liked to call them) who had been busy strengthening their ties with the Third World, expanding trade with Latin America in the mid-1950s and announcing technical assistance programmes to the impoverished countries in the US's backyard. A descendant of the missionary tradition originated by Christian Europeans (Cobbs, 1996), the Peace Corps sought not specialists but "representative Americans" who could transmit values by example and create a catalytic effect by introducing ideas from a supposedly higher point on the developmental arc. Although initially set up as a means of countering growing Soviet commitments to decolonisation and foreign aid, the Peace Corps also played a role in domestic nation-building within the US and helped to project a non-opportunistic image of the US as seeking not to "dominate" but only to help (see Figures 3.13 and 3.14). To an extent it often framed development around a North–South axis (rather than East–West) through various means, including a refusal to send volunteers to Vietnam, elaborate provisions to avoid CIA infiltration and a "rhetoric of universalism rather than anticommunism" (Cobbs, 1996: 105). As a foreign policy initiative, it was one of the most successful strategies of the

FIGURE 3.11 President John F Kennedy, accompanied by Deputy Special Assistant for National Security Affairs Walt Rostow (fourth from the left, behind two men), meets with members of the Parliament of Ceylon at the Oval Office in the White House, Washington, DC on June 14th, 1961. Photo by Abbie Rowe. White House Photographs. John F Kennedy Presidential Library and Museum, Boston.

post-World War II period for making friends for the US and by 1965 there were some 13,248 volunteers in the field (Cobbs, 1996: 89). Many other countries followed suit based on their own larger national policies and as a way of showing solidarity with the US and the wider Atlantic alliance. Between 1958 and 1965, Britain, Australia, Canada, Japan, Israel, France, Germany, Denmark, The Netherlands, Norway, Sweden, Italy, Belgium and even Liechtenstein started volunteer programmes to spread the message of economic development and international goodwill. For Geidel (2015), however, the development ventures of the Peace Corps legitimated the (violent) exercise of American power around the world, contributed to the destruction of indigenous ways of life and embodied and disseminated a particularly heroic and compelling iteration of modernisation

FIGURE 3.12 USAID agronomist Emory Howard stops to talk with an Afghan farmer during work on irrigation and hydroelectric power facilities in the Helmand Valley in the 1960s. Courtesy of the US National Archives Still Picture Unit.

FIGURE 3.13 Peace Corps Volunteer Jane Keiser, 26, of Minneapolis, works as part of a tuberculosis team training Afghan counterparts in Baghlan hospital. Photo courtesy of Peace Corps.

FIGURE 3.14 Peace Corps volunteer Mary Jean Grubber administers a tuberculosis skin test to a villager in Nsanje, Malawi in 1965. Granger Historical Picture archive/Alamy.

that allowed the US to maintain global hegemony in the face of widespread decolonisation struggles by placing modernity, rather than independence or economic justice, at the endpoint of those struggles.

One of the hallmarks of the modernisation school (and the various attempts to apply it to "Third World" contexts) was the idea of replicating and mimicking the development of others, particularly the US. Modernisation theorists separated the contents of modernity from its origins as a description of specific periods of Euro-American history and stylised them "into a spatio-temporally neutral model for the process of social development in general" (Habermas, 1987: 2). The "model of the modern" on offer was an image of the US writ large (Luke, 1991) and a key component of modernisation was its theory of convergence. The dependency scholar André Gunder Frank once referred to this as a kind of "Sinatra Doctrine":

> Do it my way, what is good for General Motors is good for the country, and what is good for the United States is good for the world, and especially for those who wish to "develop like we did" (Frank, 1997: 13)

The Dependency approach pioneered by Gunder Frank in *The Development of Underdevelopment* (Frank, 1966), as a counter-analysis to the modernisation school, first emerged in Latin America and initially took root in Brazil, Chile and Colombia (and the US) before later opening out into a variety of regions including Africa, the Caribbean and the Middle East. The promised altitude that was the fifth stage of Rostow's model envisaged an urban-based, Western lifestyle of consumption, but the dependency scholars argued that the planes of the South had been stalled on the runway by unequal relations and a history of colonialism, denying them a chance of ever being airborne or "industrialised" (Power, 2003). Just as the modernisation approach was adopted by international institutes and bilateral donors, the dependency school was made up of all those opposed to US policy and became a focal point for Third Worldism. Its message about the manipulation of the periphery by the core was very timely in the context of proliferating state socialisms and radicalisations of the Bandung project. Dependency was an "insurgent theory" (Slater, 2008) that constituted an important attempt to "theorise back" from the Third World and provided a valuable counter-analysis to the assumptions about growth and the diffusion of "progress" by modernisation theorists. In doing so the dependency scholars constructed and deployed an important alternative geopolitical imagination which sought to prioritise the objectives of autonomy and difference and to "break the subordinating effects of metropolis-satellite relations" (Slater, 1993: 430).

Both modernisation and dependency approaches were "inextricably intertwined in one another's assumptions" (Munck, 1999: 203) (including key binarisms such as modernity vs tradition, core vs periphery) and seemed in the end "to checkmate each other" (Schuurman, 2001: 6) in that the main propositions of the dependency scholars seemed counterpoised "point by point to Rostow's theory" (Rist, 1997: 110). Ironically, given the role philanthropic foundations played in creating the conditions for the growth and diffusion of modernisation thinking, much of the research on what became dependency theory in Latin America was funded by the Ford Foundation, which had helped to create networks of researchers concerned with development in the Third World. The independence of philanthropic foundations like Ford and Rockefeller from the US government enabled them to initiate research in areas wary of US geopolitical ambitions and bring them into US circles of debate, by appealing to a nebulous concept of global development free of political interest (Arnove, 1982).

CONCLUSIONS: THE GHOSTS OF COLD WAR MODERNISATION

> The various hells that postcolonial countries from Indonesia to Iraq to Colombia have entered in the last thirty years were almost always preceded and justified by well-intentioned modernizers, both liberal and Communist, who believed that they knew what was best for these lands. (Gilman, 2003: 20)

Cold War geopolitical discourses constructed and invented a "Third World" as a strategically vital third global space in need of modernisation and development. As an ideological project, the Third World took multiple forms and unfolded across various sites, nodes and networks and its boundaries and "revolutionary orientation" (Sauvy, 1952) were performed and enacted across a number of diplomatic sites and spaces in Asia, Africa and Latin America. Although it enabled a series of "revolutionary myths" of socialism and national liberation (Chaliand, 1977) as various political leaders claimed to represent the entirety of its people and to speak on its behalf, the Third World also gave rise to a number of radical developmentalisms and an important "counter-geopolitics" (Amin, 2014: 131) that sought to push back against imperialist geopolitical ambitions. Ironically, although the US set out to contain communist insurgencies, many of its interventions did as much to radicalise the Third World as any form of socialism. The apparent successes of socialist regimes in offering an alternative to capitalism and an alternative geopolitical alliance certainly helped constitute and radicalise the Third World but in seeking to confront revolution, US policy inadvertently created a strong and coordinated bloc of opposition and solidarity and strengthened the desire to find alternatives. It was the US then that, by 1970, "had done much to create the Third World as an entity both in a positive and negative sense" (Westad, 2006: 157). Interventions against radicalism in places like Iran and Guatemala, interference in the Congo in the early 1960s, support for Israel and a laissez-faire economic system that effectively kept much of the Third World in poverty, all served to alienate those who had supported the agenda of the non-aligned movement (Westad, 2006: 157). Whilst the (geo) politics of Third Worldism took distinctly different routes in various national contexts, the 1950s and 1960s was an age of "Third World

Euphoria", a period "when it seemed that First World leftists and Third World guerrillas would walk arm-in-arm toward global revolution" (Shohat and Stam, 1994: 26). The activist aura once enjoyed by the Third World may have evaporated and the post-war configuration that justified Third World identity and non-alignment may no longer exist but the "imaginary cartography that justified the Third World still does" (Grovogui, 2011: 178).

Modernisation was a global project in character and scope, taking on multiple forms (including its Soviet and Chinese varieties) with "high-modernist faith" being found across the political spectrum (Scott, 1998), especially among those that wanted to use state power to bring about huge changes to people's work habits, living patterns, moral conduct and world view. For both Moscow and Washington, the key objectives were "control and improvement" (Westad, 2006: 5) and right from the off modernisation was framed around insurgency, either as a means of fomenting and spreading revolution and anti-imperialism or as a means to confront and counter insurrection. Decolonisation had increased the threat of collectivist ideologies gaining the upper hand in the Third World and the transnational practice of development, as it emerged in the 1950s, was about managing these insurgencies and responding to the threat posed by popular mobilisation, enabling the US to develop a set of rules for a "proper revolution" (Latham, 2011): it must be a solemn affair, with minimum disorder, led by respectable citizens with moderate goals (Hunt, 1987: 116). In the US, modernisation was inextricably linked then not just to the transference of Western historical experiences, democratic ideals and values but also to the maintenance of political order and stability. It offered not only a meta-narrative that provided a sense of the "meaning" of post-war geopolitical uncertainties but also "an implicit set of directives for how to effect positive change in that dissilient world" (Gilman, 2003: 5).

For various US administrations, the peoples of the Third World needed to be defended against communism even when they had themselves elected a communist government. Modernisation would provide the necessary inoculation against the communist contagion, this opportunistic virus that took out "infant nations" not yet blessed with constitutional maturity during the difficult and dangerous transitional stages of economic development. Defending "freedom" took on the paternalistic role of the US knowing what such countries *really* wanted, better than the people themselves – a classic colonial attitude that led the US to support anti-communist forces in Vietnam and Korea, colonial autocracies such as the apartheid government in South Africa and various neo-colonial regimes in Latin America (Young, 2005). US policy towards the Third World included support for authoritarian regimes, the use of covert operations and aid payments to ensure political stability (Kolko, 1988) and saw material assistance provided to a variety of regional allies like Brazil, Indonesia, Iran and Zaire to manage local conflicts and contain Soviet influence without direct American military intervention. Both military and civilian programmes of intervention were developed, each aiming to create (within the recipients of US aid) modern army and police forces capable of countering insurgencies. The US alone provided grants amounting to over US$90 billion in military equipment and training to some 120 countries before the end of the Cold War and licensed approximately US$240 billion in arms sales to more than 100 countries (Kuzmarov, 2017). Experiences with countering insurgency also fed back into and framed wider US approaches to foreign aid. The foundations of USAID, for example, were closely shaped by counter-insurgency experiences and practices as its founding members, returning from the Vietnam War, transferred the counter-insurgency tactic of "population control" to USAID's strategy (Connelly, 2008) (see chapters 4 and 6).

Typically, writings on modernisation on both the left and right "characteristically focus on strategies rather than on the developers who devise them or the historical circumstances in which they arise" (Cullather, 2001: 243) as usually "the origin of the intention to develop is omitted from discussion" (Cowen and Shenton, 1996: 440). Modernisation was shaped by particular localities (from Addis Ababa to Washington) but also by particular projects (from the Virgin Lands programme and the TVA to the CIS, Project Troy and UNESCO) and particular individuals (such as Rostow). The passage of modernisation programmes was rarely smooth, however, and neither were they unilaterally "imposed" on subject populations. Even as US policymakers tried to promote modernisation and direct its course in countries like Egypt, India and Ghana, "foreign actors embraced, modified and reformulated the ideology to fit their own purposes" (Latham, 2011: 5). In Iran, for example, the Shah was adept at exploiting American fears of communist subversion during the JFK, LBJ and Nixon administrations, presenting himself and his vision of modernity as the only viable option for ensuring Iranian security (Offiler, 2015) and refocusing modernisation on Iran's military to secure the Pahlavi dynasty.

Similarly, in Indonesia President Suharto was able to manipulate US fears about Indonesia becoming a communist bulwark of Moscow or Beijing in Southeast Asia to promote his own "army-led modernisation" and consolidate his authoritarian regime (Simpson, 2008).

Further, despite the intensity of Cold War confrontation, it was not the case that Third World leaders had a simple choice between the two camps, or that their interventions were always entirely separate or diametrically opposed. Post-colonial leaders were willing to experiment and combined elements of both American and Soviet experience in order to generate more rapid progress (Latham, 2011). Programmes of development intervention are often pulled together from various sources in a "bricolage" (Li, 2007: 6) and by the late 1960s, American and Soviet aid advisers were collaborating on the ground in India, Afghanistan and the Middle East; by the early 1970s, even the more stubborn philosophical differences, such as over birth control, had begun to dissipate (Connelly, 2008). While the US was able to shape the overriding architecture of international development, many African countries had technical cadres trained in the Soviet Union, Eastern Europe and China, and implemented mixed development plans combining elements of state planning, cooperatives and state farms, with US models of rural development, community development and the land grant system of agricultural extension (Amanor, 2013).

As Karabell (1999: 12) has argued, US interventions in the Third World depended (to varying degrees) upon a "convergence of interest" with local elites, whose own actions were often dictated by their personal domestic concerns. It is important therefore to consider the ways in which modernisation discourses and practices were engaged with and appropriated locally in various contexts. Ethiopia was one example where millions of dollars of US foreign assistance failed to transform Haile Selassie into a democratic leader and to create a democratic state. US leaders seemed to overlook the possibility of a nation importing America's economic and scientific advances while rejecting its political ones (McVety, 2008). Further, many twentieth-century Third World leaders valued modernisation as a critical means of consolidating their own power over a state that they could now "know" and, in consequence, more effectively dominate (Scott, 1998: 77). Despite JFK's idealistic calls to spread "the disease of liberty" to communist nations, his administration's determination to fight the Soviets at all costs meant that it proved as willing as

Eisenhower's to overlook the reality that many of its allies were steadily building up their immunity (Ambrose and Brinkley, 1997: 173; McVety, 2008).

The Swedish economist Gunnar Myrdal, himself a key contributor to modernisation thinking who had been involved with the Marshall Plan, worked with the Rockefeller Foundation and chaired the Swedish Committee for Vietnam (Scott, 2009), once noted that international civil servants working in UN agencies filled roles only recently vacated by colonial officials and told a gathering of agricultural experts in 1966 (Myrdal, 1966: 666) that they were the "inheritors of the imperial *mission civilisatrice*". Indeed, many of the "apostles of modernity" approached the cause with a missionary, messianic, proselytising zeal and a sense of self-righteousness, personified and embodied in the work of key figures like Walt Rostow. Development itself increasingly constituted a kind of modernist messianism (Rist, 1997) that, at least in the US, promised to exorcise the secular demons of the post-war world – poverty, communism and colonialism – and became so popular in post-war America that it "approximated a civil religion championed by liberal Cold warriors" (Nashel, 2000: 134).

Modernisation resonated with Americans' cultural understandings of themselves and their country's "manifest destiny" and echoed and amplified American sentiments about the condition of modernity at home. With the "idea of America as an insecure space at the heart of a Cold War map" (Farish, 2010: xxii), modernisation discourses were as much about defining America as attacking communism and were as much shaped by events that occurred within the country as by those outside of it. Rostow saw his work as giving "fresh meaning and vitality to the historic American sense of mission – a mission to see the principles of national independence and human liberty extended on the world scene" (Millikan and Rostow, 1957: 2–8). Further, in the context of the persistent and pervasive militarisation that existed in the US in the 1950s and 1960s, Rostow and other social scientists tried to "weaponise" their understanding of modernisation (Gilman, 2003; Latham, 2011), "like nuclear scientists who applied their knowledge of the atom to the task of making bombs" (Danforth, 2015: 478). The result was often a form of modernisation discourse that justified military intervention and authoritarian rule and violence in countries like Indonesia and Vietnam on the grounds that only strong states could implement the social and economic programmes necessary to change traditional societies. Modernisation was thus an ideology before it became a theory of development.

Rostow subtitled *Stages* (1960) as "A non-communist manifesto" and made it his "life mission to offer the world a better alternative to Karl Marx" (Frank, 1997: 2), seeking to supplant Marx as the inspiration for revolutionary intellectuals and to "reclaim Marx from the communism of the Soviet Union" (Gilman, 2003: 201). Marxism inspired a variety of radical Third Worldists and the movements and states they tried to form but also the alternative geopolitical imagination and counter-analyses to modernisation that dependency scholars pioneered in seeking to "theorise back" from the Third World.

For some, the promises of modernisation were impossible, even unreal, goals because as a process it could not be imposed or invited and efforts to do so were doomed to fail (Escobar, 1995). Similarly, Mitchell (2002) suggests that modernisation was purely a discourse used by Western powers to justify exploitative imperial or neo-imperial relationships with Third World peoples. In this view, modernisation discourse only existed as rhetoric for "the natives", while in private "canny 'experts' knew better than to fully believe their own schemes" (Danforth, 2015: 491). US officials may have sought to conceal their strategic geopolitical interests and motivations behind multilateral institutions and technical interventions and found in multilateral development a "constructive" alternative to the clumsy, contentious manoeuvres of national diplomacy (Staples, 2006), but the peoples and political leaders of the Third World were able to negotiate and remake modernisation discourses in a variety of ways and to negotiate the multiple modernities and political geographies of the Cold War (White, 2003) in attempting to create a new geopolitics. Modernisation was discredited in the wake of US failures in Vietnam and Iran but many of its fundamental assumptions are still present today in the way US interventions in Afghanistan and Iraq, for example, have linked development to the enhancement of security in confronting insurgency in the South, as the "ghosts of modernisation" (Latham, 2011: 186) continue to haunt the crumbling Development edifice.

Chapter 4

Cold War geopolitics and foreign aid

INTRODUCTION: COLD WAR FOREIGN AID AND THE BATTLE FOR THE THIRD WORLD

IN October 1975, on the eve of Angolan independence, *El Vietnam Heroico*, one of three improvised Cuban ships, put in to Puerto Amboim, just south of the capital city Luanda. It had been sent as part of a Cuban solidarity action in Angola entitled *Operation Carlota*, framed as a "people's war" and named after a Cuban slave called Black Carlota who, working on the Triunvirato plantation in the Matanzas region in 1843, led a slave rebellion in which she lost her life. The Cuban ships were subjected to all kinds of provocation by US destroyers, which followed them for days on end, and by warplanes that menacingly buzzed above and photographed them (Marquez, 1977b). At the time, the US was just emerging from the Vietnam debacle and the Watergate scandal with an unelected president and a government keen to avoid appearing as the ally of racist South Africa in the eyes of both the majority of African countries and the black population of the US itself. The fall of South Vietnam, however, along with the painful reassessment that it led to in the US, made possible the subsequent Soviet–Cuban intervention in Angola – the first large-scale deployment of Cuban troops on behalf of a Soviet client state in a Third World country (some 36,000 troops, half of whom were black, were deployed during Operation Carlota) (Mallin, 1987; Adams, 2012).

As the Portuguese colonial forces were due to abandon Angola on the day of independence (November 11th, 1975), nationalist movements backed by a wide variety of external sponsors (including the US, South Africa, China, North Korea, Zaire, Romania, the

Soviet Union, Cuba, the German Democratic Republic, France, and Yugoslavia) jockeyed for position as they sought to seize strategic points and infrastructures, particularly around Luanda. Angola was in many ways "an improbable locus for a superpower collision" (Marcum, 1976) but as international economic and military assistance flooded in it quickly became a hotly contested Cold War battleground. The real cost of this foreign aid was huge, however – during the period of conventional warfare between 1975 and 1976, 50,000 Angolans, the majority of them non-combatants, were killed or died from starvation and disease caused by the war (Clodfelter, 2002: 625–626). Upon independence the post-colonial Angolan state was beset by an insurgency that was to last almost three decades, plunging the country into a devastating civil war.

This chapter examines the ways in which foreign aid practices were shaped by and folded into the wider Cold War enframing and imagination of modernisation and development and seeks to show how aid and overseas development assistance became a key site of Cold War geopolitical competition and rivalry with important implications for donors and recipients alike. Foreign aid, oriented towards both civilian- and military-focused interventions in recipient countries, includes a wide range of resources transferred by donors to recipients such as physical goods, skills and technical know-how, financial grants (gifts) or loans (at concessional rates) (Riddell, 2007: 17). The establishment of domestic welfare states in the 1930s and 1940s in Europe and North America had paved the way for the setting up of the foreign aid regime and the willingness to consider governmental programmes of assistance to people overseas (Lumsdaine, 1993), which had also been helped by the proliferation of private and philanthropic humanitarian assistance – by 1910 over 300 NGOs based in the "developed world" were active abroad (Smith, 1990). The Cold War provided an important rationale for foreign aid and from the very beginning it was linked to political conditionalities of various kinds that sought to create particular kinds of states or to advance the donor's security and strategic objectives. As such, through the Cold War aid became "a key part of the architecture of international relations" (Riddell, 2007: 22). At the height of the Cold War some even argued that aid was imperialism (Hayter, 1971) whilst scholars like Andre Gunder Frank argued that it was a form of neo-colonialism that had created only dependency. Foreign aid enabled a variety of countries to create spheres of strategic influence and alliance during the Cold War, to project their economic, political and cultural power, to pursue their foreign policy objectives and to sponsor (or counter) various kinds of revolution. It also enabled donors to "train" and "discipline" client states and to promote particular kinds of political formations and politics. In several cases, recipient countries (such as Afghanistan and Angola) became the site of "tournaments of modernisation" (Cullather, 2002: 530), as competing global powers sought to enact their different conceptions of revolution, progress and development in a bid to win hearts and minds.

In addition to its significance in the Cold War confrontation between the US and the USSR, Angola was also the focus of a vitriolic propaganda battle between the USSR and China. Having initially sided with the Marxist–Leninist MPLA (People's Movement for the Liberation of Angola), the first African liberation movement with which China had contact, Chinese policy later favoured the FNLA (National Liberation Front of Angola) and UNITA (National Union for the Total Independence of Angola) simply to preclude a Soviet victory and not because of the ideological merits of the organisations themselves (Jackson, 1995: 405). The Chinese armed anti-MPLA groups to show Africa to be "populist, peasant-oriented, anti-Soviet and rural" (Jackson, 1995: 396), which suited their wider agenda for the continent. For its part, Cuba had been playing an active role in fomenting and supporting anti-colonial and anti-imperial revolutions (Gleijeses, 2002, 2006). The momentous victory of the "26th of July Movement" led by Fidel Castro over the Cuban bourgeoisie (and their corporate American allies) was regarded as only the start of an international proletariat revolution which would spread across the periphery, with Cuba committed to facilitating the outbreak of "another Vietnam" somewhere in the Third World (Robbins, 1979).

Cuba's first informal contacts with the MPLA were made in the 1950s and MPLA guerrillas received their first training from Cubans in Algiers in 1963 before the MPLA's leader, Agostinho Neto, met with Che Guevara and Fidel Castro (in 1965 and 1966, respectively), to raise the possibility of wider forms of aid (Marcum, 1969). Castro sent troops to Angola without informing the Soviet Union and deployed them at Cuba's own expense from November 1975 to January 1976 (when Moscow eventually agreed to a degree of support by arranging for a maximum of 10 flights from Cuba to Angola) (Gleijeses, 2002). Castro himself came to see the troops off as they departed on *El Vietnam Heroico* – as he did with every contingent that left for Angola (Marquez, 1977b) – and kept himself informed of the minutest details of the war, quoting any statistic relating to Angola "as if it were Cuba itself… as if he had lived there all his life.... there was not a single dot on the map of Angola that he was unable to identify, nor any feature of the land that he did not know by heart" (Marquez, 1977b: 134).

Cuba's motivation to enter the war in Angola arose independent of the USSR (Kessler, 1990; George, 1999) and the decision to escalate to a full-scale intervention appears to have been purely Cuban (George, 2005), although there is no denying that the timing of the arrival of Soviet heavy arms was critical or that the Cubans were, in some respects, a "proxy" force for the Soviets in Africa. Cuba would maintain a presence in Angola for the next 16 years with close to 500,000 Cubans serving in various positions from military personnel to humanitarian aid workers. Estimates suggest that around 10,000 Cubans died in the Angolan campaign from 1978 to 1980 (Horowitz and Suchlicki, 2003; see Figure 4.1) and by 1988 there were over 65,000 Cuban troops in Angola, proportionally four times the American commitment to Vietnam (George, 2005). Moscow also rewarded Cuba financially for its role in Angola, replenishing Cuba's military inventory and offering to purchase Cuban goods that could not be sold on the world

FIGURE 4.1 Angolan President Jose Eduardo dos Santos (first from left) and his Cuban counterpart Fidel Castro (second from left) pay their respects to the mortal remains of Cuban soldiers who fell in combat in Angola, at the Cacahual Mausoleum in Havana on December 7th, 1989. RAFAEL PEREZ/AFP/Getty Images.

market (Porter, 1984). Almost immediately after Angola the Cubans and Soviets followed up with a similar large-scale operation in the Horn of Africa, where they supported the Marxist government of Haile-Marian Mengistu of Ethiopia in a war against their former clients in Somalia over the Ogaden in 1977–78. Success in Angola led Soviet officials to believe "the world was turning in our direction" (Westad, 2006: 241).

The chapter builds on the discussion of the geopolitics of Third World modernisation in the previous chapter by exploring the ways in which three countries that offered some of the most compelling models of modernity in the Third World, the US, China and the USSR (along with its Eastern European allies), used foreign aid to advance their geopolitical strategies and particular visions of "development". It examines some of the different discourses, institutions, practices and imaginaries involved and also how these were in some cases resisted, reworked and remade in particular places by recipients.

FROM THE PERIPHERY TO THE PERIPHERY: THE USSR AND FOREIGN AID

The role of the Soviet Union in shaping the coordinates of "Third World" geopolitics is a neglected theme in the literature and the instrumentalisation of the Soviet model in development practice has not been well documented (Laïdi, 1988). After 1945 and as a result of the war with Nazi Germany, the USSR had vastly expanded military power and was left with huge stocks of surplus armaments along with large and well-trained military forces as it began to use foreign aid to spread socialism in the Third World (see Figures 4.2–4.4), to build strategic alliances and to create a bloc in opposition to the West (Bervoets, 2011). Key to this was Khrushchev's rediscovery of Lenin's argument that the peoples of the colonial world represented de facto allies of the proletariat and of the first proletarian state, the Soviet Union (Kanet, 2006), whose duty it would be to lead this international alliance of the proletariat and the fight of the world's "progressive" forces against imperialism. To underline this commitment the Soviets frequently voiced their condemnation of Western colonialism and pushed in the UN Trusteeship Council and in other international fora for the immediate dismantling of the remnants of Western colonialism (cf. Pearson, 2017).

Several organisations were established by the USSR that sought to coordinate actions between communist parties under

FIGURE 4.2 Soviet propaganda artwork entitled "Human rights" (1977) by Kukryniksy Art Group. Courtesy of Yevgeniy Fiks/Wayland Rudd collection.

FIGURE 4.3 "Africa is Fighting, Africa Will Win!" Soviet propaganda poster by Viktor Koretsky (1971). Courtesy of Wayland Rudd collection.

FIGURE 4.4 "Great Lenin showed us the way!" Soviet propaganda poster by V. Boldyrev (1969). Courtesy of Wayland Rudd collection.

Soviet direction (Berliner, 1958; Goldman, 1967), including the International Department of the Central Committee of the Communist Party (set up in 1943) and the Council for Mutual Economic Assistance (CMEA or Comecon) set up in 1949 in response to the Marshall Plan and the founding of the Organisation for European Economic Co-operation (OEEC, later the OECD). Comecon originally consisted of the Soviet satellite states of Eastern Europe but was later joined by Mongolia (1962), Cuba (1972) and Vietnam (1978) and began offering subsidised oil, technical assistance, grants and loans (all tied to the purchase of Soviet goods and services) along with weaponry and military training. China stopped participating in Comecon in 1961 following the Sino-Soviet split but several other Third World countries had observer status including Angola, Mozambique, Laos, Nicaragua, Afghanistan, Mexico, Iraq and Ethiopia. Comecon did not however play a dominant or multilateral role in aid policy or programming and served as little more than a forum for donor consultation and coordination (Kanet, Miner and Resler, 1992). Attempts to maintain and develop relations with communist parties and various left-wing organisations in capitalist countries were managed through the Communist International (Comintern) and later the International Department whilst the Soviets organised conferences, seminars, and cultural exchanges to promote personal contacts with Third World intellectuals and "advertise" the socialist experiment. Organisations such as the Soviet Afro-Asian Solidarity Committee (SKSSAA), founded in 1956 and a member of AAPSO, mobilised solidarity efforts directed towards national liberation movements from Moscow and functioned in practice as a Soviet semi-official foreign policy organ (Weinberger, 1986). There was also the USSR's Communist Youth League (Komsomol), which worked to advance the Soviet cause among Third World youth during the Cold War (Hornsby, 2016) using methods from cultural propaganda through to attempts at political subversion.

By 1960, and as decolonisation gathered pace in Africa, the Soviets had established cordial relationships, based on commitments of economic and military support, with the West African governments of Ghana, Guinea and Mali, all of which had strained relations with their former colonial powers (Kanet, 1967; Andrew and Mitrokhin, 2005). Initially the USSR was not able to provide emerging anti-colonial movements with much more than verbal support (Allison, 1988) but by the 1950s and early 1960s the Soviets were focused both on expanding the community of socialist states that represented the growing global attraction of the Soviet model and on forging relations with "bourgeois" states such as India that, while important for Soviet economic or security interests, were not viewed as likely to establish Marxist–Leninist regimes (Kanet, 2006: 339). Africa became increasingly important in these efforts and in March 1958 the Secretariat of the Soviet Communist party enacted a series of measures to expand the Soviet reach on the continent, instructing various Soviet ministries to magnify radio and print propaganda efforts, to expand the number of scholarships for African students and to train a network of Soviet Africanists (Telepneva, 2017). The popular geopolitics of the Soviet campaign in Africa included countless propaganda posters (see Figures 4.5 and 4.6), intended to represent the Communist party's supposedly more enlightened perspective on race and other social issues (Nash, 2016). Cinema in particular was regarded as an important tool through which the Soviet Union would extend itself by developing "a *cinegeography* of socialist friendship" (Cummings, 2012, emphasis in original) as the Soviets saw in Africa a parallel with the establishment of their own cinema (Woll, 2004), arguing that the transition from tsarism to communism was akin to the change from colonialism to independence, which chimed well with emerging African filmmakers (Cummings, 2012).

According to Valkenier (1983), the initial motives for the Soviets' increased ties with developing countries were apolitical in that Soviet authorities were anticipating overproduction in capital goods and

FIGURE 4.5 Russian propaganda poster entitled "Long Live the World October" (1933) by Gustav Klutsis. Courtesy of Wayland Rudd collection.

FIGURE 4.6 Soviet propaganda poster entitled "For Solidarity of Women of the World" (1973) by V Rybakov. Courtesy of Wayland Rudd collection.

wanted to find markets in exchange for raw material supplies. Soviet policy towards the Third World from the mid-1950s was initially concerned with the provision of modest amounts of political, economic and military support to potential allies and clients, before shifting to extensive involvement in regional conflicts through the supply of military capabilities and the use of proxies (e.g. Cuba) and finally to direct Soviet military intervention (Kanet, 2006). The total amount of Soviet weapons exports to nonaligned countries between 1954 and 1964 exceeded US$2.7 billion with nearly 80% going to non-Communist nations (Porter, 1984: 19). Egypt, Syria and Iraq were all viewed by Moscow as progressive regimes that had broken with Western imperialism and were in the process of laying the foundations for a socialist socioeconomic orientation (Kanet, 2006). From 1954, providing military and economic assistance to Nasser's Egypt was seen as a way to undermine Western political dominance in the Middle East and South Asia and increase Soviet influence (Nation and Kauppi, 1984). Egypt became the focus of the USSR's first massive military efforts on behalf of a non-communist client at war since its efforts to back the Kuomintang in China in the 1920s (Porter, 1984), with total arms transfers to Egypt in the six years after the Six-Day War of June 1967 amounting to more than US$4.4 billion (Porter, 1984: 23). Under Khrushchev, US$1 billion worth of arms were also provided to Indonesia to back Sukarno between 1958 and 1965 in three successive conflicts (Porter, 1984: 19) and a further US$4 billion was provided to Ethiopia by 1984 in support of the Marxist–Leninist Mengistu regime, along with thousands of Soviet military advisors.

Direct participation of Soviet personnel in combat or combat support was initially avoided but this changed with the war of attrition in the Middle East between Egypt and Israel (1969–70) over territory Israel had taken in the Sinai Peninsula during the Six-Day War. Following billions in arms transfers to various parts of the Third World (reaching a value of roughly US$16.5 billion from 1973 to 1977 – Porter, 1984: 30), pro-Soviet revolutionary regimes soon came to power as a result of violent conflicts or coups in South Vietnam, Laos, Angola, Ethiopia, Afghanistan, South Yemen and Cambodia. Soviet military aid to North Vietnam increased massively after 1965, following the US decision to increase its ground forces and in the context of intensifying Sino-Soviet rivalry, with over US$3 billion worth of weaponry provided to North Vietnam between 1965 and 1972 (Porter, 1984). Whereas Soviet involvement in the late 1950s and early 1960s was based on limited but strategically important alliances with nationalist forces, some of the new relations forged between Moscow and the Third World from 1970 onwards were intended to be more comprehensive and pervasive. In particular, the fall of South Vietnam in 1975 marked the beginning of a highly successful strategic offensive by the USSR in the Third World (Korbonski and Fukuyama, 1987), buoyed not only by the Vietnam experience but by the reduced dangers of an American nuclear response – the Nixon administration had pursued Strategic Arms Limitation Talks (SALT) as part of a process of "linkage", whereby the Soviets were to be rewarded with nuclear arms control in return for behaving themselves in the Third World (Young, 2006).

Soviet aid to Third World countries foregrounded the provision of equipment for infrastructure and industrial development projects, such as the Aswan Dam in Egypt, the Bihar Steel Mill in India, and various smaller projects across the Middle East and Africa, as well as technical assistance and the education and training of local people to build the foundations of modern industrial and agricultural enterprises. Soviet aid almost always went for large and visible projects in the state sector that were expected to increase the productive capacities of the recipient country, to create an "independent" industrial base and reduce dependence on the capitalist West. In Indonesia, for example, this model of modernisation was evident in Soviet-financed industrial plants such as the steel mill at Cilegon on the coast of western Java which, according to Simpson (2008: 49), was promoted by Soviet officials "as a harbinger of rapid industrialisation". Such projects were supported by an army of Soviet economic advisors despatched to various parts of the Third World to facilitate them (Kanet, 2010). Both the Soviets and the Americans also had models of large-scale hydroelectric development that they were keen to share with countries of the Third World who often saw them as "highways" to modernisation. The Soviets in particular approached dam building in the Third World with real gusto. Engineers from both countries offered their expertise in "a sort of hydroelectric diplomacy geared toward winning the hearts and minds of the Third World" (Roe, 2015: 308). When the US withdrew funding for the Aswan High Dam in Egypt the Soviets stepped in, providing US$1.12 billion in 1956 at 2% interest for the construction of the dam (which was designed by the Soviet Hydroproject Institute) along with technicians and heavy machinery (Roe, 2015).

Although Soviet secrecy around its foreign aid donations means figures are uncertain, Soviet development assistance was just under US$2 billion in 1970, reached US$2.25 billion in 1980 and peaked at US$4.2 billion in 1988 (Chekhutov, Ushakova and Zevin, 1991). Expenses for development aid surpassed even the Soviet budgets for science and internal security by more than 20% during the 1980s (Bowles, 1992: 70). This assistance was usually offered in the form of credits (in contrast to the grants given by the US and Western European states), available at low interest rates, for the acquisition of equipment and technical assistance from the USSR. In the early stages of Soviet outreach in the Third World, the model of development that clients were expected to follow was that of the Soviet Union itself (and to a lesser extent the other Marxist–Leninist countries of Europe) with a focus on state control of the economy, heavy industrial projects, import substitution, reduction of ties with the capitalist West and closer integration with the socialist states (Valkenier, 1983). Many Soviet projects proved however to be unsuitable for the needs of the recipient countries and both the theory and practice of Soviet aid were typically focused on politics and political institutions, paying relatively little attention to the mechanics of development. Given the chronic dearth of spare parts throughout the Soviet economy, serious problems and costs emerged in those countries that depended on Soviet development projects whilst the quality of Soviet technical assistance was highly variable (Valkenier, 1983; Brun and Hersh, 1990).

Alongside aid from the USSR there were also the related flows of assistance from Eastern Europe to the Third World which reached around US$514–537 million annually for most of the 1980s (Rudner, 1996). Taken together, it is estimated that Soviet and Eastern European communist aid represented nearly 10% of total world overseas development assistance (ODA) disbursements during that period (Rudner, 1996). Czechoslovakia, for example, established important commercial relations with Third World countries and during the late 1940s and 1950s, when the Soviet Union was itself little interested, the Czechs managed to restore and extend their economic relations with the Third World, especially through sales of capital goods and equipment (Pechota, 1981). In many ways Czechoslovakia was the earliest and most ambitious Eastern European aid donor, its involvement in Africa pre-dating World War II, and it had significantly more autonomy from the USSR in conducting its foreign relations with Africa than has previously been assumed (Muehlenbeck, 2016). As a smaller Eastern bloc country Czechoslovakia was not regarded in the same light as its superpower ally and exploited the economic and cultural space this differentiation afforded (Muehlenbeck, 2016).

Czechoslovakia's status as the leading Eastern European donor was gradually eclipsed, however, by the emergence of the German Democratic Republic (GDR) whose foreign policy in general, and in Africa in particular, was a strategy seeking both closer ties with the Soviet Union and greater international recognition and legitimacy (Winrow, 2009; see Figure 4.7). Hungary, and to a lesser extent Poland, Bulgaria and Romania, also enhanced their own aid profiles in the Third World during the Cold War (Rudner, 1996). Many of these donors used project aid (often tied to donor sources of procurement) for heavy industry and capital infrastructure projects and typically involved themselves in the design, construction and equipping of the complete plant. Industrial joint ventures were also common

FIGURE 4.7 Angolan President Eduardo dos Santos (fifth from left) visits the Berlin Wall in front of the Brandenburg Gate in East Berlin on August 14th, 1981. Photo by Horst Sturm/Bundesarchiv.

along with licensing arrangements for the co-production (usually assembly) of Eastern European products by firms from recipient countries whilst, as with Soviet aid, agricultural investment typically favoured large-scale state-run projects, particularly irrigation schemes. Support was also provided for political institutions intended to enhance recipient-country capabilities in the administration of state enterprises (see Figure 4.8). Experiences of involvement in these aid programmes served to broaden the professional horizons of Eastern European technical specialists, most notably in agriculture, geology, industry, medicine and transportation, but also to augment the donor's knowledge base concerning recipient countries (Rudner, 1996).

While Americans feared a communist takeover of the continent, the relationships the USSR forged in Africa did not last long (Bervoets, 2011). Hasty and careless evaluation of potential socialist states, a limited understanding of African cultures and political economies, the heavy emphasis on military cooperation and inadequate aid, all prevented the Soviet Union from developing "anything more than, at best, friendly associations with countries that would eventually align themselves with the West" (Bervoets, 2011: 1). Many African countries cut off economic and diplomatic relations with the Soviet Union during the period from 1965 to 1974, seeing relations with Western Europe and the US as more advantageous (Hosmer and Wolfe, 1983: 27). Under

FIGURE 4.8 Skilled workers from Angola receive vocational training in communist East Germany on May 16th, 1983. The Angolan participants were taking part in a six-month course at the Central Institute for Industrial Safety in Dresden. Photo by Ulrich Häßler/Bundesarchiv.

Khrushchev the Soviets did not get the results they had hoped for in Africa as Soviet partners on the continent struggled with the process of transitioning to socialism. Failed Soviet attempts to incite revolution in Egypt, Ghana, Mali and Sudan alienated the leaders of those countries and created credibility issues for the USSR in Africa (Nation and Kauppi, 1984: 2) but also led the Soviet leadership to think more carefully about which countries were most suited to socialism. When an alarming number of African countries began pursuing a capitalist path of development in the mid-1960s, the Soviets under Leonid Brezhnev revisited their strategy for the region, concluding that economic aid was a limited policy tool and that Africa's transition to socialism would take longer than initially expected (Hosmer and Wolfe, 1983).

By the end of the 1970s, Angola had become the focus of the Soviets' efforts to spread socialism in Africa, seen as an opportunity to maintain the global strategic and diplomatic momentum gained with the communist victory in Vietnam. The Soviets initially acted in Angola to counter the growing influence of the Chinese, not the US, who had held ties with all three of the Angolan liberation movements.

Their support was focused on the MPLA in its fight for independence and in the years after (Ogunbadejo, 1981; Guan-Fu, 1983). Agostinho Neto visited Moscow in 1964 and 1973 whilst MPLA cadres travelled to Moscow for military training in December 1974 (Porter, 1984). After independence hundreds of Soviet advisors and technicians began arriving in the Angolan capital (Falk, 1988). More generally, the USSR's involvement in Angola helped foster a diplomatic climate in the Third World that was more conducive to Soviet political and military initiatives, proving that the fall of South Vietnam was not an anomaly and that the USSR would be the "natural ally" of the Third World. It also gave the Soviets direct access to SWAPO guerrillas in Namibia, increased Soviet influence in the Congo and Zambia (Guimarães, 1998) and enhanced Soviet prestige with another former Portuguese colony that turned to Marxism–Leninism after independence, Mozambique (Nolutshungu, 1985).

To receive Soviet aid, the Angolan government had to show significant progress in industrialising its economy, nationalising its industries, instituting land ownership reforms, developing readiness among its people to support a cultural revolution, and establishing a vanguard party in alliance with countries of similar political ideology (Nation and Kauppi, 1984: 32). Considering the political instability of Angola at the time, achieving these objectives proved difficult and consequently military assistance initially constituted the majority of Soviet aid to the country (Guan-Fu, 1983), amounting to an estimated US$4 billion in support between 1975 and 1985 along with the provision of 1500–1700 advisors (Central Intelligence Agency, 1985; see Figure 4.9) and reaching a total of nearly US$15 billion by 1988 (George, 2005: 281). Neither Soviet nor Angolan leaders however were deeply invested in developing a strong alliance with each other as pragmatism and national interest often came before ideology (Ogunbadejo, 1981). Although the MPLA leadership had a sense of obligation to the USSR after independence for its support during the anti-colonial war, relations with the Soviets were seen primarily as a means to consolidate the MPLA's hold on power, rather than transitioning to a socialist state (Ogunbadejo, 1981). As Soviet diplomats realised in the mid-1980s that their efforts to transform Angola into a socialist state had been futile, they began to de-escalate relations and commenced a wider withdrawal from the continent, conceding that the socioeconomic conditions had not been developed for the implementation of socialism and that class struggle was not the most important struggle for Africa, even advising African states to prudently cultivate relations with foreign capital and the world market (Brun and Hersh, 1990; Bervoets, 2011). From 1974 to 1979, Africa received 33% of Soviet

FIGURE 4.9 A Soviet military expert poses with Angolan officers in Lucusse, Angola in the spring of 1986. Source: Sergey Sheremet, http://www.vvkure.com/

credits (a total of US$ 2.7 billion) of which states of socialist orientation received only US$300 million (Brun and Hersh, 1990: 151) as the Soviets increasingly confined their offers of assistance to modest programmes in agriculture, irrigation, technical training and geological surveys.

Aside from Angola, Afghanistan also figured prominently in the list of Soviet foreign aid destinations. The USSR was the world's largest donor of economic and technical assistance to Afghanistan between 1954 and 1991, beginning around 270 major construction projects, 142 of which were completed (Robinson and Dixon, 2010), and providing approximately US$1.265 billion of credit between 1955 and 1979 (Porter, 1984: 610). Projects included dams, roads, electrical power stations and power lines, irrigation canals, factories, housing, grain elevators, bakeries, automotive repair plants, airports and educational institutions. A joint Soviet–Afghan committee decided on projects to be funded by the US$100 million loan granted in 1955 when Afghanistan became the first target of Khrushchev's "economic offensive", the Soviet Union's first venture in foreign aid, which financed a fleet of taxis and buses and paid for Soviet engineers to construct airports, a cement factory, a mechanised bakery, a five-lane highway from their own border to Kabul, and, of course, those archetypal temples of progress and modernity, dams (Cullather, 2002). At the peak of their influence in the 1980s, Soviet projects produced over half the country's power, three-quarters of its factory output and almost all the government's tax income (Robinson and Dixon, 2013). The Soviets also made a significant effort to improve Afghanistan's human capital, providing on-site training to up to 42,000 Afghans by 1969 (mostly in basic skills such as the use of industrial equipment), and gave thousands of tons of humanitarian assistance in the form of food and medical aid.

A closer look at the impact of Soviet development aid on Afghanistan reveals that expertise and aid were often transferred from one periphery (the internal Soviet/Russian one) to the other (Third World contexts like Afghanistan) (Nunan, 2016) as Soviet aid efforts focused on building up the state sector of the national economy (especially in industry and agriculture). Soviet experts set about building dams and state farms, in particular around Jalalabad in eastern Afghanistan, yet much of the expertise deployed to these locations emerged from internal peripheries within the Soviet Union itself as in the case of Soviet Azerbaijani botanists trained in developing olive cultures for the Caucasus who were sent to the state farms in Afghanistan. Nunan (2016) frames this as a story of "South–South" transfers mediated from one (Soviet) periphery to its Afghan cousin. Another example is the advisors dispatched by the Soviet Communist Youth League (Komsomol) to Afghanistan following the Soviet occupation. Komsomol advisors focused on the construction of a mass socialist party in Afghanistan but also worked closely with Afghan orphanages to, in effect, traffic Afghan youngsters to Soviet orphanages and educational institutions as a means to inculcate socialist values, as many of the Afghan teenage orphans sent abroad ended up in *internats* (boarding school-like institutions) in the Soviet periphery (Nunan, 2016). Through Afro-Asian solidarity groups like the Moscow branch of the SKSSAA the Soviets also lobbied the Third World by focusing on its own peripheries, the Soviet republics, their ethnic status and their experiences since the October revolution, offering them as models for wider Third World development (Amos, 2012).

CHINA IN AFRICA: ADVANCING A "SUBALTERN GLOBALISM"?

From the very beginning the Chinese Communist Party was explicitly internationalist in premise and in promise (Kirby, 2006). The

Chinese increasingly came to believe that Africa in particular seemed to be "re-enacting" their own recent past and that it was China's duty to "explain" such historical parallels to Africans (Power, Mohan and Tan-Mullins, 2012). In the 1950s, for example, when Cameroonians were fighting French colonialism, Mao is said to have presented one of their leaders with a copy of his work *Problems of Strategy in the Guerrilla War against Japan*, inscribed with the greeting: "In this book you can read everything which is now going to happen in the Cameroons" (Snow, 1988: 82). Africa dominated in the list of destinations for Chinese foreign assistance during the Cold War – between 1956 and 1973, out of the total US $3.38 billion in aid granted by China, almost half (US$1.73 billion) went to African countries (Brautigam, 2009). At the height of the Cultural Revolution at the end of the 1960s, the Chinese media regularly portrayed African national liberation movements as using Maoist thought as their primary ideological tool for liberation and revolution. This was a vital part of the strategic doctrine of "People's War" which held that the advanced industrial nations constituted the "cities" of the world, and the poor nations of Asia, Africa and Latin America were the "countryside" (Jackson, 1995). By fomenting revolution in the various "rural" areas of the world, eventually the liberation movements would surround and overrun the urban areas, just as they had in China during its civil war.

Guided by Mao's three worlds theory as it sought to spread the gospels of nationalism and independence and in the good works delivered by its aid projects, China would be the "champion" (Guimarães, 1998: 154) of a Third World alliance that would counterbalance the Cold War superpowers and wealthy advanced economies of the First World. China's aid was strictly bilateral and was usually given as a grant, or as an interest-free loan, in contrast to the Soviet model where interest was charged at 2.5% (Snow, 1988). Chinese aid workers were also unique in their approach, not being permitted to "loll in hotel suites and run up expenses as other expatriates did [having to] content themselves with the same standard of living as the ordinary Africans they worked with" (Snow, 1988: 146). Official discourses on China's foreign assistance would often accentuate how distant the West's historical experiences in achieving "development" were from its own and those of the Third World more generally and emphasised the struggle between the developed and underdeveloped worlds. China has also regularly underlined its historical credentials by drawing attention to its long-standing connections with the Third World (in the diaspora as well as through trade and diplomacy) and to its own experiences of being colonised and an aid recipient. Like the Soviets, China made use of propaganda to emphasise its anti-imperialism and anti-racism as well as its closer proximity and affinity to Third World peoples and issues (see Figure 4.10). Maoism seemingly offered the Third World a vision of "subaltern globalism", spreading a message of revolt against the white Europeans and Americans who had exploited them (Friedman, 2015).

A key influence on Chinese foreign policy towards the Third World was its relationship with the Soviet Union. The PRC was "born pro-Soviet" and the Soviet Union had helped create the CCP (Kirby, 2006). By 1949 the shadow of a "Soviet model" of state-led industrialisation and foreign trade was already evident in China whilst Mao initially had a strong desire to transplant Soviet experience to China, taking the country along a clearly Stalinist path. Politically, economically and militarily China became a client of the Soviet Union, guided by Soviet advisors and sustained by Soviet aid (Snow, 1988). By the mid-1950s however, China was increasingly uncomfortable with being confined to the Soviet camp, seeing a chance to establish a separate identity in the world of newly independent Asian and African

FIGURE 4.10 Chinese propaganda poster (1963) entitled "Awakened peoples, you will certainly attain the ultimate victory!" Source: Stefan R Landsberger collection, International Institute of Social History, Amsterdam.

states (Snow, 1988). These ambitions were buoyed by the Bandung conference, by the idea of "Afro-Asia" as a viable political concept that it gave rise to and by the creation of solidarity organisations like AAPSO. Although China focused on working through Third Worldist fora concerned with Afro-Asian solidarity (such as AAPSO and Bandung), Chinese wishes were often stubbornly and effectively resisted within these organisations and by no means did China fully control them (Neuhauser, 1968; Larkin, 1971).

In 1960 the Soviets withdrew all economic aid and technical expertise from China, leading to the Sino-Soviet split. The bitter ideological dispute that ensued between the two led to an intensified competition for dominance over various organisations of Afro-Asian solidarity and the non-aligned countries (Ismael, 1971) as China sought to mobilise the Third World to seize the revolutionary mantle from the Soviet Union. When Chinese premier Zhou Enlai began a tour of African countries in 1963, considerable alarm was raised in the West (Large, 2008). The Sino-Soviet split is best understood not in terms of personal clashes between Mao and Khrushchev, or even "objective" differences of geopolitical interest, but rather as a fundamental difference in conceptions of revolution (Friedman, 2015). For the Soviets, revolution was fundamentally about rolling back exploitative capitalism whereas for China it was imperialism that became the core enemy in the Maoist diagnosis of global injustice (Friedman, 2015). In seeking to add a touch of Third Worldist solidarity, President Nkrumah of Ghana even offered his services as a mediator between China and the USSR, whilst the Angolan MPLA brought Chinese and Russian officials together at a conference in Tanzania to promote dialogue (Snow, 1994).

Between 1949 and 1976, during the most ideological phase of China's revolution, assistance to Africa was scripted as a "heroic endeavour", with the continent becoming the "object of a philanthropic crusade" (Snow, 1988: 146) as China sought to discharge its "missionary duty of setting Africa free" (153). China's relations with its "partners" and "poor friends" in the Third World were often "either thin or troubled through much of the Maoist period" (Harding, 1994: 394) as it refused to join key organisations like the G77 or the Non-Aligned Movement. Chinese aid went to various sectors such as industry, transport, agriculture, water control and irrigation, public health, power and communications, sports and cultural complexes. Cotton textile mills were particularly popular. Like the USSR, China funded state-owned factories in Africa, where skilled technicians from Shanghai's pre-war industries trained Africans to manufacture substitutes for exports (Brautigam, 2009). Chinese teams constructed bridges, roads, power plants and ports but China also seemed happy to work on projects that were effectively inessential monuments to the glory of the African regimes they worked with such as conference halls, sports stadia and party headquarters and on projects that "seemed calculated less to promote the development of a country's economy than to win for Peking the favour of the [recipient] regime" (Snow, 1988: 156). Teams of doctors and medics were dispatched to the Third World and there were educational and cultural exchange programmes whilst each Chinese province also had a foreign aid bureau and each was twinned with a province in Africa to help the aid transfer process. China also specialised in training Africans in the arts of guerrilla warfare, initially at the Military academy in the city of Nanjing but also in Africa itself at a series of remote training camps in countries like Tanzania and Ghana (Snow, 1988).

For its first official aid project in sub-Saharan Africa (a cigarette and match factory just outside Conakry completed in 1964) China chose Guinea, which had been the first sub-Saharan African country to recognise China in 1959. Chinese aid was announced shortly after President Sékou Touré had expelled French diplomats and French aid had come to an end. Soviet aid personnel left Guinea shortly after, with Chinese officials declaring that Soviet assistance had been "imperialist aid" (Copper, 2016: 4). By the start of the 1970s Chinese teams were building close to 100 different turnkey aid projects around the world and had committed aid to seven countries in Asia, three in Latin America, six in the Middle East and 29 in Africa by 1973 (Brautigam, 2009: 41). Between 1967 and 1976 China's aid reached an average of 5% of government expenditure and by the end of the 1970s the level of aid and the number of costly projects initiated during the Cultural Revolution "had far outpaced China's capacity" (Brautigam, 2009: 51). Zhou Enlai and Mao Tse-Tung had embraced 19 enormous "100 million RMB" projects (each worth about US$50 million in the 1970s) during their terms in power. Chinese teams had completed some 470 projects in the Third World between 1970 and 1977 and as the Cultural Revolution wound down, China's foreign policy began to lose its strong ideological inflection, although it did continue to make active efforts to export its domestic experience to foreign clients (Harding, 1994). China's own "four modernizations", announced by Zhou Enlai in the mid-1970s, required enormous resources, leaving little extra for overseas aid. In March 1978 China announced an ambitious ten-year plan that focused on 120 key modernisation projects domestically, including 30 electric power stations, seven trunk railroads, eight coal mines, ten new steel plants, five harbours and ten new oil and gas fields (Brautigam, 2009: 45).

This experience was to prove highly influential in the infrastructurally oriented foreign aid programmes China has since begun to roll out in contemporary Africa (see chapter 7).

China often supported schemes which the West had rejected on narrowly economic grounds or which were important to African states for political or psychological reasons and made a point of "doing something" for districts which the Europeans had been content to leave as "backwaters" (Snow, 1988). The principles for aid and cooperation reflected China's own experience as an aid recipient itself over the preceding 60 years where the Chinese had not appreciated their "client" status (Snow, 1988). Chinese aid would serve not as "a kind of unilateral alms but [rather] something mutual" (MFA, 2000; see Figure 4.11). Projects would use high-quality materials, have quick results and boost self-reliance whilst Chinese experts would transfer their expertise "fully" and live at the standard of local counterparts (Power, Mohan and Tan-Mullins, 2012). Aid also became an important geopolitical tool for China in the contest with Taiwan (itself an aid donor) where it was used to restrict the

FIGURE 4.11 Chinese propaganda poster (1972) entitled "The feelings of friendship between the peoples of China and Africa are deep". Source: Stefan R Landsberger collection, International Institute of Social History, Amsterdam.

international recognition of Taiwan's sovereignty. China saw its aid as a way to expose the limits of its opponents, both Soviet and Western, and was often reluctant to coordinate efforts with other foreign powers, having a tendency to "go it alone" (sometimes resulting in active hostility to other aid personnel) or to "launch spectacular rescue operations" (Snow, 1994: 288). China came forward, for example, to build the massive Tanzania–Zambia Railway Authority (TAZARA) project and funded it with a US$450m loan after Western donors, including the World Bank and the UN, rejected initial approaches to back it. TAZARA was built between 1967 and 1975 at a cost of over US$600 million (more than the Aswan Dam had cost the USSR) with the labour of 50,000 Tanzanian and 25,000 Chinese workers (Monson, 2009). The news was greeted with alarm in the West as the *Wall Street Journal* warned that "the prospect of hundreds and perhaps thousands of Red Guards descending upon an already troubled Africa is a chilling one for the West". One US Congressman even described it as a "great steel arm of China thrusting its way into the African interior" (United States Congress, 1973: 286). The construction, use and operations of the TAZARA line however mirrored those of colonial administrations, despite TAZARA being framed as an anti-colonial project (Monson, 2009).

Following readmission to the UN in autumn 1971 (which had much to do with the diplomatic relations it had forged in the Third World), China supported the emergent "subaltern nationalism" of the New International Economic Order (NIEO) and continued to present itself as the natural leader of the Third World. By 1978 some 74 countries were receiving aid from China, the largest group of which were in Africa and by then China had aid programmes in more African countries than the US. By the early 1980s nearly 150,000 Chinese aid technicians had been sent to Africa "in a call to Chinese national glory and sacrifice" (Power, Mohan and Tan-Mullins, 2012: 47). In 50 years of cooperation China built some 900 infrastructure and public benefit projects in Africa alone, many of them small or medium sized (Brautigam, 2009). Throughout the 1980s and well into the 1990s, however, China focused the bulk of its aid on rehabilitating the dozens of former aid projects that had collapsed or were barely limping along and began developing ways to make their initial benefits more sustainable. For every new project launched during this period, three were being consolidated (repaired, renovated or restored) (Brautigam, 2009). Further, many Chinese projects struggled with issues around skills and technology transfer and long-term sustainability and there were cases where China failed to cooperate sufficiently closely with African countries about the detail of the planning processes involved (Brautigam, 1998). PRC leaders also failed to grasp the significance of regional antagonisms and cultural and historical differences between the various countries while trying to apply a general model of revolution to all African liberation movements (Neuhauser, 1968). In Angola, Mozambique and South Africa, for example, China "backed the wrong horse in all three cases" (Cheng and Shi, 2009: 89). Similarly, Snow (1994) argues that the Chinese were not especially interested in domestic developments in African countries, let alone in actively propagating communism there. Issues of race, ethnicity and assimilation together with the regional role of South Africa were key in Angola, for example, but China failed to adequately grasp these as "no analogous controversies had existed in their own revolution" (Jackson, 1995: 392).

US FOREIGN AID AND THE COUNTERING OF INSURGENCY

> Fundamentally, A.I.D.'s purpose is national security. By national security, we also mean a world of independent nations capable of making economic and social progress through free institutions. Economically, we're not aiming for standards of living; we're aiming for internal dynamics, self-sustaining growth. (David Bell, USAID administrator, 1962–66, quoted in Norris, 2014)

In order to administer and implement the "bold new programme" for the "underdeveloped world" that Point Four had announced, in the early 1950s President Truman sought to consolidate the management of US aid programmes which had been scattered across several agencies. This included the Economic Cooperation Administration (ECA), set up in 1948 to administer the Marshall Plan but increasingly involved in an expanding number of programmes in the Far East, the Technical Cooperation Administration (TCA) established within the Department of State in 1950 and the International Development Advisory Board established in the same year (initially chaired by Nelson Rockerfeller) to consult with the Secretary of State. Military assistance and related economic aid were administered by the Department of State and the Department of Defense (DoD) with the former trying to coordinate the efforts of the different agencies to ensure overall coherence.

Under the auspices of the TCA, American experts and advisors were dispatched to various countries around the globe where they implemented, in collaboration with NGOs and local counterparts, a diversity of technical projects (Berger and Borer, 2007). By the end of 1952 TCA development projects were underway in 35 countries and had a budget of US$153 million (Butterfield, 2004: 37). Most of the projects were in Central and South America or in Asia with only Libya, Liberia, Egypt and Ethiopia represented in Africa, much of which was still under colonial rule. The formation of the Mutual Security Agency (MSA), set up under the Mutual Security Act of 1951, sought to bring America's military assistance programmes together with the economic aid being provided to "underdeveloped areas" under the umbrella of Point Four (Berger and Borer, 2007), making explicit the link between the two (Beall, Goodfellow and Putzel, 2006). The MSA administered economic and technical aid programmes in countries that were also receiving US military aid whilst the TCA administered aid to countries receiving only technical and economic assistance. The creation of the MSA heralded a major programme of American foreign aid between 1951 and 1961, to numerous countries, aimed initially at shoring up Western Europe as the Cold War developed but, as the 1950s wore on, intended to secure a growing number of East Asian "partners" thought to be on the frontlines of the Cold War (Setzekorn, 2017).

Some 95% of all US aid in 1954 was military and between 1949 and 1960 the US provided nearly US$24 billion worth of military aid to more than 40 nations around the world (Birtle, 2006). By 1956, the US was also supervising the training of nearly 200 military divisions worldwide whilst 20% of all US Army officers had served as military advisers to foreign forces (Kuzmarov, 2017). Initially, much of this effort focused on creating conventional armed forces to resist Soviet or other external aggression but "internal" subversion, either indigenously generated or assisted by external communist forces, soon became "equally as menacing" (Birtle, 2006: 23). US soldiers first began to contemplate measures to combat socialist revolutionaries as early as the 1880s and classical counter-insurgency techniques were subsequently tried out by the US in the Chinese Civil War (1945–49), during insurgencies in Greece (1945–49) and the Philippines (1945–55), the Indochina war (1945–54) and the Korean civil war (1945–54), in Vietnam (1965–73), Thailand (1962) and the Dominican Republic (1965). Counter-insurgency tactics originally developed to suppress earlier anti-colonial insurrections were also seen as readily adaptable (Pimlott, 1985), particularly the British and French experiences. During the decade and a half following World War II, Britain had become enmeshed in a succession of counter-insurgency campaigns involving different environments (urban and rural as well as jungle and mountain) against a variety of opponents (anti-colonialists, communist revolutionaries and ethno-nationalist separatists) in places as diverse as Palestine, Malaya, Kenya and Cyprus (see Figure 4.12). In Malaya, the British first coined the term "hearts and minds" to describe the process of winning the loyalty of the people through developmental initiatives (Thompson, 1966). French experiences with countering insurgency in Algeria and Indochina also provided important reference points, as did their model of the "agroville", a forerunner of the strategic hamlets later used by the US in the Vietnam war (see Figure 4.13).

In addition to deploying 85 counter-guerrilla mobile training teams to 14 countries between 1955 and 1960 to help foreign governments plan and organise counter-insurgency efforts, the Eisenhower administration set up the top-secret "1290-d" project to train foreign police in counter-subversion and established a

FIGURE 4.12 Troops of A Company, 3rd (Kenya) Battalion, King's African Rifles, search an abandoned hut in Malaya for signs of terrorists during the Malayan Emergency (1948–60). Ministry of Information Second World War official collection, Courtesy of Imperial War Museum.

FIGURE 4.13 A mocked-up Vietnamese hamlet at the US Army Training Camp at Fort Gordon, Georgia. Viet Cong suspects are rounded up for questioning during a simulated search and destroy operation in 1966. Granger Historical Picture Archive/Alamy.

police aid programme, the Civil Police Administration (CPA), in 1954, guided by the belief that police units were the best mechanism (rather than employing direct military force) for defeating local insurgencies and addressing problems of internal security. The CPA began operating in Guatemala, for example, after the 1954 CIA-backed coup that removed the democratically elected government of Jacobo Arbenz. Eisenhower's policies for Latin America, codified in 1953, sought to ensure Latin American support for the US at the UN, to work to protect the hemisphere from communist invasion, to continue to have the region produce raw materials and to eliminate the "menace of internal communist or other anti-US subversion" (Taffet, 2007: 14) although they did not underwrite a major economic aid programme to promote Latin America's economic development and were primarily focused on inter-American military cooperation (Rabe, 1988, 1999).

Between 1952 and 1968 the US spent US$687 million in military assistance to Latin America in addition to arms sales valued at US$187 million (Kuzmarov, 2017). Thousands of Latin American military officers (including several future dictators such as Manuel Noriega) were also trained at the US Army School of the Americas (SOA) founded in 1946 in Panama before relocation in 1984 to Fort Benning in Georgia, with a curriculum centred on counter-insurgency. This approach began to be questioned, however, following Khrushchev's "economic offensive" in 1955 which appeared to involve a more sophisticated Cold War strategic use of development assistance. In August 1958, the Senate Foreign Relations Committee declared US military and development assistance to be inadequate given world realities and argued that the US needed to stop the Soviets from supplanting it as the primary aid donor throughout the world (Taffet, 2007). Through the CIS, Walt Rostow, Max Millikan and others increasingly underlined the importance of economic aid programmes to the pursuit of US foreign policy objectives (Millikan and Rostow, 1957), with Rostow's *Stages of Economic Growth* (1960) arguing that a well-funded economic aid programme could provide a sharp stimulus and quickly usher Latin American countries into the take-off phase.

The MSA was replaced first by the International Cooperation Administration (ICA) created in 1955 and eventually by USAID in 1961, created by the Foreign Assistance Act which had called for a separation of military and non-military foreign aid programmes and emphasised wide-ranging economic and social assistance efforts in the "underdeveloped world". USAID would bring together the work of various agencies and programmes including the Marshall Plan, the MSA and ICA, the Foreign Operations Administration (1953), the Food for Peace programme (1954) and the Development Loan Fund (1957). JFK had outlined how "collapse" in developing countries would be "disastrous to our national security" and "harmful to our comparative prosperity" (quoted in McMahon, 2006) and USAID explicitly intended to use aid to fight the Cold War and pursue its emerging modernisation agenda as another weapon in America's strategic arsenal. In the words of David Bell (USAID administrator from 1962 to 1966), fundamentally the purpose of aid "is national security" (cited in Norris, 2014), meaning both the national security of the US but also the international security of recipient states. The 1959 Mutual Security Act specifically encouraged military involvement in nation building and in May 1960 the Eisenhower administration gave the Army limited authority to promote "civic action" programmes overseas. In the same year the National Security Council directed the DoD to develop a new counter-insurgency doctrine and the Departments of State and Defence reaffirmed this decision by informing all US embassies, unified commands and military assistance groups that US policy "was to encourage foreign military and paramilitary organizations to promote economic development" (Birtle, 2006: 161). Accordingly, the global geopolitical imaginations of the American national security state played a central role in "identifying regions and resources suitable for programmes of technological intervention" (Sneddon and Fox, 2011: 452), creating new geographies of development in the process.

For JFK, only the diffusion of technology and know-how could help bring the Third World across the "dangerous" period of uncertainty in which communism threatened. Further, the receptivity of Third World countries to US technology would imply an acceptance of the Americans' leading role in the global drive towards modernity. In this sense, as Westad (2006: 31) notes, US aid sought in part "to make the world safe for capitalism" with assistance linked to recipients' acceptance of market access and export of profits, administrative restructuring and the exclusion of communists and left-wing socialists from government. US interventions were thus aimed at creating states that could be both successful in their own capitalist development *and* a part of American containment policies against the Soviet Union and its allies. When JFK took office in 1961 the threat of armed revolution became the main concern and the basis for military aid shifted from "hemispheric defence" to internal security and protection against Castro-communist guerrilla warfare (Kuzmarov, 2017). At Rostow's suggestion, JFK's administration set up a counter-guerrilla operations and tactics course at Fort Bragg in 1961 (Birtle, 2006) and according to Roger Hilsman, who was responsible for counter-insurgency policy at the State Department during the Kennedy administration, from the time that JFK took office in January 1961, he was preoccupied not just with counter-insurgency but with its "critical political element", or with the new political tactics required to counter communist revolution (Hoffman, 2006; Nagl, 2002).

JFK's administration moved to build on the work of the 1290-d programme and established a successor agency to the CPA within USAID, known as the Office of Public Safety (OPS), based on a template that originated in occupied Japan (Roy, Schrader and Crane, 2015; Kuzmarov, 2017). The OPS mission was clearly ideological; its director was from the CIA and CIA personnel worked within many of its aid programmes, its purpose was the development of covert intelligence networks and it taught subjects such as communist tactics and ideology. It also provided equipment that was subsequently used for torture and abuse, mainly in Central America (Hills, 2006). In 1962, a Special Group on Counterinsurgency was established within the OPS at USAID with the intention that it would "develop the civilian police component of internal security forces in underdeveloped states" and "identify early the symptoms of an incipient subversive situation" (quoted in Kuzmarov, 2009: 201). The OPS provided aid to police agencies in around 50 countries during its 12-year existence, spending US$300 million on training, weaponry, telecoms and other equipment (Spillane and Wolcott, 2012). The goal of these US "public safety" programmes was to unify a country's police and military under a central command, overseen by OPS advisors.

Within a few years, officers were operating out of US embassies, police headquarters and safe houses in 15 Latin American countries (Kuzmarov, 2017) and the OPS remit was soon extended beyond the hemisphere into Asia and Africa. Across the Third World hundreds of retired and active US police officers travelled to undertake the training of tens of thousands of police officers in administration, riot and traffic control, interrogation, surveillance and intelligence whilst thousands of police officers from various Third World countries travelled to Washington to receive training (Nadelmann, 1993). In practice, US development aid, military aid and counter-insurgency were often indistinguishable (Simpson, 2008) as in the case of Indonesia in the 1960s where USAID, the State Department, the CIA, US-based capitalists and a host of other official and private actors tacitly and even overtly supported the violent purge of communists and ousting of Sukarno in favour of General Suharto's New Order military dictatorship. Aid directed to the training and support of domestic police forces in Indonesia as well as "civic action programs" supporting military-led construction of rural infrastructure "fudged the line between COIN [counter-insurgency] and civic action" (Simpson, 2008: 80), with USAID and the CIA working closely together.

Six mobile teams of area-oriented, linguistically trained experts called Special Action Forces (SAFs) were also created (one each for Europe, Latin America, Asia, Africa, and the Middle East plus a reserve) in 1962 to supplement more conventionally oriented military assistance advisory groups in helping foreign armies perform unconventional warfare, counter-insurgency, civic action and nation-building activities (Birtle, 2006: 199). Each consisted of a Special Forces military group augmented by engineering, civil affairs, psychological warfare, military police, medical and intelligence detachments. US military forces also widely used psychological operations (psyops) during situations short of war to win popular support for intervention forces, "to counter hostile propaganda and to convince the indigenous population that America's actions were legal, that its intentions were benevolent, and that its presence would be temporary" (Birtle, 2006: 196). By 1965, 13 Latin American countries had US-funded civic action projects as across the length and breadth of the continent the US army cleared jungles, installed sewage systems, repaired roads and built schools, whilst army doctors provided free medical care and paratroopers assisted civilian aid agencies in distributing food and clothing (Birtle, 2006). Army psyops specialists also supported these activities through millions of printed propaganda items and thousands of hours of loudspeaker and radio broadcasts.

From the beginning, USAID and counter-insurgency were inextricably linked (Essex, 2013). In Latin America throughout the 1960s and 1970s, the agency's efforts were directed towards social and economic development to "combat communism and 'Castroism' across the region" (Essex, 2013: 46). In South Vietnam, the agency played a pivotal role in American pacification efforts, including addressing the "conditions that facilitated the successes of the Viet Cong" (USAID, 2007: 5, quoted in Essex, 2013: 48). Under pressure to develop a civilian economic and social component to the counter-insurgency effort in Vietnam, USAID spearheaded the "other" war for rural "hearts and minds" through two distinct, yet related suites of spatial interventions (Phillips, 2008). In order to pacify insurgent South Vietnam, USAID produced spaces that were designed to physically and psychologically separate rural populations from National Liberation Front (NLF) fighters. This archipelago of counter-insurgency spaces was secured through a policing effort that "sought to assay and manage the diverse flows of bodies, commodities, and capital constantly traversing the countryside of South Vietnam" (Attewell, 2015: 2259). To this end, under USAID, Vietnam became the site of an OPS programme that sought to turn the National Police "into a modern professional law enforcement organization capable of maintaining law and order and of constituting an effective first line of defence against subversion, insurgency and guerrilla activities" (OPS, 1966: 16). USAID advisors provided the South Vietnamese police forces with training and technical assistance in areas such as population control, prison management and the neutralization of insurgents and civilian populations (Attewell, 2017).

The relocation of rural populations in "supervised" centres with schools, clinics and other facilities (ostensibly to promote their progress and "development" but also to restrict communist influence) has long been a strategy used in counter-insurgency operations. It was implemented by the US and the government of South Vietnam to combat insurgency during the Vietnam war, beginning with the Rural Community Development Programme introduced by President Diem in Vietnam in 1959, which modelled itself after the success of a similar programme used by the British in Malaya (1940–60) and the "agrovilles" used by the French in the First Indochina war (1946–54) in a policy called "pacification by prosperity". The plan was closely shaped by Vietnamese actors and events and was not simply the product of US interventions (Stewart, 2017). It first emerged in the mid-1950s when Diem's advisers proposed a civil service project in the villages and hamlets aimed at pacification which was formalised as Civic Action in 1955. This became the foundation for the establishment of model villages in which cadres recruited and dispatched from Saigon would assist local development projects (Stewart, 2017). Unwilling to compromise national sovereignty by inviting greater US military involvement to counter the communists, Diem first doubled-down on the model village concept by forming "agrovilles", but the formation of the communist NLF in 1960 raised new challenges which Diem's government attempted to meet by transforming the Civic Action plan yet again, this time into the Strategic Hamlet Programme. Strategic hamlets required a greater US military presence which further alienated peasants, nudging them closer to the NLF and channelling "the forces of modernization in Vietnam in a direction more conducive to wider American Cold War interests" (Stewart, 2017: 218).

Pacification included local security efforts and a process of restating governmental influence and control in areas beset by insurgents but also efforts to distribute food and medical supplies along with wider reforms like land redistribution. More than just providing security, their goal was to facilitate "improvement" and reform. This represented a tactical shift away from merely securing territory to securing *and winning* populations (cf. Ansorge, 2010). By July 1963 some 8.5 million people had been settled in 7,205 hamlets according to figures given by the Vietnam press (Osborne, 1965: 25). The various efforts of US agencies around security and development in Vietnam were brought together in 1967 in a further pacification programme called Civil Operations and Revolutionary [later Rural] Development Support (CORDS) (cf. Belcher, 2012). CORDS represented President Lyndon B Johnson's attempt to define the Vietnam conflict as a progressive expression of the Cold War through modernisation theory (Fisher, 2006) where pacification had become the handmaiden of progress and he filled his speeches with the vision of creating security through development (Fisher, 2006: 31). CORDS was hailed by many as a successful integration of civilian and military efforts to combat the insurgency, despite being heavily contested and offering only the illusion of progress (Fisher, 2006: 31).

Beyond Vietnam, in the 1960s aid programmes in Latin America "were a top, if not *the* top, US concern" (Taffet, 2007: 2). For JFK, Latin America had become the "most dangerous area in the world" (Rabe, 1999: 91) with the Cuban Revolution in 1959 triggering a sharp increase in US aid to the region (Thérien, 2002). The main vehicle for this was the Alliance for Progress (AfP), a ten-year US $20 billion foreign aid programme that drew heavily on modernisation theory and aimed at promoting economic growth and political reform as a means of containing communism. Funding would allow port facilities, hospitals, roads, housing, power plants and schools to be built and in return Latin American governments would commit to instituting tax reform, promoting land redistribution and extending democracy and political freedom. The AfP was not solely an economic programme, however, it was also "a political program designed to create certain types of political outcomes" (Taffet, 2007: 10). In addition to mobilising a response to Castro's Cuba, the AfP was also about building alliances and spreading the positive vision at the heart of US democracy, fuelled by the belief that ideas inherent in the foundation of the nation could and should be exported and that money and technical expertise could solve Latin America's problems.

To create enthusiasm in Latin America, the United States Information Agency (USIA) tried to promote Kennedy's grand vision for the AfP and narrate the progress generated using several media including radio, TV, film, pamphlets and travelling speakers as well as planting news stories to explain how the programme would help ordinary people (Taffet, 2007: 43). Some AfP comics ran stories about how the programme helped communities, focusing on workers such as agronomists trained with US funds that were helping people produce more food. Others were designed to scare Latin Americans about the Cuban revolution. The USIA didn't only focus on the AfP in their efforts in Latin America, producing a stream of anti-Soviet propaganda that circulated across the Third World as well as material on US culture and society (Cull, 2008). The AfP struggled to meet its objectives, however, in part due to "an inherent conflict between lofty humanitarian goals and a desire to fight the Cold War" (Taffet, 2007: 5). AfP funding was used in countries like Chile and Brazil not to advance progress around development but to stop the spread of communist political parties. There was little connection then between aid distribution and levels of poverty in recipient countries, with almost 60% of all funding going to just four countries: Chile, Brazil, the Dominican Republic and Colombia (Taffet, 2007: 7).

Beyond Latin America, the US also made southern Afghanistan a showcase of nation-building, providing approximately US $600 million between 1955 and 1979 (Williams et al., 1988; see Figure 4.14) and making the country one of the largest recipients of US Cold War foreign aid (Attewell, 2017). A significant

FIGURE 4.14 Afghan Ambassador Mohammed Kabir Ludin signs an agreement in 1954 providing US$18.5 million of credit with Glen E Edgerton, President of the US Export-Import Bank. The money was used primarily for investment in irrigation and power projects in Helmand Valley. Courtesy of the US National Archives Still Picture Unit. Photograph by Joseph O'Donnell.

proportion of this went to the flagship Helmand Valley Project (HVP), a dazzling top-down integrated rural development programme intended to "reclaim" and modernise a swathe of territory comprising roughly half the country (Cullather, 2002; see Figure 4.15). In part, this was inspired by domestic infrastructure projects like the Salt River Project in Arizona (Caudill, 1969: 1) and the belief that irrigation canals would enable farms to produce food surpluses that could be exported for profit. It was also a direct response to the Soviets' development assistance offensive in 1955 when over US$100 million in credits paid for Soviet engineers to construct airports, roads, factories, dams and irrigation canals. New schools, modern hospitals and recreation centres would rise from the sand along with factories fed by electricity from a generator at a dam upriver, funded by a US Export-Import Bank loan of US$80 million (Cullather, 2002). There would also be model towns built from scratch like Lashkar Gah (known locally as "little America") that would serve as an example of a modern community, one that could be replicated, whilst a 45-minute colour motion picture produced by the USIA

FIGURE 4.15 Afghans view photographs of Helmand Valley Authority (HVA) initiatives at the 1956 Jeshyn Fair. The James Cudney Collection: Photograph by James A Cudney.

featured the HVP prominently in globally narrating the benefits of the economic development enabled by USAID (Cullather, 2002).

By the 1960s, Afghanistan had Soviet, Chinese and West German dam projects underway (Nunan, 2016) and was receiving one of the highest levels of development aid per capita of any nation in the world. *The Atlantic* magazine called it a "show window for competitive coexistence" (*The Atlantic*, 1962) whilst *U.S. News* described it as a "strange kind of cold war", fought with money and technicians, instead of spies and bombs. Publicly, US officials said this was "the kind of Cold War they wanted, just a chance to show what the different systems could do in a neutral contest" (Cullather, 2002: 530). Afghanistan had "become a new kind of buffer, a neutral arena for a tournament of modernization" (Cullather, 2002: 530). The country was also a key destination for Peace Corps volunteers, with 1,652 arriving in the period 1962–79 (Meisler, 2011) before the post was closed (along with other US aid efforts) following the Soviet invasion in 1979. Much of this grand development venture failed, however – the valley never became Afghanistan's breadbasket (although it did become the world's largest grower of opium-producing poppies). In Soviet-occupied Afghanistan, US policy focused instead on the provision of extensive weaponry to anti-Soviet insurgents with USAID providing operational support to the *mujahideen* resistance through the construction of a transnational logistics network designed to channel development and humanitarian assistance to rural populations. By 1987 the US was aiding the *mujahideen* to the tune of US$700 million per year in military assistance whilst US covert aid to the Afghan resistance virtually doubled each year from 1983 to 1987 (Cordovez and Harrison, 1995: 157) and during the 1980s cost American taxpayers over US$3 billion (Hartman, 2002). By the mid-1980s under President Reagan the US was ready to work with anyone who opposed Soviet adventurism in the Third World: the *mujahideen* in Afghanistan, the contra guerrillas in Nicaragua, Pol Pot and the Khmer Rouge in Cambodia, UNITA in Angola and other insurgents challenging leftist regimes.

CONCLUSIONS: AN EMERGING GOVERNMENTAL RATIONALITY OF DEVELOPMENT

It is in many ways impossible to separate foreign policy from the Cold War foreign assistance given by the US, the USSR and China as they were inextricably connected. As the realist scholar George Liska noted in 1960, foreign aid has served as an "instrument of political power" (Liska, 1960: 14) and it was widely used to fight the Cold War. Within the JFK administration, for example, it was regularly framed (particularly by Rostow who later played a key role in determining how USAID would operate) as another "weapon" in America's strategic arsenal. It would be used to win friends and strategic alliances, to create an enabling environment for the pursuit of wider foreign policy objectives and to sponsor (or counter) different kinds of revolution. It enabled donors to advance their quest for international recognition but also, in the case of China's relations with Taiwan, to restrict it.

Vast sums were spent in the name of foreign assistance and modernisation – at the end of the twentieth century the World Bank (1998: 2) noted that since the late 1940s "developed countries" had allocated around US$1 trillion to development cooperation (to say nothing of emerging South–South flows of development assistance). The establishment of domestic welfare states in the 1930s and 1940s paved the way for the foreign aid regime and the willingness to consider governmental programmes of assistance to people overseas but the idea that aid can be used as a foreign policy tool to create a particular kind of world can partly be traced to perceptions of the success of the Marshall Plan and the idea that aid led to stability, which inspired many leaders to recreate the programmes elsewhere. Aid allowed the US government to express a set of Judeo-Christian ideas held by most Americans about the moral responsibilities the rich have to the poor and demonstrated "that the country is not simply a powerful nation, but a powerful nation committed to a higher purpose" (Taffet, 2007: 3). In this sense, foreign aid also provides a valuable "window into the national political soul" (Taffet, 2007: 4).

The US government had long been interested in the improvement of living conditions in poor countries and the creation of effective economic systems around the world, but that interest was not spread evenly. Cold War development assistance had very particular political geographies then as "[s]ome poverty was regarded as more important than other poverty" (Taffet, 2007: 2). Aid sought to address "internal [political] dynamics" and became, as in the case of JFK's AfP, a set of political programmes designed to create particular kinds of geopolitical outcomes. It quickly became a tool, however, to reward allies and prevent threats emerging and in the pursuit of wider geopolitical aims of dominating, pacifying, protecting, strengthening or transforming recipient countries. It also put in place important networks and connections, linking donors and recipients, which for some critics meant that it was little more than imperialism and would create only dependency. The AfP, for example, was not an alliance and "was not even always about economic *progress*" (Taffet, 2007: 196, emphasis in original) but it did enable the US to manipulate its relations with Latin America to its own advantage as "decisions about economic aid programs were part of every major US action in the region" (Taffet, 2007: 197).

US policymakers believed that communism would be unable to threaten countries with healthy economies and as such the simple "task" of development for many US modernisation theorists was to provide "an ethos and system of values which can compete successfully with the attraction exercised by Communism" (Watnick, 1952–53: 36). In addition to countering communism US aid also sought to "make the world safe for capitalism" (Westad, 2006: 31). Aid was the primary tool through which the US offered what JFK promised Vietnam in 1956: "a political, economic and social revolution far superior than anything the Soviets had to offer" (JFK cited in McVety, 2012: 164). This competitive logic behind foreign aid meant that many recipient countries became "tournaments of modernisation" (Cullather, 2002: 530) as competing global powers sought to enact their different conceptions of revolution, progress and development, often through large-scale "showcase" projects (such as the HVP in Afghanistan or the TAZARA railway in Africa). In particular, the Third World witnessed a rapid proliferation of river basin development projects during the Cold War and even though they were often represented as purely technical exercises, they were nonetheless "rife with geopolitical calculations" (Sneddon and Fox, 2011; Sneddon, 2015). Along with a surfeit of Cold war dam projects there were also a huge number of infrastructure projects focused on roads, railways and airports, schools and hospitals, along with a raft of factories, textile mills and state farms (many inspired by similar schemes developed by donors domestically). Each would seek to show how the donor's particular combinations of technology and know-how would vanquish underdevelopment in the Third World.

In 1982 a report by the US National Intelligence Council examined the question of what makes superpowers attractive to "less developed countries" (LDCs) and noted that "[m]ost LDCs will seek support wherever they can best meet their needs, regardless of ideology" (National Intelligence Council, 1982: 4). Reviewing 30 years of US–Soviet competition in the Third World, the report argued that the Soviets held several "advantages" including continuity, faster arms delivery with fewer strings, assistance from allies like Cuba, Moscow's ability to identify itself with widely held positions on the key issues of self-determination for Palestinian Arabs and black majority rule in Southern Africa, the Soviets' use of subversion and military intimidation to "force" LDCs to cooperate and their freedom from parliamentary and public opinion constraints. The report also noted the perception that the US was linked to the colonial policies of its Western European allies and that the Soviets' centralised, authoritarian political model was regarded as more suitable to the Third World, even though the USSR had "mismanaged" its relations with LDCs. Two years later a CIA report on the USSR and the Third World (Central Intelligence Agency, 1984: 3) scripts the latter as a "volatile arena of US-Soviet political struggle" where the Soviets' advance had bolstered their claim to be a global power, noting that the "inherent instability" of the Third World and the "inhibitions of US policy in the post-Vietnam period" had created tempting opportunities for Soviet gains.

Around this time there were concerns expressed in several popular geopolitical texts that the Soviets were running rampant in the Third World, which solidified support for increased military spending in the US. The film *Red Scorpion* (1989), for example, tells the story of Nikolai, a Soviet Spetznaz operative played by Dolph Lundgren, who is sent to a Marxist Southwest African state called Mombaka (a country modelled on Angola) to kill an African resistance leader named Sundata (loosely modelled on UNITA leader Jonas Savimbi) who switches sides and helps the rebels as he cannot accept the brutal tactics used by the Russians and Cubans against the local people. Discarded by his communist bosses for failing in his mission, he is rescued in the desert by bushmen who nurse him back to health and initiate him into the group, turning the Hollywood anti-Soviet action film into a conversion narrative (Robinson, 2007). During the Cold War, foreign aid was all about conversion – transforming recipient states into a model of the modern fashioned in the donor's own image, persuading them of the pitfalls of communism or capitalism and transforming their economies, infrastructures and state institutions in the process. The pursuit of modernisation through foreign aid was also all about convergence as industrialisation and the transfer of technology and modern values would pull all recipient nations towards a common point (modern, industrial societies), enabling them to "catch up".

The Cold War was highly cartographic and numerous maps were produced that articulated spatial ideologies of containment and liberation, mapping the Third World but also projecting the power of the US or Soviets within it (Barney, 2015). Vast swathes of the Third World were coloured red for communism as alarmism about the "dangers" of its diffusion was commonplace, but the reality was a far more complex and dynamic patchwork of allegiances that were never as simple, constant, uniform or unilateral as the projections of power and influence created by Cold War cartographers implied. The "Communist world" in particular was never as unified and cohesive as the mapmakers suggested (Nunan, 2014; Friedman, 2015). Cold War geopolitical alignments were often fluid and dynamic (especially in the case of Afghanistan which regularly shifted allegiances between the US and the USSR) and there was also lots of discontinuity in Chinese, American and Soviet engagements with the Third World as their interest peaked and troughed at different historical moments and was always subject to a wide range of national, regional and global geopolitical dynamics.

Foreign assistance did not simply reproduce a single model of political economy around the world – in the case of US aid to Taiwan, often touted as an exemplar of non-communist development practice, Cullather (2001) has documented how US aid and advice often responded creatively and in an experimental fashion to particular threats, local crises and opportunities. To an extent, aid projects were also demand led and were not simply driven by superpowers seeking to export their wares (although aid was in various cases tied to the use of donor services, companies, technicians etc.). Neither was it always a straightforward core–periphery transfer of aid – Soviet assistance to the Third World, for example, was often mobilised by Moscow from parts of the Soviet Union's internal periphery (e.g. from Azerbaijan to Afghanistan). In practice, modernisation projects funded by foreign assistance also frequently struggled with issues of implementation and were often resisted and reworked by recipients and intended beneficiaries whilst their reception "on the ground" was far more complex than anticipated by the planners and developers who dreamt them up.

Although many aid projects made a significant difference in recipient countries, there were a catalogue of expensive failures and white elephants which often revealed how little donors understood about the cultural, economic and political contexts in which they operated. The British politician Shirley Williams claims in her autobiography, for example, that the Soviet Union sent snowploughs to Ghana along with roadmaking equipment (Williams, 2010). Similarly, in Angola the Soviets found it difficult applying a revolutionary ideology based on class to a revolution whose protagonists saw their struggle largely in terms of race, whilst the Chinese struggled to comprehend the sheer complexity of liberation struggles in Southern Africa. Promoting anti-capitalism "had been one thing when Moscow's allies were in Prague, Warsaw, and Budapest. It was another thing entirely when they were in Luanda, Addis Ababa, and Kabul" (Nunan, 2014: 6). Further, the "experts" that most often came with the aid were also often resented because "they created a social sphere over which the recipient country had little control, even when they came from countries with which the regimes had close relations, such as the Soviets in Angola or the Americans in Iran" (Westad, 2006: 96).

Around the world, whilst presenting itself as the antithesis of empire, the US "lavished billions on corrupt regimes that ignored poverty" during the Cold War (Engardio, 2001: 58). The 2011 revolution in Egypt drew attention to the historic role that US aid had played in Third World state-building and in helping a recipient government to win and maintain legitimacy, propping up unsavoury administrations like the illegitimate Mubarak regime. During the uprising Egyptian riot police were photographed in the streets of Cairo hurling teargas canisters labelled "Made in the USA", whilst tanks, rifles, helicopters and fighter jets funded by US aid and manufactured by US companies were used by the Mubarak regime to quell the protests. As Mitchell (1995) has argued with respect to USAID's operations in Egypt, the "forgetting" of certain forms of engagement (such as these kinds of military aid) is as significant to the framing of development as the assertion of "new" (aid) relationships. One of the main consequences of Cold War foreign assistance was a massive militarisation of the Third World. Over time the Soviets increasingly prioritised military over economic assistance and the US provided grants amounting to over US$90 billion in military equipment and training to some 120 countries before the end of the Cold War (Kuzmarov, 2017). Vietnam alone received US$16 billion in military assistance from the US between 1955 and 1975 (Kuzmarov, 2017).

A "strange kind of Cold War" unfolded, fought not just with military hardware but with competing models of the modern and different ideas of the correct pathways to development, with loans and grants, with technicians and planners. There was also a propaganda war for hearts and minds where donors sought to narrate the "progress" their assistance was enabling through popular geopolitical texts including films, radio programmes, comics, posters etc. and which created significant affective communities, as in the idea of "Red Africa" created in Soviet propaganda (Nash, 2016). This was also not a just a simple confrontation between the US and USSR: Soviet attempts to organise the Third World existed in contention with projects organised out of places like Belgrade, Havana, Beijing and Algiers (amongst others). Shubin (2008) argues that Soviet support to the Third World was not determined by a competitive rivalry with the US, but by Moscow's desire to respond to Third World demands for emancipation, but so much of what China, the USSR and the US sought to do in the Third World through foreign assistance was shaped by their interactions with each other and by the competition for strategic influence. China, for example, entered Africa locked in combat with the US (and later the USSR). Similarly, Soviet support certainly influenced ZAPU intelligence cadres in Zimbabwe in the 1960s, but it did so in negotiated, pragmatic and, at times, surprising ways, which were shaped by interactions with many other foreign hosts (Alexander, 2017).

US Cold War foreign aid was, in particular, heavily focused on the countering of communist insurgency and consequently development became increasingly central to the goal of not just securing territory but securing and *winning* populations as pacification was directly linked to prosperity. Counter-insurgency was increasingly framed as what David Kilcullen (a senior US counter-insurgency advisor in Iraq) has termed "armed social work". During the Cold War development increasingly came to be fashioned as a set of biopolitical compensatory and ameliorative technologies of security (Duffield, 2006b), that sought to "accomplish or attempt to accomplish, stable rule through certain sorts of governable subjects and governable objects" (Watts, 2003: 12). In this way development has historically served as a locus of disciplinary (bio)power whose primary concern is with violently fixing potentially insurgent populations within an extended archipelago of highly securitised space (Duffield, 2007a, 2008, 2010a). This emerging governmental rationality of development is particularly evident in the work of USAID which was centrally involved in the project of countering this communist "contagion", championing increasingly "total" forms of rural development as the key to transforming populations of potential insurgents into "governable subjects" (Attewell, 2017). In a range of contexts such as Vietnam, Palestine or Afghanistan, USAID increasingly practiced development as a form of governmentality, its interventions becoming ever more concerned with the government and management of restless rural populations (Attewell, 2017). This question of the emerging governmental rationalities of development and the management and countering of insurgency is now taken up in the next chapter, which considers the state as a key point of entry for thinking about geopolitics and development.

Chapter 5

The State and Development

INTRODUCTION: THE STATE IS DEAD, LONG LIVE THE STATE

ALTHOUGH many discussions of the history of development begin with President Truman's Cold War announcement of a "bold new programme" for the underdeveloped world, the origins of development as an idea can also be traced to other, earlier periods of military conflict, such as in the power struggle that began to unfold among emerging states in a turbulent and industrialising Europe during the nineteenth century (Cowen and Shenton, 1996). Security-motivated state intervention has historically played a large and significant role in European history, both in periods of industrialisation and welfare and in periods of economic depression and war (Hettne, 2010). As Foucault noted in his lectures on governmentality, the idea of security emerged in a period of transition in the history of French statecraft, as "old" approaches to rule based on repression and extraction (criticised for their tendency to create periodic shortages and disruptive crises) were replaced with tactics designed to foster productivity and trade (Dwyer, 2014). In the nineteenth century, heavy industries were regarded as a means to achieve national strength and security, one that required a strong interventionist state (Hettne, 2010) whilst economic doctrines viewed wealth creation as directly linked to order and state security. Indeed, with its in-built sense of design, "development" has always represented forms of mobilisation associated with order and security (Hettne and Odén, 2002) and the state has consistently been regarded as central to that. The general aim has remained that of attempting to reconcile the inevitable disruption of progress with the need for order (Cowen and Shenton, 1995: 27–43); an objective that it has failed to achieve (Duffield, 2001). In many ways state sovereignty "constitutes a precondition for Development" (Tickner, 2003a: 318) and it has often been argued that the inverse is no less true; economic well-being and wealth are important enabling factors for the realisation of sovereignty (Tickner, 2003a: 318).

During the Cold War, the primary focus for both the main superpowers was on supporting and allying with Third World states and particularly in the 1950s and 1960s, both development and security were primarily framed as inter-state affairs. Foreign aid, inspired by the dream of modernising the Third World, centred on enhancing security by strengthening the state apparatus and creating and securing a "steady state" in the periphery as a means of promoting development and making the world safe for capitalism. In "theorising back" (Slater, 1993) against the grain of modernisation discourses during the 1960s and 1970s, the dependency scholars also placed the state at the forefront of their analyses in arguing that the periphery's global insertion had harnessed state-building processes to global capitalist dynamics in ways that had hampered the consolidation of the state's basis for internal legitimacy and that active state intervention would be necessary in the country's economy to refocus the industrialisation process. Theories of modernisation implied that, in order to "catch up" or to capture the secret of (or mimic) the success of wealthier states, countries of the Third World should internalise within their domains "one or other of the features of the wealthier countries, such as industrialisation and urbanization" (Arrighi, 1991: 40). State institutions in the Western world were also offered up as models that the newly emergent states of the Third World should seek to internalise and replicate. By the 1980s many countries of the Third World had "internalized elements of the social structure of wealthier countries through 'modernization' but had not succeeded in internalizing their wealth" (Arrighi, 1991: 40). Consequently, many lacked the means of fulfilling the expectations and accommodating the demands of the social forces that they had brought into existence through modernisation and "as these forces rebel a general crisis of developmentalist practices and ideologies begins to unfold" (Arrighi, 1991: 40). This chapter seeks to examine the historically central role of the state in development theories and practices, arguing that states and the crises, resistances and insurgencies that they engender lie at the very heart of the geopolitics–development nexus.

One group of states that managed to internalise many features of the wealthier Western countries in the name of modernisation were the "developmental states" of East Asia (such as Japan, South Korea, Taiwan and Singapore) that provided strategic partners in the Cold War balance of power. Drawing on classical statist political economy, the literature on "developmental states" (Amsden, 1989; Evans, 1995; Wade, 1990; Johnson, 1999; Woo-Cumings, 1999) emphasised how, contrary to prevailing free-market explanations, the "catch-up" development of much of East Asia was more the result of a high level of state intervention. The goal of a developmental state was conceived as building markets but then "deliberately distorting them to serve specific national development objectives through the judicious use of incentives, tariffs, subsidies and especially control of finance" (Bishop and Payne, 2017: 1). The label itself is now variously applied, sometimes very loosely, to a vast array of countries as diverse as Indonesia, Brazil, Mozambique, Mauritius and Botswana (amongst many others) (Bishop and Payne, 2017; Ovadia and Wolf, 2018). Even some Western states that claimed to have developed via open markets, did precisely the opposite and were (and in some cases still are) to an extent "developmental" states themselves (Chang, 2003). Further, neoliberal actors often try to "retro-fit" their theory to the South Korean context whilst typically ignoring the historical importance of the "development state" approach in achieving growth (Chang, 2003).

Typically, however, there have been far too few efforts to understand the East Asian developmental states within the geopolitical and historical contexts in which they emerged (Berger, 2004; Yeung, 2017). In particular, examining the historical intersections of the Cold War, development and security in the making of East Asian developmental states can be richly suggestive (Glassman and Choi, 2014). The US (and, later, Japan) invested substantial resources into these East Asian economies in an attempt to shape the development of their political institutions and create alliances with "strong" states. It can even be argued that US aid indirectly helped create the actual developmental state apparatus since its financial contribution and market support to East Asian development, as indispensable parts of the US-led Cold War imperative, were necessary conditions for underwriting the initial emergence of the developmental state (Yeung, 2017). The modern post-colonial Korean state was in many respects "no less an externally created structure than the colonial state it replaced" (Pirie, 2008: 62) as its new institutions were constructed around the old structures of the colonial state and its "fiscal base" was US aid whilst the first Korean president was a "US placeman" (Pirie, 2008: 62). Perceived geopolitical threats in the East Asia region spurred the emergence of developmental states led by "authoritarian strongmen" (Yeung, 2017), several of whom struggled to maintain security within their own borders (Doner, Ritchie and Slater, 2005) and by the late 1980s, major social and political movements had prompted the decline of the authoritarian state in countries like South Korea and Taiwan (Yeung, 2017).

As the Cold War ended, development discourses that envisaged the "engineering" of states in the South diminished and there was a paradigm shift towards a more non-interventionist approach (see chapter 6), guided by a "post-national" logic with a humanitarian focus on "human security" (Hettne, 2010), incorporating a transnational assumption of responsibility, as if one could no longer rely on states to fulfil their duties for their citizens (Hettne, 2010: 34). Some scholars also implicitly or explicitly posit the decreasing relevance of states in driving development processes and in shaping development outcomes by reference to the increasing import of "horizontal" and relational processes of connectivity, such as transnational commodity chains or development programmes, along with migration, governance or NGO networks, philanthropic foundations (like those of Bill Gates and George Soros) and wider informal social and economic networks. These are seen to possess logics, causalities or mechanisms of coordination that cut across and to an extent "transcend" states. The murky concept of "failed states" also became prominent during this period (Call, 2008). This chapter argues that discourses around the weakness, fragility and failure of states in the South need to be understood in historical and geopolitical context and despite forecasts of its demise the state remains the primary site of development interventions. States remain the main institutional realm for addressing political and economic grievances and claims, as they legislate and have obligations towards their citizens and under international legal frameworks. States also play a vital role in the wider diplomacy around development whilst international governmental organisations along with a range of non-state actors and para-diplomatic entities (e.g., governments in exile, secessionist movements and sub-national or regional governments) can also "mimic" conventional state diplomacy (McConnell, Moreau and Dittmer, 2012; Mamadouh et al., 2015). However permeable to global and transnational social processes, state institutions "still decisively filter the flows of commodities, people and ideas unfolding across and beyond them, and states continue to be central vis-à-vis the specific ways in which global development interventions are defined and implemented" (Novak, 2016: 491). Contrary to neoliberal development paradigms which promised to roll it back, the state was, to an extent, "smuggled back in" again through the "security and development" rubric (Luckham, 2009) (see chapter 6).

James Scott's (1998) pivotal work, *Seeing like a State*, is hugely relevant here and has been foundational in shaping my own understanding of states in the South. Scott provides a host of examples, from state forestry to villagisation and other "grand schemes" for infrastructure and large-scale or "high modernist" environmental-developmental changes that could be considered as definitional to states. In this sense, development serves as a key means through which states seek to create a "legible" society to be governed (Scott, 1998), both through measuring and codifying a population (e.g. their landholdings, their harvests, wealth, volumes of commerce etc.) but also in creating standard grids through which this could be centrally recorded and monitored. Scott (1998) articulates these practices (establishing territorial boundaries, controlling movement across them, policing, border control etc.) as the "state simplifications" and "schematic categories" needed for grasping a "large and complex reality". These grids and grand schemes are key to geopolitical imaginations of development. As Luke (1996) has argued, states have often sought to establish their power by *in-state-ing* themselves in space, imprinting a mark of their territorial presence (cf. Call, 2008). One of the ways they do so is through infrastructure and large-scale rearrangements of socio-natural environments (e.g. through dam projects) enabling states to demonstrate and project their strength and power. There is thus a continuing need to detail how state practices transform environments and how the state is consolidated, and constituted, in relation to "nature" (Mitchell, 2011) since the "process of mapping, bounding, containing and controlling nature and citizenry are what make a state a state" (Neumann, 2004: 202). A key part of the way in which states come into being is through these claims and assertions of control over territory, resources and people. Infrastructure development has also been at the heart of recent South–South cooperation initiatives led by China and other (re)emerging powers in Africa (see chapter 7) and is central to China's "One Belt, One Road" (OBOR) initiative launched in 2014, with US$1 trillion worth of infrastructure works planned or already underway. Infrastructures thus represent an important political terrain, the critical analysis of which provides a useful point of entry into further understanding the geopolitics–development nexus.

The chapter also considers the ways in which neoliberalism has impacted on conceptions of the role of the state in development but also the many forms of insurgency and resistance that the pursuit of neoliberal state-led models of development has given rise to. Neoliberal development discourses advocated giving resources to governments to make markets work so as to reduce poverty or disaggregate and marketise the state (Craig and Porter, 2006), breaking up existing forms of state rule (corrupt, patrimonial) and then "using markets to replace and reconstruct the institutions of governance" (9, 100), whilst re-embedding markets in regulatory and constitutional frameworks such as the rule of law or freedom of information (Mosse, 2011). As a term neoliberalism is "promiscuously pervasive, yet inconsistently defined, empirically imprecise and frequently contested" (Brenner, Peck and Theodore, 2010: 184). Rather than being a static homogeneous ideology with a uniform outcome, it results from complex and dynamic processes of "neoliberalisation" (Peck and Tickell, 2002) which are "contextual and contingent" (England and Ward, 2007: 250). Similarly, simplistic binary distinctions are often drawn between "neoliberalism" and "developmentalism", yet development almost always embodies in practice, even if not in theory or ideology, a complex mix of state–market interactions (Bishop and Payne, 2017). Neoliberal deepening was itself a state-led process, just as developmentalism relies on market mechanisms and countries can and do exhibit characteristics of both simultaneously (Bishop and Payne, 2017). States are central to the reforms prescribed by neoliberal development policy (such as the "shock therapy" or privatisation and market deregulation required by Structural Adjustment Programmes (SAPs) and the more recent Poverty Reduction Strategy Papers or PRSPs which seek "country ownership" of neoliberal prescribed pathways to poverty reduction) and rather than "hollowing out" the state, such interventions exemplify the use of state power to impose market imperatives across societies.

Brazil's current socioeconomic policy regime, for example, is a hybrid of economically *liberal* policy goals and instruments associated with the Washington Consensus and more developmentalist objectives (e.g. to eradicate hunger and extreme poverty) that have their roots in the so-called "national-developmentalist" paradigm of the 1930s and 1940s and the work of the Economic

Commission for Latin America (CEPAL) (Ban, 2013; Dauvergne and Farias, 2012). This "neo-developmentalism" entails a new form of state activism and rather than rolling back its interventions in leading sectors of the economy, the Brazilian state has consolidated its presence not only as a regulator, but also as owner and investor (Ban, 2013). More generally, Latin America has also seen a recent rise in "post-neoliberal" developmental states (such as Bolivia and Ecuador) that are today heralded by some observers as contributing alternative practices around development with potentially positive impacts on state–society relations.

The development processes that unfold within as well as across and beyond states intersect and entangle with each other around borders, which constitute important points of contact, division and articulation between various geographies of development (Novak, 2016). While borders attempt to shape space along state-centred scales of discourse and practice they are today increasingly the focus of a series of intense and wide-ranging interventions by state and transnational actors, with several neoliberal agencies arguing that integrating national economies into world markets means that borders need to be restructured or "thinned" through multilateral institutional alignment, the reduction of tariffs and restrictions, the liberalisation of capital markets and the establishment of new governance authorities and bodies (Novak, 2016). Similarly, de- and re-bordering strategies, whereby territorial jurisdictions are redefined in regional, cross-border and/or sub-national terms have also been aggressively promoted, leading to a proliferation of "growth corridors", Free Trade Areas (FTAs) and Special Economic Zones (SEZs). By creating territorial governance units among, across and within states, these "political technologies" (Elden, 2013) are increasingly seen as important in fostering FDI, regional cooperation and trade-led growth. In this sense it is useful to examine the "borders–development nexus" since a variety of disparate neoliberal development policies can be reinterpreted as instances of border management interventions (Novak, 2016).

As the devastation caused by neoliberalisation has proliferated, so the increase in the number of social movements has gathered pace. As Paudel (2016: 1047) notes, "[s]ubaltern political possibilities are always present and inherent in the historical material processes of development" and it is important to examine how subaltern struggles have contested hegemonic conceptions of state, citizenship and society or sought to defend places and ecologies, how they have generated alternative projects, practices and imaginaries of development and challenged the anti-politics of the neoliberal development regime. As states have retreated in the wake of neoliberalism the number of NGOs has also grown significantly. Founded largely on values of "world citizenship and universal human kinship" that were also promoted by the labour movement (Lumsdaine, 1993), NGOs were initially focused on charitable giving and short-term humanitarian aid, but their remit gradually extended during the inter-war years to include long-term development projects in the areas of health, education and agriculture (Duffield, 2007a). During the Cold War, NGOs were at pains to position themselves as operationally outside of states; their selling point was the promotion of "bottom-up" community-based development as opposed to the bureaucratic "top-down" efforts practiced by states and multilateral development agencies (Jones, 1965). NGOs are now less likely to be outside of states looking in, but rather on the inside looking out (Hulme and Edwards, 1997), even building or extending state capacity or acting as its surrogate (Duffield, 2007a). There are concerns however that NGOs often depoliticise and blunt the edges of resistance, interfering with self-reliant social movements whilst turning confrontation into negotiation and defusing political anger (Roy, 2014b).

This chapter is divided into four sections. The first sets out a theoretical framework for conceptualising the state and its relations to development and examines the concept of governmentality before considering the theorisation of African states in particular. The second argues that infrastructures are key sites around which the meanings of development are contested and that they help create, destroy, expand or limit the contours of the state. It also argues that a focus on resource geographies and the "resource–state nexus" (Bridge, 2014) can reveal a great deal about the role of states in shaping the political geographies of development. The chapter then considers how post-colonial state formation in the South has often been beset with insurgencies of various kinds (and explores some of the strategies used by states to counter them) before a final section examines the ways in which state power, imaginaries and narrations of development have been contested by a range of social movements.

THEORISING THE STATE

> It [the state] should be examined not as an actual structure, but as the powerful, metaphysical effect of practices that

> make such structures appear to exist. In fact, the nation state is arguably the paramount structural effect of the modern social world.... By approaching the state as an effect, one can both acknowledge the power of the political arrangements that we call the state and at the same time account for their elusiveness. (Mitchell, 1991b: 95)

Despite long-standing calls to rethink the state "as a social relation", reified understandings that view the state as a differentiated institutional realm separate from civil society have persisted in academic and political debate (Painter, 2006). Steinmetz (2008) and Ferguson and Gupta (1997) have shown how the state, insofar as "it" can be said to "exist", is a mythologised, contradictory and constantly challenged entity (Manchanda, 2017). Despite the almost unavoidable tendency to speak of the state as an "it", the domain we call the state is "not a thing, a system or subject, but a significantly unbounded terrain of powers and techniques, an ensemble of discourses, rules, and practices" (Brown, 1995: 174). Feminist, anthropological and poststructural approaches to the state have steadily reoriented attention away from a focus on formal state institutions toward the more socially embedded processes through which ideas of the state are reproduced. Rather than conceptualising the state as a self-generated and governing stable structure, Mitchell (1991b), for example, advocates rethinking the state as a "rhetorical effect". States do not simply exist; rather, they are accomplishments reified and reformulated through a variety of prosaic and quotidian activities and mundane practices (Painter, 2006). Such an approach reveals the "heterogeneous, constructed, porous, uneven, processual and relational character" of states (Painter, 2006: 754) and can bring to light the uneven geographies of state power with greater complexity and subtlety, which is particularly useful in the global South context where state power can be fragile and contingent. The state emerges as an imagined collective actor "partly through the telling of stories of statehood and the production of narrative accounts of state power" (Painter, 2006: 761) and the ideas, imaginaries and practices of development play a key role in the telling of these stories and in the wider narration of state power. The focus on "statisation" can help us then in understanding the perpetual process through which the state comes into being by attending to the mundane social and material practices of "development" through which stateness is actualised.

Foucault's concept of governmentality is also useful here in illuminating the diverse political rationalities of government, its "technologies" and the bringing into being of the things, people and processes to be governed through development programmes (Watts, 2003; Li, 2007). In exploring questions of governmentality and practices of rule in development, attention focuses on how state power operates to shape the conduct of conduct, as tactics and technologies are deployed to control, subdue and oppress the citizenry. Here the state no longer stands as an unquestioned *source* of power, but, rather, as its *effect*. Despite concerns in some quarters about the applicability of Foucault's thought, Death (2011) argues to the contrary that it is precisely in spaces of contragovernmentality, ungovernability, anarchical governance or in the borderlands of global politics that a governmentality approach can provide illuminating insights into the contemporary operation of politics at various scales. Africa in particular has increasingly been identified by the US military as having many such spaces of ungovernability and disorder, seen as having the potential to fuel Islamist insurgencies (see chapter 6). The governmentality approach thus has the potential to map the fragmented, uneven, heterogeneous, overlapping and fractured spaces of global politics as well as to make sense of the various attempts to counter insurgency there. Another example of spaces of contragovernmentality is the *favelas* of Brazil where state institutions, in preparation for a series of mega-events culminating in the 2016 Summer Olympics, tried and repeatedly failed to reclaim these spaces to ensure security and social order (see Figure 5.1). The Brazilian state has attempted to retake territories that have been controlled by drug gangs and regarded as largely off-limits for as long as 30 years through a series of aggressive interventions involving military occupation, infrastructure provision, beautification and selected removal (Freeman, 2014). The practices of "legibility and simplification" associated with these programmes are often justified as something that is to an extent "necessary" for residents to fully participate in modern society but these measures also expose residents to predatory aspects of the state and capital (Freeman, 2014). Such urban peripheries of devastating poverty and inequality have also increasingly become spaces of insurgent citizenship (Garmany, 2009; Holston, 2009).

The concept of transnational governmentality (Ferguson and Gupta, 2002) is also useful here in capturing how the state is composed of multiple relations within and beyond national borders and enables a better grasp therefore of the multiple movements and complex spatialities of governance and how they are shaped by

FIGURE 5.1 A Brazilian police pacification unit (*Força de Pacificação*) on patrol in Favela Mare in Rio de Janeiro on July 12th, 2014. Courtesy of ITA-TASS news agency/Alamy.

transnational actors (e.g. corporations and the IFIs), alliances and networks. This focus on governmentality can also be very productive if supplemented by assemblage thinking. Abrahamsen (2017: 134) notes that many analyses of the state typically try to "apply" theory to Africa, seeking to fit its institutions and practices into an already existing model "and constantly finding it wanting". Africa here is only ever acted upon. Assemblage thinking however usefully draws attention to the multiple forms and sources of agency, and the different forms of power, resources, and capacities that different actors and actants possess. It also involves approaching the state and social orders not as fixed or static, but as something to be discovered, as contingent and evolving. Since no two assemblages are the same this gets away from the need for an explicit or implicit comparison to a Western state norm. Some recent anthropological scholarship on post-colonial African statehood similarly sees the state not as a finite entity – an apparatus of borders, personnel, budgets and bureaucracies – but rather as always emerging, becoming, incomplete or "an always-emergent form of power and control identifiable at multiple societal levels" (Bertelsen, 2016: 3).

Due to their lack of internal legitimacy, Third World states were typically regarded as having been constructed from the outside, by means of international recognition of their sovereign status (cf. Jackson, 1993; Jackson and Rosberg, 1982, 1986). This is particularly true of how African states have been regarded in some academic literatures. The debate about the state in Africa remains a highly contested domain, however, as conceptions and theories of African states present a series of questions that have "haunted the continent's place in IR" (Harman and Brown, 2013: 73), given a perceived lack of "fit" between IR's theoretical constructs and African realities (Harman and Brown, 2013: 73). African states are typically depicted as public façades behind which power operates through clientelistic networks (Bayart, 1993), as marginal actors in the international economic and political order, as violent and corrupt and as having economies and societies that have been profoundly and predominantly shaped by external interventions and resources. Many contemporary scholars have claimed that African states are governed by a pervasive "patrimonial" logic, which encourages clientelism, corruption and economic stagnation, blurring the lines between party and state. These range from "developmental patrimonial" states (Kelsall and Booth, 2010) to those reliant on violent political bargaining (de Waal, 2010; Booysen, 2011) or the "instrumentalisation of disorder" (Chabal and Daloz, 1999). As Bach (2013: 6) notes, the "standard depiction" is of African countries as quasi-states "devoid of the empirical components of statehood", where the essentials of statehood are missing, or it is often "juridical" (i.e., *de jure* recognised by international actors) rather than "empirical" (a *de facto* ability to exercise sovereignty).

Africa is also typically represented as home to an assortment of "weak", "failed" and "fragile" states as if there are states in the world that conform to an "ideal-type" when in reality nowhere does (Brown, 2006). The grammar around "failure" and "lack" in IR has been somewhat unreflexive (Manchanda, 2017) – weak states are typically described as those in which a solid national identity, or an "idea of state", is absent, or contested by a diverse array of societal actors; socio-political cohesion is especially weak; consensus on the "rules of the game" is low; institutional capabilities in terms of the provision of order, security and well-being are limited; and the state is highly personalised (Tickner, 2003a: 314). Talk of state failure is, as Manchanda (2017: 388) demonstrates in relation to Afghanistan, "often laden with the same normative assumptions that accompanied the more explicit racial biases and ethnocentric baggage intrinsic to colonial propaganda and conceptualisations of world order". In a variety of ways then, imperial hierarchy and Eurocentric modes of thought underpin notions of "statehood" (Grovogui, 1996; Hobson and Sharman, 2005) and colonialism has a direct bearing on today's international relations since "fragile state discourse reproduces some of the key assumptions and relations of colonial bureaucracy, in particular the liberal practice of indirect rule or Native Administration" (Duffield, 2009: 116). Indeed, many of the fundamental assumptions that fortify discourses of statehood, state failure and good governance have a fundamentally orientalist make-up (Hill, 2006; Gruffydd-Jones, 2014). The specific *post-colonial* circumstances of African statehood are also important here (and something that many theorisations of the state in Africa have struggled to adequately comprehend). African states are a "hybrid" blend of informal and formal institutions, institutional and patrimonial forms of rule, democratic/authoritarian predispositions and modern and charismatic sources of political authority (de Waal and Ibreck, 2013). It is thus necessary to "re-examine the utility of the state concept itself, rather than attributing its limited applicability to the shortcomings of the periphery" (Tickner, 2003a: 315).

STATES, INFRASTRUCTURES AND RESOURCE GEOGRAPHIES

Transforming the logic of research on state–society relations from seeking to understand the effects of the state on society to considering the processes and technologies through which the concept of the state emerges and appears as a discrete object (Mitchell, 2011) means tracing the production of state power to a variety of different everyday spaces and practices where it is enacted (and contested). As Meehan (2014: 216) argues, "the everyday spaces of state power are products of an entangled thicket of objects and practices: from dams to discourses, meters to mapmaking, pixels to bureaucrats". Whilst work on everyday and capillary experiences of states, on the microphysics of power and on banal nationalisms has usefully shown how symbolic acts and gestures are important to, and constitutive of, states, it is also important to focus on large-scale changes to landscapes and economies and on grand schemes of social engineering, centralisation or territorialisation and their connections to states and stateness (Harris, 2012). For the state to take on meaning and importance in the lives of rural residents, "it not only has to be enacted through festivals or parades but must also command the attention of citizens through large-scale [infrastructural] efforts as part of what characterises the state as something that stands apart from society" (Harris, 2012: 39). Harris (2012) locates the "sites" and "spaces" of the Turkish state in the differentiated experiences, narrations and spatialities of large-scale infrastructural works in Turkey's waterscape which are read as pathways through which the Turkish state emerges as a "socio-natural" effect. Large-scale infrastructural works (such as the building of centralised electricity networks and hydropower dams) constitute an important way in which states demonstrate and project their strength and power and *perform* and *narrate* the state's presence and role, with important implications for state–society relations and distinctions.

In a variety of ways then, infrastructure "helps create, destroy, expand or limit the contours of what we call the state" (Meehan, 2014: 216). Infrastructures have played a significant role in processes of nation-building, modernisation and development (Calvert, 2016) and in the production and reproduction of geopolitical imaginaries of territory, nationhood and sovereignty (Huber, 2015). State power is in part the capacity of infrastructures to order, arrange and make legible (Scott, 1998) and they are key to "seeing like a state". Similarly, for Chatterjee (2004) the postcolonial state deals with its people primarily as governed populations (rather than as full rights-bearing citizens), or as subjects to be constantly divided and rearranged by government as targets of policy and one of the key ways this mode of operation is reinforced is through programmes of "development" focused on infrastructure. Consequently, access to infrastructures is often a key issue in struggles for inclusive citizenship and political rights. Accordingly, it is important to attend to the ways in which infrastructures (such as those around electricity and water) are institutionalised within biopolitical missions like "development" (Winther, 2008; Bakker, 2013). In Africa, electricity infrastructures, for example, enable the state to extend the power and reach of state institutions, particularly in contested peripheries (Power and Kirshner, 2018). Electrification (see Figure 5.2) is part of the "will to improve" underpinning development interventions and exemplifies the practice of "rendering technical" (Li, 2007), where energy access is depicted as a series of technical "problems" responsive only to a "development" intervention, while rendered non-political as experts concerned with improvement exclude political-economic relations from their diagnoses and prescriptions through a subliminal and routine "anti-politics" (Ferguson, 1999).

FIGURE 5.2 Electricity infrastructures overloaded with illegal and informal connections in Puttaparthi, Andhra Pradesh, India. Tim Gainey/Alamy.

For Ferguson (1999), the deterioration of the electricity system in Zambia was an icon of how people's expectations of modernity came to falter, which he includes alongside the examples of the "disconnection" that accompanied the collapse of the national airline (Zambia Airways) and the declining demand for copper-wired telephone cables in the age of fibre optics and satellite communications, as metaphors for the wider decline and disjuncture of the Zambian Copperbelt, of which so much had been expected at the zenith of modernisation (see Figure 5.3). Infrastructures have become key sites around which the meanings of development are contested, with multiple social movements and justice-seeking groups in the South naming it as an object of struggle and a crucible in their work towards socially just and environmentally sustainable futures (Cowen, 2017). Mobilisation around oil pipeline projects, dam construction and toxic water, for instance, has continually challenged corporate and state conceptions of development (see Figure 5.4). In many parts of the South universal provision of public services (like electricity or water) is a panacea and consequently the under-provision of (and exclusion from) public services plays an important role in shaping contemporary geographies of development. The idea of a single national space in which state power is

FIGURE 5.3 An image from Christina de Middel's project "The Afronauts" which blends fact and fiction in remembering some of Zambia's early post-colonial dreams. In 1964 (the year of Zambia independence) a Zambian science teacher named Edward Makuka decided to train the first African crew to travel to the moon and then Mars using an aluminium rocket to put a woman, two cats and a missionary into space using a catapult system. Makuka founded the Zambia National Academy of Science, Space Research and Astronomical Research to train his "Afronauts" near Lusaka. Courtesy of Christina de Middel/DMB creatives.

exercised and rights are enjoyed in a consistent and homogeneous way by all residents has been shattered by a variety of insurgencies, leading to a kaleidoscopic array of micro-sovereignties and diverse enclaves of authority and social service provision (Gilman, 2016).

Across the North–South divide reports of infrastructural obsolescence, failure, crisis and struggle are a mainstay of the daily news and mark the volatility and vulnerability of the socio-technical systems

FIGURE 5.4 Medha Patkar (leader of the Narmada Bachao Andolan movement, NBA) and Indian writer Arundhati Roy visit Bhil communities at Domkhedi village along the Narmada river in Madhya Pradesh in September 2000. The village was submerged as part of the construction of the Sardar Sarovar dam project. Joerg Boethling/Alamy.

upon which people come to depend (Cowen, 2017). Access to vital public services such as water, electricity, health, housing and transport is also often spatially fragmented, particularly in urban spaces where populations experience water supply services, for example, as elite "urban archipelagos" rather than homogeneous networks (Bakker, 2013). Access to "public" goods and services is thus usually incomplete and a plethora of alternative strategies of service provision arise, including illegally tapping into existing networks, as with the "midnight plumbers" and "comrade electricians" operating in South Africa's townships (Bakker, 2013), and unregulated private alternatives or services provided by NGOs. In some cases, communities come together to provide themselves with the services that their governments are unable or unwilling to provide such as cooperative water supply systems and sewerage networks or self-built housing (Bakker, 2013).

Some of the literature on infrastructure and biopolitics has examined state interventions to ensure the health of populations, resources and the economy and the ways in which these were predicated upon, for example, techniques of surveillance, statistical representation and the discursive mediation of individual subjectivities (Cooper and Stoler, 1989; Corbridge et al., 2005; Legg, 2007, 2008; McFarlane and Rutherford, 2008; Bakker, 2013). This work views governmentality as simultaneously material and discursive, inscribed in physical space as well as in social relations and given the partiality and fragmentation of infrastructures (and access to them) it has shown how the extension of biopolitics through the "public" (governmental) sphere is necessarily uneven. Access to services such as water thus becomes the subject of continuous political negotiation or "a question of political struggle, often enacted by self-organised community groups operating outside the bounds of traditional pathways of democratic representation in permanent negotiation with governments over demands for political recognition via the provision of public services" (Bakker, 2013: 283). The resistance of "ungovernable" populations implies nuanced negotiations between the "governed" and "governing", which Li (1999) argues result in "compromised governmentalities" rather than the "mastery of territory".

Alongside a focus on states and infrastructures, the resource–state nexus (Bridge, 2014) can also be highly instructive in analyses of geopolitics/development. Work on resource geographies (Hayter, Barnes and Bradshaw, 2003; Bridge, 2010; Sheppard, 2013) has highlighted the ways in which states have sought to mobilise and manage natural resources or enabled various forms of accumulation. This is particularly timely given that extractive industry is expanding into new resource peripheries and unconventional locations (Bridge and Le Billon, 2013), including "ungoverned" or "undergoverned" and marginal spaces and conflict zones (Magrin and Perrier-Bruslé, 2011). Resource peripheries however have been treated not only as peripheral places but as peripheral to disciplinary thinking in Economic Geography and other social sciences (Hayter, Barnes and Bradshaw, 2003: 16). Sheppard (2013) thus raises the possibility of "theorising back" on the relationship between resources and economic geography from places outside global capitalism's core. More broadly, extractive economies have been associated with a proliferation of enclave spaces which represent important "spaces of postdevelopment" (Sidaway, 2007) and such intersections of enclosure and enclavisation demand and reward careful scrutiny (Sidaway, 2012; Kirshner and Power, 2015), especially given the numerous social conflicts over land and territory that lie at the heart of the extractive model (Bebbington, 2012).

Words like "scramble" (Carmody, 2011) and "grab" are narrative figures that loom large in recent research on the political economy of natural resources (Bridge, 2014), especially in relation to the global South and where the "rising powers" and "emerging economies" (e.g. China, India) are concerned following recent macroshifts in the geography of accumulation. Africa is one of the "epicentres" for this (Bridge, 2014). Some work on resources, however, particularly that on "land grabs", paints a picture of national landscapes throughout the global South being reimagined as "needed" resources for the rest of the world, in ways that downplay the significance of place, history and local context. Further, sovereignty is often imagined as the exclusive control of national states over internal natural resources in opposition to foreign capital, yet sovereignty must be understood in relational terms so as to consider the global geography of non-state actors that shape access to and control over natural resources (Emel, Huber and Makene, 2011: 70). In their work on gold mining in Tanzania, Emel, Huber and Makene (2011) adopt such a relational approach and raise important questions about the role of the colonial state but also about the post-colonial places of dispossession that specific sites of resource extraction create. Similarly, in a study of Chinese agribusiness in Laos, Dwyer (2014) argues that the political-economic approach exemplified by the neo-primitive accumulation literature on resource and land grabs (in which enclosure, displacement and dispossession loom large) could usefully be complemented with historically grounded, place-specific ethnographic investigation. This focus attends to questions of "micro-geopolitics" in considering local histories (rather than just international relations) and the Cold War geopolitical legacies that have shaped state–community relations in particular places (Dwyer, 2014).

THE STATE AND INSURGENCY

The impulse to classify, to standardise and to arrange populations as objects of development can be traced to the 1860s when "colonial bureaucracies became concerned with classifying people and their attributes, with censuses, surveys, and ethnographies, with recording transactions, marking space, establishing routines, and standardising practices" (Copper and Stoler, 1989: 611). This took on renewed importance in the 1940s, when imperial powers began seeking to make their empires more productive and orderly and to justify their rule on the basis of the "development" they were bringing to colonial peoples. A raft of colonial development and welfare legislation followed as colonial states were pressed into service as arbiters of national progress and development, shaping resource geographies and raising expectations. Colonial development interventions spanned a wide register of fields, from health and education, via community development and resettlement schemes, to agricultural modernisation (Worby, 2000; Beusekom, 1997; Smyth, 2004). Along with the education of nationalist elites (which played a key role in fomenting insurgency in Africa and the struggle to take control of the state), this "reformist urge" may have sown the seeds of its own destruction, hastening its own demise as:

> state-controlled development projects, making new demands on people, reallocating resources and access to power, and creating expectations on which colonial authorities could not deliver, fostered the resistance such policies were intended to avoid and focused it on the question of control of the state. (Cooper and Stoler, 1989: 619)

As an idea development quickly became central to colonial and Cold War expressions of what Routledge (1998: 245) calls "anti-geopolitics", a kind of "permanent assertion of independence from the state" or an "ethical, political and cultural force within civil society.... [that] challenges both the material (economic and military) geopolitical power of states and global institutions". These expressions of "geopolitics from below" emanate from subaltern (i.e. dominated) positions and Routledge (2003) includes within this a wide variety of anti-colonial national liberation struggles but also struggles against US imperialism (as in Vietnam) and a series of peasant guerrilla movements in Central and Latin America fighting against authoritarian rule. Development was central to many of these struggles but also subsequently to the articulations of post-colonial statehood pursued in many areas of the global South. Unlike other justifications for empire, "development came to have as strong an appeal to nationalist elites as to colonisers" (Cooper, 1997: 64). The challenges involved in transforming an inherited colonial state were numerous and complex, however. Nationalism had to transform from a "restive and reactive, what-we-are-not, separationist ideology to a persuasive image of a natural, organic, what-we-are, historic community, ready for deals, development, and practical alliances" (Geertz, 2010: 240). How better then to define "what-we-are" as a community than through a collective, national pursuit of progress and development? There were a multiplicity of challenges the "prototypical new state", just emerging from a colonial past into a world of intense great-power conflict, had to address from "a standing start":

> It had to organize, or reorganize, a weak and disrupted, "underdeveloped" economic system: attract aid, stimulate growth, and set policies on everything from trade and land reform to factory employment and fiscal policy. It had to construct, or reconstruct, a set of popular (at least ostensibly), culturally comprehensible political institutions – a presidency or prime ministership, a parliament, parties, ministries, elections. It had to work out a language policy, mark out the domains and jurisdictions of local administration, elicit a general sense of citizenship – a public identity and a peoplehood – out of a swirl of ethnic, religious, regional, and racial particularisms. It had to define, however delicately, the relations between religion, the state, and secular life; train, equip, and manage professional security forces; consolidate and codify a thoroughly pluralised, custom-bound legal order; develop a broadly accessible system of primary education. It had to attack illiteracy, urban sprawl, and poverty; manage population growth and movement; modernise health care; administer prisons; collect customs; build roads; shepherd a press. And that was just for starters. A foreign policy needed to be established. A voice in the expanding and proliferating system of trans-, super-, and extra-national institutions needed to be secured... (Geertz, 2010: 240)

The process of post-colonial state formation has also, in many cases, been beset by the challenge of maintaining security and containing insurgencies of various kinds. Governance and state capacity are key issues in the onset of insurgencies (Jones and Johnston, 2013) which themselves often develop in rugged, difficult-to-govern areas or in situations when a government's capacity to tamp down or co-opt opposition groups is in decline. During the Cold War and fuelled by superpower rivalry and the massive militarisation that ensued, the number of civil wars in the Third World rose inexorably to peak at around 50 or so in the early 1990s (HSC, 2005). As a result, the erstwhile Third World was increasingly "remapped" as a series of "borderland" spaces (Duffield, 2002) and a view of state failure leading to a breakdown in development, conflict, criminality and international insecurity began to take shape among metropolitan actors. That underdevelopment increases the risk of conflict while development reduces it was the basic rationale here in ways that are highly reminiscent of Cold War geopolitics as underdevelopment had once again become dangerous. Conflict in these "borderland" spaces is depicted as the result of a "regressive developmental malaise" (Duffield, 2002: 1066) characterised by illiberal and often corrupt or criminalised economies and state structures. Poverty, resource competition, environmental collapse, population growth and so on, in the context of failed or predatory state institutions, are now widely seen as fomenting non-conventional internal, regionalised and criminalised forms of conflict and insurgency. Such discourses, reminiscent of imperial claims to "civilise" spaces of "barbarity", carry a series of implicit "them" and "us" dichotomies establishing a formative contrast between "borderland traits of barbarity, excess and irrationality" with metropolitan "characteristics of civility, restraint and rationality" (Duffield, 2002: 1052).

Having once been regarded as justifiable on the basis of national liberation, the political violence of civil wars was, from the 1990s on, increasingly delegitimated and reinterpreted in terms of irrationality, the breakdown of order and the growth of criminality in the South (Duffield, 2008). This has been accompanied by a resurgence of state-led humanitarian, development and peace interventionism in spaces of crisis and conflict as aid actors have since expanded into a wide range of demobilisation, reintegration and reconstruction activities (Duffield, 2008). Here, development agencies reposition themselves "as a defence against the borderland forces of chaos and anarchy" (Duffield, 2002: 1064), with aid practice itself redefined as a strategic tool of conflict resolution and social reconstruction (Duffield, 1999), as a means to alter the balance of power between social groups in the interests of harmony (Uvin, 1999) or to produce desired political outcomes. Some of the liberal practices of development that have traditionally been associated with NGOs have also increasingly been rediscovered as forms of counter-insurgency (Slim, 2004). In this context and faced with the crisis of state-based security, development itself has been "rediscovered" and "reinvigorated", providing a new sense of purpose, a "second chance to make modernity work" (Duffield, 2002: 1064). Duffield (2002: 1065) writes that development has "come to acquire a new strategic role" but, as we have seen in the previous two chapters, development had long since acquired this kind of strategic, geopolitical significance, particularly during the Cold War and the links between development and counter-insurgency are far from novel.

States within the global political economy today face a "twin insurgency", from both above and below (Gilman, 2016). This includes both a *plutocratic insurgency*, in which globalised elites seek to disengage from traditional national obligations and responsibilities, and *criminal insurgencies* in which drug cartels, human traffickers, computer hackers, counterfeiters, arms dealers and others resist, co-opt and "route around" states as they seek ways to empower and enrich themselves in the shadows of the global economy by exploiting the loopholes, exceptions and failures of governance institutions. Unlike classic twentieth-century insurgents, who sought control over the state apparatus to implement reforms, criminal and plutocratic insurgents do not seek to take over the state but rather to carve out de facto zones of autonomy for themselves by crippling the state's ability to constrain their freedom of (economic) action:

> what both insurgencies represent is the replacement of the liberal ideal of uniform authority and rights within national spaces by a kaleidoscopic array of *de facto* and *de jure* microsovereignties. Rather than a single national space in which power is exercised and rights are enjoyed in a consistent and homogeneous way by all residents, the cartography of the dual insurgency represents diverse enclaves of political authority and of social service provisioning arrangements. (Gilman, 2016: 54)

This has led, Gilman argues, to a multiplication of various forms of authority between the full-blown modern state and outright anarchy, symbolised by the blurring lines between police, military and private security contractors which are themselves plugged into wider global security assemblages (Abrahamsen, 2017). In the space of the dual insurgency, citizenship "no longer signifies the liberal ideal of an identical package of rights for all, but instead means very different things depending on where individuals are in physical and social space" (Gilman, 2016: 54). This proliferation of exceptional and unique microsovereignties is typical of contemporary spaces of development (Ong, 2000; Sidaway, 2007), as are the enclavic spaces of social service provision that Gilman refers to. The paradigmatic case for plutocratic spatial segregation and secession is the so-called gated community but the enclaves of criminal insurgents are more precarious, temporary and less secure spaces (like Brazil's favelas) in which:

> some notionally social modernist state may claim authority, but in which true power is wielded by warlords, gangsters, or

> other kinds of organized criminals, who take de facto control over local security and whatever meager social service provisioning may be on offer... (Gilman, 2016: 55)

In contrast to an earlier inter-state norm, wars in the South today weave back and forth across borders to form regionalised systems of instability; they are not state-based wars in the traditional sense (Duffield, 2002). Major wars for state power in sub-Saharan Africa are now rare but there has been an increase in low-level insurgencies involving mobile rebel groups which work across national borders like the Lord's Resistance Army (LRA) (which operates in Uganda, South Sudan, the DRC and the Central African Republic), Al-Qaeda in the Islamic Maghreb (AQIM) (across the Sahel region south of the Sahara Desert), Boko Haram (Northeast Nigeria, Niger, Chad and Cameroon) and Al-Shabaab (Kenya, Djibouti, Ethiopia and Somalia, Tanzania and Uganda) (see chapter 6). There is no simple dichotomy between war and peace here and instead it is more productive to conceptualise war as an:

> ongoing, more or less "permanent" process of militarized power and conflict, varying in intensity over space and time from banal occupation and low-level warfare to sudden and terrifying bombardments and displacements, but premised ideologically on a distinct horizon of peace, and the closure of war as a discrete event. (Kirsch and Flint, 2011: 5)

The way in which different assemblages of intervention (including those of the state) are mobilised in war reveals a great deal about the contemporary operation of power (Bachmann, Bell and Holmqvist, 2015) and is important in understanding the geopolitics–development nexus.

One of the ways in which states have historically sought to counter insurgencies is through the deployment of "development" projects involving electrification, road and dam building, villagisation and population resettlement. In contemporary Asia this includes, for example, the use of transnational economic zones to promote cross-border trade among geographically remote population groups with transnational links, and the establishment of interstate highways, railways and commercial sea transportation links intended to promote trade, encourage economic growth and discourage the discontent that may give rise to insurgencies (Odgaard and Nielsen, 2014). For example, in the Greater Mekong, SEZs and infrastructure projects have been created to promote development in continental Southeast

Asian countries such as Vietnam, Cambodia, Myanmar and Thailand. After the civil war in Laos fought between the Communist Pathet Lao and the Royal Lao government (1959–75), "population management work" was conjured to meet the strategic tasks of defending the country and building socialism (Dwyer, 2014). Wrestling with the ghosts of Soviet collectivisation, China's Great Leap Forward, and Cambodia's "Year Zero", the population management led by Laos's Council of Ministers emphasised the need to control the activities of Laos's citizenry, especially its upland population, through a series of state simplifications and schematic categories (Scott, 1995) such as issuing identification cards, recording and monitoring overall population statistics, organising population relocation and arranging domicile patterns (Dwyer, 2014). State officials and foreign development experts also turned to "focal sites" in an effort to corral a shifting-cultivation-dependent population out of economically and politically sensitive forest areas without driving them over to the anti-government resistance, which was active at the time:

> Focal sites embodied the counterinsurgency-style panopticism embodied in strategic hamlets and socialist villagization schemes, but also emphasised the provision of state services like roads, schools and health clinics, as well as, eventually, formal land allocation to households. (Dwyer, 2014: 388)

Spatial tactics like focal site development were designed to achieve political and economic security. Resettlement of the population from mountainous and/or remote areas was commonplace during the Laotian civil war of the 1960s and 1970s. After the victory of the Laotian People's Revolutionary Party (LPRP) in 1975, some upland and highland populations who lived in sensitive areas were relocated into "safe" areas to prevent them providing any support to the Royalist movement and further internal resettlement programmes were implemented between the mid-1990s and early 2000s, again heavily motivated by security concerns (La-Orngplew, 2012). Indeed, in many parts of Southeast Asia, agricultural expansion has often been part of the state's geopolitical strategy, enabling control and consolidation of both frontier spaces and populations (De Koninck and Déry, 1997: 2). Upland areas in particular have often been constructed as spaces where levels of "civilisation" and "development" are still low and agricultural practices there (e.g. various forms of shifting cultivation) have regularly been used to justify the labelling of such people and places as "backward" or "primitive" (McElwee, 2004;

Duncan, 2004a; Cramb et al., 2009), requiring "civilisation" from the core (Hirsch, 1989; Li, 2007). Schemes claiming to eliminate the "backwardness" of frontier spaces often enabled a "soft" control of them through agricultural expansion and intensification, aimed at a state-led process of "civilising the margins" (Duncan, 2004b) and those people and places "on the outer boundaries of modernity" (Hirsch, 2009: 125). In the 1980s large-scale plantations were rehabilitated in a southern region of West Java, for example, where the promotion of commercial crops to peasants was strongly made in those areas that had experienced uprisings against the government (White, 1999: 237). In Thailand the clearing of new forestland for commercial crops in mountainous areas was largely driven by security concerns and a desire to prevent people from joining the communist movement (Uhlig, 1988: 15) whilst road networks were also developed in remote areas of Thailand, helping both to facilitate the penetration of the market and to maintain national security near its border with Laos and Myanmar (Thomas et al., 2008; Fox et al., 2009).

In recent years, insurgencies relating to a variety of religious, ethnic and nationalist concerns have been common in several parts of Asia including China, India, Laos, Myanmar, Thailand and the Philippines. In Buddhist Myanmar's Rakhine state the Rohingya community, stripped of their citizenship in 1982 and categorised as "nonnationals", as stateless people, have waged an insurgency whilst in response the military has attempted "ethnic cleansing" of the Rohingya Muslims in 1992, 2012 and 2016–18, with many killed and displaced as a result (cf. Cheesman, 2018; see Figure 5.5). The conflict has its genesis in control of the country's natural resources, namely the land occupied by the Rohingya where over the past few decades and in the name of "development", the military junta has been acquiring land from small-scale farmers without due compensation. Myanmar has been a key focal point of China–India rivalry (who both have significant natural resource interests there) and so the state has a strong vested interest in clearing land to prepare for further "development". China itself has faced violent social unrest in Tibet and Xinjiang and has been developing counter-insurgency strategies in response (drawing upon a range of experiences, including that of the US in Afghanistan), with some observers suggesting that Chinese interests in unstable areas may lead it to co-opt insurgents, militias and other sub-state actors who control territory (Odgaard and Nielsen, 2014). China's OBOR initiative also has to be understood in the context of its domestic objectives which, in addition to advancing China's "Go

FIGURE 5.5 Rohingya refugees fleeing a military operation in Myanmar's Rakhine state take shelter in Cox's Bazar, Bangladesh on September 7th, 2017. Rehman Asad/Alamy.

West" initiative focused on the development of provinces such as Gansu, Guangxi, Ningxia, Shanxi, Yunnan and Xinjiang, aim to help produce a more favourable regional/domestic security environment in China's western areas that confront the challenges of religious extremism, separatism and terrorism (Blanchard and Flint, 2017).

In tracing the emergence of the Maoist revolution in Nepal in the 1990s, Paudel (2016) argues that in some peasant societies of the South the geographies of developmental empowerment and subaltern rebellion have unexpectedly overlapped and expanded rapidly in recent years. The uprising emerged from the Rapti region of Nepal, the focus of international development interventions since the 1950s and a series of financial and technical investments aimed at showcasing it as a model of development. Given the country's strategic geopolitical location between China and India, the Cold War had a strong influence on Nepali politics and development (Levi, 1954; Mihaly, 1965) with fears that peasants in the Rapti region were particularly vulnerable to communism because of their geographical remoteness and relatively autonomous economic practices. In 1952 the US Government's "Mutual Security Program" noted that "under current conditions, the agrarian sector offers a major target for communist subversion in South Asia" (cited in Mihaly, 1965: 32). In response, the Nepali state first strengthened the bureaucratic and police presence in the region and implemented integrated rural development programmes with the goal of systematic transformation of peasant communities (Fujikura, 2013). Development through agrarian transformations was conceived by the West as the only way to achieve economic and political stability in the region in order "to resist internal and external communist aggression" (Mihaly, 1965: 31). With funding from USAID, the Rapti Integrated Development Project (RIDP) was implemented, beginning in the 1970s, mobilising hundreds of rural villages for the intensive implementation of development activities such as adult literacy, rural infrastructure and community forestry, ostensibly to contain an upsurge in rebellious sentiments. Development projects can have unintended consequences, however, as Paudel (2016: 1030) shows how these increasingly consolidated movements at local and district levels provided a foundation for the emergence of the Maoist insurgency, as peasants in the Rapti region "enrolled themselves as development subjects in the 1980s and emerged as a revolutionary force in the 1990s" (see Figure 5.6).

FIGURE 5.6 Nepalese troops from the Armed Police Force pause for a break by a school building during counter-insurgency operations in Kathmandu, Nepal on December 19th, 2013. Photo by De Visu/Shutterstock.

CONTESTING STATE POWER: SOCIAL MOVEMENTS

Social movements, operating at a variety of scales, have also been important in contesting hegemonic conceptions of state, citizenship and society or in mobilising to defend particular places and ecologies from interventions planned in the name of development. Social movements in the global South have mobilised a wide array of subaltern groups (including, for example, peasants and landless workers, women, informal sector and unemployed workers, slum dwellers, indigenous peoples, and marginalised youth) around sets of radical claims and practices that have challenged the anti-politics of the neoliberal development regime in significant ways (see Motta and Nilsen, 2011; Nilsen, 2015). At the global level, major summits (such as those for the G8 and G20) and meetings of the IFIs have increasingly become the focus of protests by social justice activists whilst transnational anti-capitalist social movements like Occupy (see Figure 5.7) have demonstrated their support for alternative futures to those promised by neoliberal globalisation, promoting alternative geographic imaginaries that contrast significantly with those at the centre of developmentalist

FIGURE 5.7 Portuguese protesters participating in the global "Occupy" protests knock down police barriers and take the marble staircase in front of the Portuguese parliament building in Lisbon on October 15th, 2011. Photo by Luis Bras/Shutterstock.

renderings of the spaces of globalisation (Nagar et al., 2002; Silvey and Rankin, 2010). In addition, as Lubeck (2000) and Watts (2003) have shown, the rise of Political Islam, understood as an "anti-systemic movement", is intimately linked with the implosion of the secular nationalist Development project (Hart, 2010).

Several studies have demonstrated the significance of particular spatialities such as place (Routledge, 1993), scale (Miller, 2000) and networks (Featherstone, 2008) for social movement mobilisation, with growing attention to the spatialities of contentious politics (Jessop, Brenner and Jones, 2008; Leitner, Sheppard and Sziarto, 2008; Nicholls, Miller and Beaumont, 2013; Halvorsen, 2017). The "material, symbolic and imaginary character of places can powerfully influence the articulation of protest" (Routledge, 2017: 24) since the "place of performance" has a significant bearing on where and how social movements arise and the particular cultural, economic and political milieu in which a protest emerges influences the character and form that resistance takes (Routledge, 2017). In Apartheid South Africa, for example, resistance took a wide variety of different forms shaped by the specific context including mass demonstrations, marches, music, alternative press and advertising, flag burning, graffiti, strikes, the wearing of symbolic clothing, "stay-aways" organised by labour groups, rent boycotts, international sanctions, divestment, hunger strikes by political prisoners and student movements. There were also the many forms of defying Apartheid regulations by sit-ins/occupying "whites only" spaces such as buses and beaches and the destruction of key infrastructures assembled by the Apartheid state (Kurtz, 2010).

Social movements are significant in understanding the geopolitics–development nexus because they have forged alternative modernities, refused teleological models of socio-spatial progress and sought to build coalitions attentive to diverse aspirations and trajectories (Andolina, Laurie and Radcliffe, 2009; Sheppard et al., 2009). They have, in many cases, rejected views of globalisation that celebrate the diffusion of market principles and advocated instead different ways of measuring and making value across non-homogeneous places (Sparke, 2005). They have also rejected the reductionism of logics of capital and commodification, defined political formations in terms that extend beyond their relationships to states and markets (Bakker and Silvey, 2008) or created transformative political projects that include space for non-capitalist and also alternative capitalist models, narratives and practices (Gibson-Graham, 2006; Sparke et al., 2005). Some scholars thus advocate taking seriously the "oppositional movement spaces" constructed by social movements "as part of a justice-oriented postdevelopment map that already exists" (Silvey and Rankin, 2010: 700). The claims made by subaltern groups frequently revolve around demands for redistribution or recognition in some form or other (often related to questions of citizenship) but their struggles often delegitimise dominant meanings of development by calling attention to the discrepancies between state ideologies and lived realities (Moore, 1998; Nilsen, 2010). Social movements in the South however do not always articulate their demands as expectations on "development" or the money economy. The women that were central to India's Chipko movement in the 1970s and early 1980s, for example, worked in subsistence agriculture in ways that have been interpreted by ecofeminists as a critique of the prevailing capitalist-, profit- and growth-oriented development paradigm but the Chipko movement sought primarily "to preserve their autonomous control over their subsistence base, their common property resources: the land, water, forests, hills" (Mies and Shiva, 1993: 303).

Post-development scholars have placed a great emphasis on social movements and the alternative projects, practices and imaginaries that they construct, arguing that a progressive future is not to be forged through the construction of "development alternatives, but in alternatives to development, that is, the rejection of the entire paradigm altogether" (Escobar, 1995: 215). In post-development literatures it is social movements that are crafting these alternatives to development since their "anti-development struggles may contribute to the formation of nuclei of problematised social relations around which novel cultural productions might emerge" (Escobar, 1995: 216). In a variety of ways, however, subaltern groups have appropriated "the rhetoric of development" (Gupta, 2000: 16) and inflected it with meanings that express their grievances, needs, interests and aspirations in order to make claims on dominant groups, not just on states but also NGOs and multilateral institutions (Nilsen, 2016). Various studies both of the micropolitics of everyday development encounters (Moore, 1998, 1999; Li, 1999; Gidwani, 2002; Shakya and Rankin, 2008) and of large-scale social movements (Rangan, 2000; Nilsen, 2010; Vergara-Camus, 2014) have established that subaltern groups do not always oppose or reject development in its entirety, but rather often seek to negotiate and change the direction and meaning of development. For Matthews (2017) post-development scholars have failed to fully grapple with this question of the desirability of development which has been all too easily dismissed simply as evidence of subaltern minds being colonised.

If development is seen simply as an apparatus of power, it ends up disavowing people's subjectivity (Kapoor, 2017) and fails to acknowledge that "development is a desiring machine ... not only an apparatus of governmentality" (Escobar, 2008: 175). In the Lacanian/Deleuzean approach advocated by De Vries (2007: 35), rather than being a rational, legal-bureaucratic and hierarchical order, the development apparatus functions "as a crazy, expansive machine, driven by its capacity to incorporate, refigure and reinvent all sorts of desires for development". Here, the focus is on the relations between the actual life-ways, dreams and aspirations of local populations and the virtual realm of development rhetoric, routines and procedures as the desire for development "fills in a certain lack in its actualisation" (De Vries, 2007: 32). Paradoxically, the very idea of development relies on the production of desires, which it cannot fulfil and as such there is a certain "excess" in the concept of development that is central to the functioning of the development *dispositif*:

> we have to scrutinise the disjuncture between the desire for development and its banalisation in practice.... Foucauldian post-structuralist theory fails to interrogate the very lack in development itself, its inability to engage with the dreams and fantasies it triggers. (De Vries, 2007: 32, 35)

In the 1990s, Latin America in particular witnessed the rise of a plethora of social movements seeking to move beyond developmentalism and neoliberalism whilst today the region's "post-neoliberal" countries are widely celebrated as offering the most visible alternative policies, programmes and mindsets which will break the mould of development (Radcliffe, 2015). From the Zapatistas, who seek non-violent ways to ignore or organise around the state, to the rise of militant land movements such as the *Movimento Sem Terra* in Brazil and a wave of struggles unleashed by the privatisation of water and other basic services, the region has become a key site for contestations of state-led development (see Figure 5.8). The tension between top-down neoliberal development and insurgent alternatives so characteristic of the 1990s has to an extent, however, now been displaced by the rise of Latin American countries that attempt to merge a strong state with grassroots development models (Radcliffe, 2012). These

FIGURE 5.8 Social movements protest at Paulista Avenue in Sao Paulo, Brazil on April 1st, 2016 against corruption within the country's political parties. Photo by Al Ribeiro/Shutterstock.

moves to a more developmental state have been called forth by electorates and widespread popular mobilisation against neoliberalism. Ecuador's post-neoliberal development agenda, for example, initially aimed to address the material and social exclusions associated with neoliberal, capitalist development but in recent years Petroamazonas (the state oil company) has been drilling for oil in the Yasuní national park, an area of significant biodiversity. Some of these countries initially attempted then to implement "post-development" but found the practices, grammars and logics of neoliberal governmentality hard to shift (Radcliffe, 2012: 248). *Buen Vivir* (living well) has emerged as an influential socio-political and identitarian concept (Altmann, 2015) lauded in many circles as the realisation of post-development agendas and widely taken up by social movements. The concept, which incorporates ideas of de-growth and a stern critique of extractivism, is generally defined as forming part of the Andean indigenous cosmology, representing a particular vision of society and relationships with nature, entailing a radical questioning of colonialism, the dominant development model and modern institutions (Acosta, 2008; Villalba, 2013). In Ecuador and Bolivia, it has recently obtained a distinct symbolic, political and also legal status.

Scholars like Mignolo (2000) and Escobar (2007b, 2010) have used the idea of *decoloniality* to understand and explain some of the changes, transformations and future horizons in Latin America. Decoloniality is not only a long-standing political and epistemological movement aimed at liberation of (ex-) colonised peoples from global coloniality but also a way of thinking, knowing and doing. It is an approach that has been particularly interested in *Buen Vivir* "due to the epistemological and ontological rupture in relation to Western epistemologies brought about by its emergence" (Florentin, 2016: 9). Due to its indigenous roots and its philosophical and spiritual underpinnings, the concept of *Buen Vivir* is regarded as an alternative to modernity and an ontological space from where alternative epistemologies (to dominant Western ones) can be developed. The primary premise here is that European colonisation and the making of the capitalist world-system have been constitutive elements of modernity in Latin America (Blaser, 2009), forming part of a "coloniality matrix", the application of which in Latin America has resulted in a capitalist, Christian, colonial and modern framework of society at the expense of alternative world views and cultural models (such as *Buen Vivir*). This does usefully draw attention to indigenous conceptions of "development" and territoriality and the ways in which they often conflict with those produced by states. Geographers have often examined territoriality as the top-down assertion of power over space but there is growing acknowledgement that territoriality is also produced from below by social movements (Routledge, 2015; Vasudevan, 2015), presenting overlapping (Agnew and Oslender, 2013) and clashing (Zibechi, 2012) territorial claims to that of the state and dominant institutions (Halvorsen, 2017). Thus, whilst it is important to examine the question of how the state produces its space through territorialisation, as Wainwright (2008: 27) does in noting the ways in which colonial practices territorialised Mayan spaces and bound together political identity with development and settlement, it is also important to attend to indigenous counter-mappings to colonial and post-colonial territorialisations of the Maya (Peluso, 1995; Wainwright and Bryan, 2009).

The study of social movements is a somewhat neglected field of research in African Studies. Not only does Africa remain largely absent from social science research using a social movement perspective but much of the theoretical literature focuses on social and political movements in Europe, North and South

America and tends to neglect the African continent (Brandes and Engels, 2011). With the notable exception of the South African struggle against Apartheid, Africa has either been neglected by theorists of non-violent political action against political repression and injustice or been misread from a perspective of "western ...expectations and norms" (Larmer, 2010: 257). Typically, social movements in Africa have been discussed in the framework of "civil society" (Eckert, 2017) but de Waal and Ibreck (2013) argue the case for specifying the relationship between social movements and the character of African states. Beginning with the so-called IMF bread riots and "austerity protests" in many parts of Africa (and Latin America) in the 1980s that were subjected to harsh stabilisation and structural adjustment measures, Africa has long been a locus of protests against neoliberal models of development. Structural adjustment, as Simone (2004b: 8) notes, was never just about restructuring the economy but also involved "the restructuring of the time and space of African lives". What followed these riots was the emergence of a new discourse of entitlement that would ultimately crystallise in the oppositional projects of the social movements that had consolidated across much of the South by the early 2000s (Nilsen, 2015). South Africa, which has heavily embraced neoliberal economic policies (Walton and Seddon, 1994; Marais, 2011; Hart, 2014), has witnessed significant mobilisation against neoliberalism since the early 2000s, with grievances arising from "atomising neoliberal pressures" (Bond and Mottiar, 2013: 284).

Writing about Johannesburg, Simone (2004a: 411) identifies the ways in which residents themselves become "infrastructure", as the inner city "has been let go and forced to reweave its connections with the larger world by making the most of its limited means", due to processes of fragmentation and state abandonment. Today, poor people's movements like the Landless People's Movement in Johannesburg, the Anti-Eviction Campaign in Cape Town and *Abahlali baseMjondolo* (a shack dweller movement in Durban and Cape Town) have mounted effective challenges to the post-Apartheid state and its neoliberal vision of development. These protests, very much driven by grassroots agency, are directed against the state's failure to deliver basic conditions associated with development (from access to flushing toilets to decent housing and jobs) and not necessarily against the notion of development itself (see Figure 5.9). They have often been framed as "service delivery" protests (Alexander, 2010) yet according to Pithouse (2011: 1), the "service delivery myth"

FIGURE 5.9 Residents of Hout Bay, South Africa protest on March 3rd, 2017 about their living conditions after broken promises to provide adequate housing and sanitation. Photo by Charles H B Mercer/Shutterstock.

presents people as "consumers or beneficiaries who just need to be plugged into the grid of serviced life by a benevolent state" and obscures the fact that these protests often represent "rebellion against service delivery as it is currently practiced rather than a demand for it to be speeded up". For example, there are instances where people living in informal settlements resist being relocated to state housing because the housing on offer is not suitable or is distant from employment opportunities or family and friends. Here, protestors' demands are less about "delivering already determined services" and more about agitating for greater democracy and for meaningful recognition of their citizenship (Mottiar, 2013: 605–606). The desire for "development" therefore is often "tangled up with the desire for equality, dignity and redress" (Matthews, 2017: 2658), something that has not always been sufficiently acknowledged by post-development theorists. While people "surely do want water and electricity for their own sake, protests about service delivery are also about what access to water, electricity, adequate sanitation and the like represent" (Matthews, 2017: 2658) which in the South African context is highly complex and contentious.

CONCLUSIONS: SPACES OF SUBALTERN STRUGGLE

This chapter has argued that a focus on states and on the different geographies of resistance that emerge in response to state-led discourses and imaginaries of "development" provides valuable insights into how geopolitics intersects with development. Deleuze and Guattari (2002 [1980]: 385) once wrote that "one of the fundamental tasks of the State is to striate the space over which it reigns" and the governmentalisation of the "objects" of development is key in tackling this task and in creating a "legible" society to be governed (Scott, 1998). States themselves are understood here as always emerging and becoming and therefore incomplete but also as located and experienced at many sites and across various scales, what Gupta (1995: 392) refers to as the "translocality of state institutions". The state is best approached then not as a fixed and static entity and "not a thing, a system or subject" but as an assemblage of powers and techniques, an ensemble of discourses, rules and practices, as something to be discovered, as contingent and evolving. States emerge as an imagined collective actor partly through the "telling of stories of statehood and the production of narrative accounts of state power" (Painter, 2006: 761) and development is a central part of the articulation of these stories and narratives.

States can usefully be read in relation to how they are understood, experienced and constituted through everyday spaces, practices and narrations. Infrastructures of various kinds have historically helped create, destroy, expand or limit the contours of the state (Meehan, 2014: 216) and represent important sites of contestation where the meanings of development and the authority of states have been contested. Infrastructure has also become a key focus of SSDC where companies from an assortment of (re-) emerging economies like China and India have delivered a number of infrastructure-related projects (e.g. roads, airports, ports, railways, water, energy and housing) and are significantly impacting on the infrastructural systems of recipient countries. Infrastructure is central to China's OBOR strategy in Africa too: this includes the flagship project, a 290-mile railway from Nairobi to Mombasa (with plans to extend that network into South Sudan, Uganda and Burundi), a 470-mile electric railway running from Addis Ababa to Djibouti port and the SEZ planned for an area south of the port city of Bagamoyo in Tanzania which is regarded by some in China as a "new Shenzhen" and that recalls the heady days of the TAZARA railway project built during the age of Maoist, Third Worldist cooperation (Van Mead, 2018).

Many states no longer even try to demonstrate concern with the welfare of their populations and the symbolic efforts of discourses on participatory governance "largely become performances deployed to attract donor interest" (Simone, 2004b: 8). Development however must enrol its subjects as always "becoming", in order to continuously legitimise the superiority of its ideology (Hart, 2002; Moore, 2005) and state-led development programmes and projects have been key to that. Yet development has a "double life" (Paudel, 2016) in that it always produces a subaltern relation between master (agents of development) and follower (subjects in need of developing). Subaltern political possibilities are thus "always present and inherent in the historical material processes of development" (Paudel, 2016: 1047), which means that its political outcomes are always complex and usually unpredictable: projects intended to deflate rural rebellion, for example, may stimulate new challenges to existing forms of hegemony (Paudel, 2016: 1047).

Bordering practices and territorialisation also play an important role in the process of becoming and seeing like a state and attention to the "borders–development nexus" (Novak, 2016) and to the ways in which territoriality is produced from the top down (by states) and the bottom up (by social movements) is instructive here given the recent proliferation of FTAs, SEZs and Growth Corridors and other re-bordering strategies characteristic of contemporary development, whereby territorial jurisdictions are redefined in regional, cross-border and/or sub-national terms. Historically, development has often played a key role in the process of laying claim to territory, particularly by states seeking to control and consolidate both frontier spaces and populations in the name of vanquishing rural "backwardness". The resource–state nexus (Bridge, 2014) is also highly significant here, particularly given the rise of (re-)emerging economies like China, India and Brazil where state-led developmentalisms (in which resources figure prominently) return centre stage. It is necessary, however, to avoid the tendency to subsume development interventions under the imperatives of capital and/or the strategic responses of states and to avoid overemphasising the role and power of states in shaping development processes and outcomes (Novak, 2016). The focus on enclosure, displacement and dispossession in the neo-primitive accumulation literature on resource and land grabs, for example, could usefully be complemented by more historically grounded, place-specific ethnographic investigation and a focus on the "micro-geopolitics" that shape how state authority is experienced and encountered, involving an emphasis on politics (rather than just economics); on local actors (rather than just investors and their sovereign backers); and on "specific landscapes and populations (rather than countries) on the receiving end of investment deals" (Dwyer, 2014: 386–387).

Duffield (2007a: viii) argues that "development has always existed in relation to a state of exception". Insecure, collapsed, weak and fragile states provide the other against which model, steady states in the periphery are imagined and constructed. This focus on state weakness and fragility has its roots in colonial discourses about governance and development but in a variety of other ways empire has significantly shaped theories and practices of statehood in the South. From the different colonial governmentalities and developmentalisms and the resistances they engendered to the post-colonial challenges of taking over and transforming colonial bureaucracies, colonial constructions and spatialisations of the state have had enduring consequences. As Manchanda (2017) has argued, colonial spatialisations still structure the ways in which we experience and think about the Afghan state today, which did not fully materialise as a "principle for reading reality" or "scheme for intelligibility", despite the multiple interventions of USAID. The labels that are variously applied to the Afghan state – "buffer", "rogue" or "failed" – along with the trope of the "frontier" which has played a formative role in defining Afghanistan as a state and space are "essential elements in a story of imperial sense-making" (Manchanda, 2017: 387) and part of Afghanistan's long lineage of "constructed deviance" (something many African states also have experience of). Indeed, many of IR's core concepts have been closely shaped by empire and require provincialising and decolonisation. Although sovereignty, for example, "carries the horrible stench of colonialism" (Barker, 2005: 26) it has also been rearticulated to mean altogether different things by indigenous peoples.

It is useful to retain the concern with the dynamics of power and resistance that has been central to the radical critique of development since the 1990s but modify it by understanding power in less absolute and unitary terms than post-development (Nilsen, 2016: 273). Although power is constantly exercised through discourses of development, it is also constantly challenged and consequently reshaped since "the power of development does not simply mould the global South in its own Eurocentric image" (Nilsen, 2016: 273). Resistance and insurgency is also not simply an assertion of otherness that rejects development but "is a practice of meaning- and claims-making that hinges on oppositional appropriations of dominant symbols and idioms" (Nilsen, 2016: 273). As Mitchell (2002: 77) has argued, the effects of policy and expertise do not arise from preformed designs imposed from outside, nor from their own logic, but are wrought through the ruptures and contradictions they effect in existing social, political and ecological systems. These ruptures and contradictions are highly significant for the study of the nexus between geopolitics and development. Following Mosse (2007) we might move away from the approach taken in some critical treatments of development discourse that privilege the products of public policy process materialised as text or focus on the intended and unintended effects of interventions on populations, regions and communities, to instead explore the internal dynamics of development's "regimes of truth" and the political communities and knowledges of development that they generate. How policy ideas, models and frameworks travel across networks and assemblages at various spatial scales is important here and, in this sense, anthropological approaches and methodologies may offer a way forward (Mosse, 2005, 2007).

As noted in previous chapters, the Third World has long been constructed as a space of insurgency. Colonial and Cold War forms of "anti-geopolitics" (Routledge, 2003) provided significant challenges to modernisation discourses of development and represent important chapters in the history of rebellion and resistance. In some ways social movements have built upon the legacies of popular mobilisation and today represent important forms of insurgency that continue to reject the normalising power of (state-led) development. It is the tensions between the "discourses of control" (articulated by states in a hegemonic inflection) and "discourses of entitlement" (in an oppositional inflection as articulated by social movements) that require further attention since the encounter "between such discourses in and through conflicting political projects in turn gives shape and form to development as a trajectory of sociohistorical change" (Nilsen, 2016: 272). Social movements have historically constituted important spaces of resistance and destabilisation of the development project (cf. Escobar, 2008) but do not always seek an outright rejection of development in its entirety, often instead seeking rather to negotiate and change its meaning and direction (Nilsen, 2016).

The spaces of subaltern struggle and the imaginaries, discourses and practices created by social movements can tell us a great deal about the contested geopolitics of contemporary development. As Silvey and Rankin (2010: 700–701) have argued, "attention to the geographies and histories of these movements can contribute to ongoing efforts to decolonise the political geographic imaginaries of critical development studies". In particular there is a need to take the different imaginaries created by social movements seriously as expressions of "the capacity to desire a different kind of society that is not yet defined" (De Vries, 2007: 27). Post-development has usefully called attention to the fact that such movements have created valuable alternatives, crafting alternative forms of collective ownership, for example, in the form of the worker-run factories organised by the *piqueteros* (picketers) of the *Movimiento de Trabajadores Desocupados* (Movement of Unemployed Workers) in Argentina or the cooperatives of the MST in Brazil. Social movements are skilfully linking localised struggles to the dynamics of global power structures and mobilising to achieve progressive changes across spatial scales but are also playing a key role in democratising development and promoting more participatory and deliberative forms of political decision-making, for example, in and through the practices of movements of the urban poor, such as the *Comités de Tierra Urbana* (Urban Land Committees) in Venezuela and *Abahlali baseMjondolo* in South Africa. There is also much we can learn from the study of post-neoliberalisms (Elwood et al., 2016) although popular struggles have often quickly been diminished and co-opted by states:

> We also need to remember that, across space and time, elites brought to power by the struggles of ordinary people have usually moved very quickly to diminish these struggles by reducing democracy from a day to day popular practice to the altogether more anorexic conception of occasionally voting for one of the elite groups contending for state power. (Pithouse, 2009: 3)

A case in point is the "Arab spring", a wave of violent and non-violent demonstrations, protests, riots, coups and civil wars in North Africa that began in Tunisia, in December 2010 with the Tunisian revolution but which spread to Libya, Egypt (see Figure 5.10), Yemen, Syria and Bahrain where the regime was either toppled or major uprisings and social violence occurred, including civil wars or insurgencies. The Egyptian and Tunisian protesters were able to generate "resonant collective action frames" linking immediate material grievances with the demand for systemic change and built upon more than a decade of protests about material issues including water and bread shortages that had

FIGURE 5.10 Thousands of protesters flock to Cairo's Tahrir Square, Egypt on November 22nd, 2011 to protest against the military junta that took power following the toppling of President Mubarak. Photo by Hang Dinh/Shutterstock.

shaped "people's consciousness and organisational capacities" before the revolution (Ali, 2012: 16). In Egypt, as foreign aid poured in, the uprisings were discursively reframed by the IFIs within a pro-market discourse not as a revolt against several decades of neoliberalism but rather as a movement against an intrusive state that had obstructed the pursuit of individual self-interest through the market and as due to the *absence* of capitalism rather than its *normal* functioning (Ali, 2012: 16). Popular demands to reclaim wealth, offer state support and services to the poor, nationalise those industries that were privatised and place restrictions on foreign investment, could be either disregarded or portrayed as "anti-democratic" (Hanieh, 2011). Another example is the Zapatista movement, hailed by post-development writers like Esteva (1999: 173) as embodying a "radical democratic" struggle focused not on "seizing state power" but on seeking to build decentralised, autonomous spaces that can create "new political relations". For Žižek, the Zapatistas became not a counterpoint to the state but its "shadowy double" (Žižek, 2012: 177) as its moralistic protest now places it in an increasingly unthreatening, symbiotic relationship with the state (and capital) (Kapoor, 2017) that could be seen, not as posing a threat to NAFTA (Žižek, 2012: 178) – the movement's main objective at NAFTA's inauguration in 1994 – but as helping to facilitate NAFTA's integration. The Zapatistas succeeded in gaining a degree of local autonomy but did "little to transform wider sociopolitical power" (Kapoor, 2017: 2673).

The chapter has also argued that a focus on the spaces of contragovernmentality, ungovernability, anarchical governance or the borderlands of global politics is productive. One of the key concerns of the counter-insurgency lobby has been that insurgencies today no longer appear directed at "taking over a functioning body politic, but dismembering or scavenging its carcass, or contesting an 'ungoverned space'" (Kilcullen, 2006a: 112). It is to the contemporary scripting of failed states and other such borderland spaces of ungovernability in the global South by the US military (and their attempts to counter insurgency through aid and development) that our attention now turns in chapter 6.

Chapter 6

The political geographies of contemporary US foreign assistance

INTRODUCTION: RECONSTRUCTION AS WAR

DURING the announcement of his US Presidential candidacy in 2016, Donald Trump pledged to put "America first" and to "stop sending foreign aid to countries that hate us and use that money to rebuild our tunnels, roads, bridges and schools" (*The Guardian*, November 13th, 2016). The US is the most generous global aid donor in absolute terms, but relative to the size of the American economy (total foreign assistance in 2017 amounted to approximately 1% of the total federal budget) it is less a case of "America first" than "America twenty-second" (Konyndyk, 2017). Trump suggested foreign aid was a waste of US tax dollars and subsequently proposed a budget prioritising "hard" vs "soft" power, including a huge uplift in defence expenditure but also massive cuts to current spending for US diplomacy and foreign aid. This would have meant preserving US$3.1 billion in security aid to Israel whilst cutting bilateral aid entirely to countries like Nicaragua and closing USAID missions, reducing funding for the UN and reducing spending on global health, humanitarian, refugee and international disaster assistance. In June 2018 Congress pushed back however on Trump's proposed cuts and passed a budget bill that maintained existing levels of funding of US$54.4 billion for foreign affairs programmes (Saldinger and Igoe, 2018).

Just over a year previously, when Trump first proposed the aid cuts as President, more than 120 retired US generals and admirals, including some of the most prominent US military officers to serve in recent decades (such as General David Petraeus, the former CIA director and commander of troops in Iraq and Afghanistan), signed a letter urging Congress to set about elevating and strengthening diplomacy and development alongside defence and to fully fund them, since they were "critical to keeping America safe":

> The State Department, USAID, Millennium Challenge Corporation, Peace Corps and other development agencies are critical to preventing conflict and reducing the need to put our men and women in uniform in harm's way.... The military will lead the fight against terrorism on the battlefield, but it needs strong civilian partners in the battle against the drivers of extremism – lack of opportunity, insecurity, injustice, and hopelessness. (United States Global Leadership Coalition, 2017)

Whilst the US is certainly not the only Western donor country considering cuts to its foreign aid budget in the context of austerity and to suggest that the efficiency of aid could be improved or that charity should begin at home, what is interesting and significant about the proposed cuts is that so many senior US military personnel were so heavily invested in defending US aid (and its importance to US diplomacy and conflict resolution) alongside defence. In order to secure aid funding, it is clearly more effective to present ODA to the US Congress as a "strategic defence system" that protects the homeland rather than as a mechanism for poverty alleviation in the far-off countries of the borderlands (Brainard, 2006). In recent decades and particularly after shifts in the global security agenda following 9/11, aid has more than ever become about the safeguarding and promotion of donor interests (Woods, 2005). This chapter explores these recent global trends in development assistance and in particular the creeping securitisation and militarisation of contemporary aid, with particular reference to US aid policy. Further, it considers the recent revival of a long-dormant interest in development-based counter-insurgency techniques ignited by US military setbacks in Iraq and Afghanistan (in which lessons on pacification were drawn from classical counter-insurgency doctrine) but also the enrolment of USAID within its contemporary application. The chapter is also concerned with the current configuration of the US military assemblage in Africa and the contemporary scripting of the continent as a "swamp of terror" infested with Islamic terrorist insurgencies along with the use of civil affairs projects, reconstruction and capacity-building "security cooperation" to counter them.

Threats to national security have often stoked a "geopolitics of fear" that less developed areas will somehow "invade" or "infect" donor countries, or that the South will "leak" into the North, if adequate aid transfers are not provided. In this sense states produce the *threat* of crises in order to authorise various enforcement measures to eliminate them (Amoore, 2006; Sparke, 2006; Hyndman, 2007). As Hyndman (2009: 874) argues, public anxiety "consequently sets in motion political demands for protection from often ill-defined, geographically diffuse threats: disease, asylum seekers, transnational crime, terrorism, all ostensibly linked through a global web of risk". In this way the global North attempts, quite unsuccessfully, to isolate itself from the South by exerting control over migration and security matters through a series of shared tactics (Amoore, 2006). Proponents of American exceptionalism "paint the world as a hostile place, an environment in which America must constantly strive to control and eliminate evildoers before their malevolent acts hit the American homeland" (Holsti, 2011: 392). This "permanent aura of exaggerated insecurity" (Holsti, 2011: 394) may also help create and sustain the role and efforts of the US "to liberate others" and guide them to the trappings of liberal democracy and capitalist development. Aid is not just about maintaining the safety and security of donors, however, it is also a tool that recipient governments use "to stabilize social relations at home and is often an important part of domestic security arrangements" (Hyndman, 2009: 875). The question of whose security is being talked about here is key, although within the liberal policy mainstream "one finds surprisingly little serious interrogation of the concept of security itself and of how, by whom and with what political agendas security issues are framed and security functions are exercised" (Luckham and Kirk, 2013: 2) (see also Buzan, 1983).

The US, the EU and other governments now regularly use the rhetoric of "opposing terrorism" as a basis on which to allocate aid whilst OECD rules governing how member states give aid have been adjusted to include terrorism prevention and a range of military activities (OECD DAC, 2003). The growing emphasis on security and countering terrorism has also enabled several states in the global South to attract and secure various forms of economic, social and military assistance, as in the case of rotating members of the UN Security Council (UNSC) who extract rents by trading their votes for political or financial favours during the two years in which they enjoy a boost to their diplomatic importance (Malone, 2000). Indeed, the US reportedly issued "promises of rich rewards" to rotating members in exchange for their support during the run-up to the 2003 invasion of Iraq (Renfrew, 2003) and UNSC membership has a large positive effect on foreign aid receipts, especially in years when the attention focused on the UNSC was greatest (Kuziemo and Werker, 2006).

Aid agencies and NGOs are however concerned that the increasing focus on security "will divert aid away from the poorest countries and communities and weaken donors' commitment to poverty reduction" (Oxfam, 2005: 49) or that this trend represents a "drifting back to the darkest days of the Cold War... when aid was just as liable to prop up dictators and their regimes, as it was to build hospitals or drill wells" (Christian Aid, 2004: 6). In this view, the period between the Cold War and the War on Terror was just a brief hiatus with US foreign aid policy seen as returning to something like its Cold War past (Buzan, 2006). During the Cold War, short-term foreign policy decisions driven by the State Department strongly influenced the US aid programme (Zimmerman, 1993) and there have been concerns that such short-termism in contemporary US foreign assistance priorities has resulted in alliances with some rather unsavoury regimes, from Colombia to Cameroon, in a manner reminiscent of the Cold War alliances struck with anti-communist dictators such as Suharto in Indonesia, Marcos in the Philippines and Mobutu in Zaire.

Although poverty reduction remains one of the stated goals of US foreign assistance, the 2006 Foreign Assistance Framework (FAF) defines the primary mission of US aid as helping to build and sustain "democratic, well governed states that respond to the needs of their people" and encouraging them to "conduct themselves responsibly in the international system" (United States Department of State, 2017). The increasing inclusion of USAID in the national-security structure along with its ever more ambitious ventures into the domestic security environments of recipient countries suggests that it has increasingly become a quasi-security agency concerned with crafting particular kinds of states and promoting political order and stability as a means of enabling economic "take-off". These long-standing geopolitical priorities related to the promotion of security, democracy and liberal peace have increasingly overlapped and intersected with the neoliberal agenda around aid effectiveness (Duffield, 2008) but also with wider projects of neoliberalisation that involve making countries "safe" for the passage of US capitalism and creating opportunities for US companies. USAID's work is very much about the roll-out of neoliberalism (Essex, 2013) where aid is allocated to countries seen to have appropriately neoliberal state institutions or to have sufficiently committed to the requisite neoliberal reform packages.

Aid is usually taken to refer to "all resources, physical goods, skills and technical know-how, financial grants (gifts) or loans (at concessional rates) – transferred by donors to recipients" (Riddell, 2007: 17). Foreign aid is not, however, a story featuring only nation states as the actors and one-way coherent flows of money as the objects of analysis (Roberts, 2014) and is about more than just resource transfers since aid also represents "a dynamic bundle of geographical relationships at the intersection of war, neoliberalism, nature and fear" (Hyndman, 2009: 867). It is important therefore to consider the various ways in which US geopolitical interests intersect with geoeconomic strategies through programmes of foreign aid and development. One useful approach to doing this that will be drawn upon here is to conceptualise aid as a set of complex networks, networked elements or assemblages, constituted by flows of capital, knowledge, influence, practices, material objects and people (Roberts, 2014). Key network or assemblage elements in US foreign assistance for development include various and multiple government agencies, contractors (firms of many different kinds, NGOs and individuals, including consultants), industry organisations and lobbyists, and actual contracts, sub-contracts, grants and procurements (Roberts, 2014).

Following interventions in Iraq and Afghanistan, the US has undertaken large-scale assistance programmes intended to stabilise both countries, rehabilitate their economic infrastructure and introduce representative government, among other objectives. Such reconstruction efforts are in part intended to enhance political stability and secure and inoculate against the contagion of insurgency but also to make Iraq and Afghanistan safe for the advance of Western capitalism. In Afghanistan, aid and development interventions have often sought to promote the advance of neoliberalism and to discipline the population "into their peripheral position within the global neoliberal economic structure through modernization and capital-driven privatization" (Fluri, 2009: 987). The American invasions of Afghanistan and Iraq indicate a significant departure from previous models of geopolitical engagement (Dahlman, 2011) in that instead of upholding a threatened state or rebuilding a failing one, the model of breaking and then remaking the state suggests a different model of militarised power, that of "reconstruction as war" (Dahlman, 2011; Kirsch and Flint, 2011).

Post-war reconstruction constitutes an important and fertile arena for pursuing a critical analysis of geopolitics but also for understanding the intertwining of geopolitics and development (see Figure 6.1).

As noted in chapters 3 and 4, the engagement of the US military in the provision of aid has a long history. Reconstruction activities were institutionalised within the US military during World War II in response to the devastation of European infrastructure and

FIGURE 6.1 "How neocolonialism works" by Andy Singer. http://www.andysinger.com

additionally became a prominent tool in Western warfare, from the decolonisation struggles to Vietnam, and most recently in Iraq and Afghanistan. This militarisation of development, where humanitarian language has increasingly been recruited to justify military operations and where military actors (and technologies like drones etc.) are increasingly involved in the design and delivery of foreign assistance, embeds processes of reconstruction within combat brigades and subsumes post-war tasks as explicit tactics for countering an insurgency (Dahlman, 2011). Here, as Patrick and Brown (2007) argue, the military is regarded as more "nimble" than civilian agencies in reconstruction and development efforts and has the ability to force projects but "often remains overly concerned with narrow, short-term security interests rather than transformational development" (Essex, 2013: 120). As a result of focusing too heavily on security as understood by the DoD and State Department, the economic and political objectives that previously formed the core of USAID strategies "have been subsumed within immediate strategic and military concerns that only contingently support development progress, and may even undermine it rather than laying the foundations for accountable governance and sustainable development" (Essex, 2008: 1633).

In Iraq and Afghanistan the DoD has operated schemes like the Commander's Emergency Response Program (CERP), which has provided immediate reconstruction and humanitarian assistance at the local level to support the work of US military commanders, providing them with on the ground "walking around money" intended to win hearts and minds and buy short-term local support through small-grant funding for infrastructure projects (e.g. for roads, wells, sanitation, railways and schools) and agricultural support or the provision of micro-grants to businesses (Tarnoff, 2009). Here, as Gilbert (2015) argues, aid money (e.g. micro-finance) becomes a "weapons system", an "entrepreneurial way of war" that promotes marketisation and neoliberalism. The US has also deployed Provincial Reconstruction Teams (PRTs) bringing together small teams of military and civilian agencies to address how the lack of security and the lack of reconstruction feed into each other to exacerbate instability. Overwhelmingly military in scope and operation (Patrick and Brown, 2007), PRTs typically consist of 80–100 soldiers, under the direct command of a military officer and are focused heavily on force protection and security assistance alongside a handful of individual representatives from the Department of State, USAID and the Department of Agriculture (USDA). The DoD has also deployed Business Task

Forces to help stimulate private sector growth and investment in both Iraq and Afghanistan (Berteau et al., 2010). In Africa, the US military is increasingly making use of related "civil affairs" operations (United States Army, 2006) that attempt to foreground development and are engineered to improve US visibility, access and influence with foreign military and civilian counterparts but also to promote the security and foreign policy interests of the US in the "ungoverned" spaces of the continent (Bachman, 2014, 2017).

In 1980 the total amount of US foreign economic and military aid stood at US$9.69 billion (of which US$2.12 billion was military) with just under half (US$4.06 billion) coming from USAID (United States Census Bureau, 2012b). By 2017 this total had reached US$45.21 billion (United States Department of State, 2018). For the most part, aid is disbursed by the DoD, USAID and the State Department along with a slew of other agencies (including the Peace Corps, the US Export-Import Bank, the Department of Homeland Security, the Millennium Challenge Corporation, the US-African Development Foundation and the USDA). In 2017, just under a third of all US aid (US$14.5 billion) was spent on peace and security. New security imperatives have increased flows of US foreign assistance to countries deemed to be of geostrategic importance and since 9/11 the revised list of countries that receive US military aid now includes several previously ineligible states, including Armenia, Azerbaijan, India, Pakistan and Tajikistan. Pakistan went from being a nuclear pariah state, ruled by an unelected military dictator and languishing on the world aid blacklist, to America's closest ally in the global War on Terror, receiving some US$3.5 billion in 2011 (Ali, 2016), up from just under US$88.5 million in 2000. In recent years, however, the US has substantially cut aid to the country and in 2017 the combined total for economic, development and security assistance was US$376.35 million (United States Department of State, 2018). In January 2018 President Trump announced further cuts in security assistance to the country, claiming it had not done enough to counter terrorism and that the US had "foolishly given Pakistan more than US$33 billion in aid over the last 15 years, and they have given us nothing but lies and deceit" (quoted in Zakaria, 2018).

In addition to Iraq, Afghanistan and Pakistan, the list of major recipients includes seven African countries (Nigeria, Ethiopia, Kenya, Tanzania, Uganda, Mozambique and Zambia) together with long-standing regional partners like Jordan and Israel. US aid flows to Africa are increasingly focused on countering the rise of Islamic insurgencies there and the US military assemblage on the continent has been considerably expanded in recent years. US assistance in Africa is of course not just about countering terrorism but also involves geoeconomic concerns with the security of US access to hydrocarbon resources and the potential commercial opportunities for US firms. The 2017 US National Security Strategy noted that "Africa contains many of the world's fastest growing economies, which represent potential new markets for US goods and services" (United States, 2017: 52), even though attempts to realise this potential have only been backed by relatively modest investment from the US Export-Import Bank (of about US$6.6 billion since 2009) (Ismail, 2018).

The geographies of US foreign assistance reflect the recent prioritisation of conflict zones where the US has been active (such as Iraq and Afghanistan) and the emerging centrality of Africa to American foreign policy priorities concerning global security and stability but also underline the continuing importance of major historical recipients like Israel, Egypt and Jordan. Israel alone receives about US$3 billion in direct foreign assistance each year and is both the largest annual recipient of direct US economic and military assistance since 1976 and the largest total recipient since World War II, with total cumulative US aid between 1949 and 2010 estimated at US$109 billion (Sharp, 2010). US Cold War foreign policy was geared towards supporting the development of oil-producing countries, maintaining a neutral stance in the Arab–Israeli conflict while supporting Israel's security, and preventing Soviet influence from gaining a foothold in Iran and Turkey. The "special relationship" between the two countries is based around the perception of shared strategic interests (including oil in the Middle East and North Africa), but also on Israel's democratic status, the idea of Israel as "Western" and the influence of the Jewish lobby in the US (Riech, 1996). In December 2017 the US Ambassador to the UN, Nikki Haley, warned General Assembly members not to vote against the US decision to recognise Jerusalem as Israel's capital and that the US "would be taking names" to learn which of its supposed allies was really on its side and still deserving of US assistance (Novak, 2017).

The chapter is divided into four main parts. The first traces the emergence of the "security–development" nexus and the (re)imagining of underdevelopment as "dangerous" and explores the intrinsic relationship between development, security and containment. The next section then investigates the role of USAID and the ways in which the agency seeks to advance neoliberalisation.

The chapter then moves on to consider the recent rediscovery of development-based counter-insurgency techniques and the enrolment of USAID within this before moving on to examine the current configuration of the US military assemblage in Africa and US attempts to counter Islamic insurgency there.

THE SECURITISATION OF DEVELOPMENT

> Rather than development leading to international security... development is part of a deepening and ultimately unwinnable global civil war. This war is not being fought between armies, it is embodied in the developmental agencies, relations and networks that seek to proselytize the attitudes and behaviour that liberal states deem to reflect acceptable as opposed to unacceptable ways of life.... From communism to terrorism, through its marginalizing effects and its ability to foster resentment and alienation among common folk, poverty has been monotonously rediscovered as a recruiting ground for the moving feast of strategic threats that constantly menaces the liberal order. (Duffield, 2010a: 57, 61)

With its promise of redemption through making full and wholesome what would otherwise remain incomplete (Duffield, 2008), development has arguably functioned as a technology of security since the dawn of industrial capitalism (Cowen and Shenton, 1996). The link between development and security is not particularly new (for some it is even intrinsic to modernity) (Watts, 1995) but since the end of the Cold War, as Duffield (2008, 2010a) argues, there has been a rediscovery of poverty as recruiting ground for the strategic threats "menacing the liberal order", this time terrorism rather than communism. There has also been greater recognition of the interdependence between poverty in the global South and insecurity in the global North and consequently a "renewed emphasis on the need for social cohesion at home while, at the same time, urging a fresh wave of intervention abroad to reconstruct weak and fragile states, or remove rogue ones" (Duffield, 2008: 162). This post-Cold War merging of developmentalism and securitisation has been legitimised in part by the blending of discourses of humanitarianism and human security with neo-colonial metaphors of tutelage and protection (Doty, 1996) and is suffused by ideas of Western economic and political liberalism.

This "development–security *nexus*" can itself be understood as a *dispositif* or "constellation of institutions, practices, and beliefs that create conditions of possibility within a particular field" (Slater, 2008: 248). It constitutes a field of development and security actors, aid agencies and professional networks, complete with their own forms of subjectivity, that call forth the conditions of need and insecurity to which they seek to provide solutions (Duffield, 2008, 2009).

For Duffield (2008: 162), what is at stake here is the West's ability to contain and manage international poverty and the pervasive security risks of globalisation's "unruly borderlands" while maintaining the ability of mass society to live and consume beyond its means. In the post-Cold War period it is now commonplace for policymakers to assert that development and security are interconnected and that you cannot have one without the other but Duffield (2008) argues this remains incomplete without the third category of containment, including all the interventions and technologies that seek to restrict or manage the circulation of "incomplete" or "threatening" life. As such, "an expanded nexus would add the proviso that you cannot have development or security without containing the mobility of underdeveloped life" (Duffield, 2008: 146). Following decolonisation, the direction and perception of global migration changed (Balibar, 1991) from a North-to-South dynamic positively associated with exploration, escape and fortune, to a South-to-North dynamic with various negative connotations (Balibar, 1991). While development has a long history as a strategic response to various threats (including those posed by migration), this role is still not widely appreciated since as a practical technology of security, development exists in the here and now and its benefits are always cast as a future yet to be realised (Easterly, 2002; Duffield, 2008). It is in this context and in response to the crisis of containment, especially as a means of "capturing and securing non-insured life", that the underdeveloped state "has once again moved to the centre of development policy" (Duffield, 2008: 160). Rather than focus on the reasons why poverty persists, however, it is important to examine the political function that its constant rediscovery serves – especially how it validates and revitalises a "liberal will to govern" (Duffield, 2010a: 61).

Humanitarian and development agencies have also been brought together with political and military actors through the UN's "multidimensional integrated stabilisation missions" currently operating in places like Mali and the Central African Republic. These integrated missions represent a new kind of humanitarianism which demands that international enforcement (rather than mechanisms of entitlement or redistribution, for example) be mobilised regardless of the consent of its "recipient" populations (Amar, 2012). Contemporary Western humanitarian intervention, state-building and development initiatives differ significantly from previous imperialisms then in "being conducted in the name of the international community", including not only powerful states but also the "entire panoply of international organisations, international financial institutions, aid agencies and global civil society organisations" (Luckham and Kirk, 2013: 4). Besides conflict resolution, the development–security nexus was initially mainly concerned with practical issues relating to the transition from humanitarian relief to sustainable development but through the medium of integrated missions eventually became integral to much more ambitious and complex programmes of disarmament, institutional reform, capacity-building, economic development, and societal and state reconstruction. In the UK, under the New Labour government (1997–2010) much of this effort was focused on the "securitisation" of Africa, as dealings with the continent moved from the realm of development and humanitarianism into one of "risk/fear/threat" (Abrahamsen, 2004: 677). Some of the literature on the "security–development" nexus has, however, yet to sufficiently address how the increasing focus on securitising development has been "adapted into a variety of national/institutional policy settings" (Friis, 2014: 6). This has at times "overplayed the coherence of the securitisation project and underestimated the conflicts and tensions between the major political, military and humanitarian players in global and national security marketplaces" (Luckham and Kirk, 2013: 4) (cf. Selby, 2013).

Despite not specifically writing on themes of colonialism or development (Stoler, 1995), Foucault's work on biopolitics (Foucault, 1975–76, 1976, 1977–78) is very useful in understanding the intrinsic relationship between development, security and containment because of the way it drew attention to the liberal idea of "government of the population and the imperatives that are derived from such an idea" (Dean, 1999: 113). Since the beginnings of modernity, "a liberal rationality of government has always taken the protection and betterment of the essential processes of life associated with population, economy and society as its object" (Duffield, 2008: 145). Today the effectiveness of modern states is now typically defined in terms of how well they support the life and well-being of their populations. Typically, development experiences non-Western peoples as somehow incomplete or lacking in the essentials for a proper existence

(Mehta, 1999; Duffield, 2008) including a lack of education, an absence of capacity, the inability to save properly or the need for greater gender awareness, all absences that mean life cannot be lived properly. As a response to this experience of inadequacy, development seeks to provide solutions in the form of a moral or educative trusteeship and the rule of experts (Mitchell, 2002), aiming to bring incomplete or underdeveloped life to its full potential and to make it wholesome through education and empowerment (Duffield, 2008, 2010a). During decolonisation, development was "reconfigured as an inter-state relation of governance"; it moved from the colonial bureaucracy into the institutions of external expertise arranged "to help and support the newly discovered underdeveloped state" (Duffield, 2008: 148).

Through this "will to improve" (Li, 2007) and the idea of bettering non-Western peoples and spaces through changing behaviour and attitudes, development has always been a way of governing others, functioning as a liberal technology of security (Duffield, 2008). This *liberal way of development* (Duffield, 2008, 2010a) privileges local and adaptive *self-reliance* and has historically favoured the encouragement of community self-reliance and local entrepreneurship based upon the small-scale ownership of land or property (Cowen and Shenton, 1996: 266–267), today recognisable as "sustainable development" (Duffield, 2008: 148) and visible, for example, in a significant recent proliferation of land titling programmes in Africa. Development has thus long served as a kind of "educative trusteeship" (Duffield, 2008: 148) and there are clear lines of continuity between its colonial and post-colonial histories. The Millennium Villages Project (2004–15) conceived by Jeffrey Sachs has received a great deal of attention in recent years as a demonstration project that claimed to use an integrated approach to rural development to achieve the Millennium Development Goals (MDGs) but the design of development around demonstrative village-level interventions can be traced back to the Baptist "free villages" of 1830s Jamaica (Hall, 2002: 120–139). In this sense, following Duffield, it is possible to trace a clear continuity from colonial missionary projects right through to today's focus on human security and integrated and sustainable development, although he makes an important distinction between this "liberal way of development" and the state-led modernisation and economic catch-up strategies pursued in the early post-colonial period.

Two factors distinguish the configuration of the development–security nexus today (Duffield, 2010a): the global outlawing of spontaneous or undocumented migration and the shift in the focus of security from states to the people living within them. There have "never been so many frontiers, checkpoints or restrictions for undocumented migrants" (Duffield, 2010a: 62) and this trend has been accompanied by the growing surveillance and policing of all forms of international circulation. While the *geopolitics* of border control identifies and seeks to neutralise the threat of unmanaged migration, Duffield argues that the *biopolitics* of development, "through self-reliance, basic needs and community support, seeks to manage in its natural habitat the risks posed by underdeveloped life" (Duffield, 2010a: 63). There has also been a "biopolitical turn within aid policy" (Duffield, 2010a: 55) where policy discourse now conceives development and underdevelopment *biopolitically* or in terms of how life is to be supported and maintained, within what limits and level of need people are required and expected to live, rather than according to economic and state-based models. As a result of these trends, the soft power of development has found a new premium as a technology of security and counter-insurgency.

REINVENTING USAID: THE SECURITY–ECONOMY NEXUS

USAID's significance as a tool of US foreign policy is certainly not new. In the 1980s, during the Reagan administration's aggressive strategy of containment in Ethiopia, for example, USAID officials used disaster relief (over US$500 million and 800,000 tons of food) as an instrument to weaken Mengistu Haile Mariam's socialist government, to discredit its troubled feeding programme and to discourage Ethiopian land reform plans that involved the resettlement of people onto collective farms (Poster, 2012). Despite this and being one of the more deeply internationalised institutions within the US state, relatively scant attention has been paid to the logics and frameworks that shape USAID's inner workings and external relations (Essex, 2013; Bhungalia, 2016). USAID initially floundered during the post-Cold War years (see Figure 6.2) as development duties splintered among a dozen federal departments and as successive administrations sidelined it. However, particularly since the Bush administration (2001–09) which played a key role in reshaping US aid policy (Lancaster, 2008), development has become a central plank of US national security and a pillar of US foreign and economic policy, receiving

FIGURE 6.2 American Marines guarding food aid distribution in Mogadishu during "Operation Restore Hope". In December 1992 US Marines landed near Mogadishu ahead of a UN peacekeeping force sent to restore order and safeguard relief supplies. The US forces withdrew in 1993 following the debacle of the infamous "Black Hawk Down" battle. Photo by Paul Lowe, Panos Pictures.

extensive treatment in the 2002 and 2006 *National Security Strategy* (NSS) (United States, 2002, 2006).

As a "soft power" asset USAID was more firmly placed inside the State Department in 2006, losing its policy division in the process and with its human resources shrinking dramatically. In 1980, USAID had 4,058 permanent American employees but by 2008 the number had dropped by 45% to 2,200, causing "a dramatic loss of expertise" (Atwood, McPherson and Natsios, 2008: 127) as there were just six engineers and sixteen agricultural experts when "it had hundreds in the 1980s" (Natsios, 2010: 25). USAID remained a relatively minor player in foreign policy, however, being framed as a crucial but definitely subsidiary partner in the "3 D's" approach (defence, diplomacy, and development) and increasingly operating in the shadow of the DoD (Hart, 2010; Roberts, 2012), its operations increasingly guided by military strategy, specifically counter-insurgency (Bachmann, 2010: Hodge, 2011). Today, USAID regularly draws direct connections between economic liberalisation, development progress and US national security, with developing states understood as sources of, and focal points for combating, global terrorism, although interest in and strategising around development policy has considerably broadened out to include multiple federal agencies and departments beyond USAID.

In 2006 there were more than 40 US government agencies involved in giving "foreign aid" of some kind (Moss, Roodham and Standley, 2005) and an interagency budgeting and strategy process has subsequently emerged focused on a more "comprehensive approach to managing aid, built around country-based strategies and long-term development planning" (Essex, 2013: 32). The DoD in particular has taken a more active and expansive role in development policy and strategy, with its share of ODA rising from 3.5% in 1998 to 22% in 2013 (Charny, 2013). The Pentagon's involvement in ODA-eligible activities spans several distinct activities and challenges, "from providing humanitarian relief to training and equipping border and customs services, and from HIV/AIDS programmes for foreign militaries to technical assistance aimed at drug interdiction and counter-narcotics programmes" (Patrick and Brown, 2007: 4). The DoD is increasingly preoccupied with addressing the roots of instability and extremism in weak and failing states and preventing their collapse into conflict (Patrick and Brown, 2007), inspiring programmes to build counterterrorism capabilities in "developing countries" and including several activities that might in principle be undertaken by the State Department, USAID and other civilian actors.

In 2004 USAID outlined a programme designed to redefine the objectives of its foreign aid programmes, reassert the connections between development, economic openness and security, and re-establish the relevance of both the agency and foreign assistance within US foreign policy (Essex, 2013: 90). This strategy (USAID, 2004) laid out "five core operational goals" for the agency and US development assistance which included: "promoting transformational development"; "strengthening fragile states"; "providing humanitarian relief"; "supporting US geostrategic interests"; and "mitigating global and transnational ills" (USAID, 2004: 5). For USAID (2004: 13) the world of developing states can be subdivided into "relatively stable developing countries" (where the basis for political will and development progress already exists and must be maintained or enlarged) and "fragile states" (where basic governance and stability are perilous and where it remains questionable whether aid can adequately promote development progress) (Essex, 2013: 117). As it has done for many years, USAID plays a normative role here in scripting exactly what form these states should take. The level of deservedness is determined by "criteria of need, commitment by the host government and/or nongovernmental actors to reform, feasibility of achieving results, and foreign policy importance" (USAID, 2004: 20). USAID's *Fragile States Strategy* has also attempted to categorise state fragility as "failing", "failed" and "recovering from conflict" (USAID, 2005).

The Foreign Assistance Framework (FAF) agreed in 2006 which guides foreign aid allocation is similarly built around a taxonomy of developing states that reconfigures aid criteria of need and deservedness and represents the emergence of what Essex (2008: 1634) calls a new "statist cartography of development" within and through the US state based on free market capitalism and a narrowly defined conception of security. Here deservedness and development remain "properties of states" (Essex, 2013: 124) in an approach that "only contingently addresses underdevelopment" (124). USAID has played a leading role in constructing this taxonomy, in which the states most likely to receive development assistance are those that are strategic relative to US economic and security interests, but this also "hinges on the particularity of the neoliberal state's forms and functions" (Essex, 2013: 118) – formal multiparty democratic elections, liberalised economic governance that facilitates capital mobility and security reforms that enhance the investment climate and promote law and order. During the Cold War USAID was centrally involved in US counter-insurgency efforts abroad and the pursuit of modernisation but in recent years has become increasingly embroiled in the machinations of the global neoliberal roll-out (Essex, 2013: 118). Through USAID there is an increasingly neoliberalised basis for aid allocation (urging societies to develop in accordance with appropriately neoliberal state institutions) that also aligns with US geopolitical objectives post-9/11. USAID's development interventions have thus sought to promote US geopolitical and security interests abroad while expanding capitalist markets across the Global South.

Enduring geopolitical concerns in Washington with containing security threats and nation building have increasingly been rethought and remade from the Cold War onward through forms of geoeconomics that emphasise global engagement, investment and partnering (Essex, 2013), all of them in turn tied to the economic forces, visions, practices and partnerships of market-led globalisation (Sparke, 2016). Essex (2013: 4–5) argues that we should not see this however as "a crude chronological progression from the geopolitics of Cold War developmentalism to the geoeconomic rhetoric of 'borderlessness' that marks contemporary neoliberal globalisation" and notes the error in any simplistic geographical partitioning of geopolitical and geoeconomic space (Sparke, 2016). Glassman (2018: 412) notes that "geopolitical economy" has frequently been more a term than a concept but if geopolitical and geoeconomic logics of power are grasped dialectically it has the potential to bring to the fore "specific – and sometimes unduly neglected – aspects of development dynamics". One way to do this is to examine how US geopolitical interests intersect with geoeconomic strategies, or the "security–economy nexus" (Coleman, 2005), in shaping American bilateral assistance, considering the inter-articulation of ideational and material. Only through close attention "to the evolution and circulation of geostrategic discourses in concert with the formation and exercise of political and economic decision-making powers... can the spatial assumptions and projects underlying and animating USAID be understood in their full geostrategic importance" (Essex, 2013: 18)

A key element of the security–economy nexus shaping the delivery of US bilateral assistance is the powerful contracting assemblage at its heart which has seen a significant proliferation of private development assistance providers in USAID's work (Roberts, 2012, 2014). In the period 2000–08 outsourcing at USAID grew 690% (Roberts, 2014) whilst the development contracting industry now handles more than US$12 billion a year from USAID (Roberts, 2014: 1035), most of which goes to for-profit US firms, some of which have origins in the Cold War and have grown to be large corporations. Poised to deliver on the promises of science and social science to execute development efficiently, the firms embodied a distinctive Cold War US liberal internationalism and an optimism about development that reflected the times and places of their origins (Wolfe, 2013). Roberts (2014) argues that the industry of contracting firms and the development contracting "market" coevolved and co-constitute one another and both lie at the centre of a "development–industrial complex assemblage" that has expanded dramatically in recent years. This contracting assemblage has a specific regional geography centred on the Beltway region in the greater Washington, DC area, a classic agglomeration that is home to many US development contractor firms (Roberts, 2014) but also includes a range of institutions that sustain and are sustained by USAID's contractor networks such as trade associations, lobbying groups, policy think tanks, research groups, professional societies and clubs, and information brokers of various types (Roberts, 2012). American for-profit firms have been major recipients of USAID funding and in a 2011 speech Rajiv Shah, Obama's USAID administrator, drew parallels between USAID's reliance on these firms and Eisenhower's warnings about a military–industrial complex, claiming that USAID was "no longer satisfied with writing big checks to big contractors and calling it development" and that development firms were more interested in keeping themselves in business than seeing countries graduate from the need for aid (Shah cited in Norris, 2012).

With the growing emphasis on security sector reform, whereby development resources are used to train and equip police in what is considered to be an appropriate democratic fashion, USAID has continued its historical focus on police training programmes and the work of the OPS (see chapter 4) to improve the capabilities of civilian police and paramilitary forces through "training, technical assistance and equipment", justified as long-term "institution building". The notion of using an organisation such as USAID to improve the counterterrorist capacity of Africa's police forces in the pursuit of US national security is, however, seriously flawed (Hills, 2006: 631). USAID's acceptance of a security-oriented remit ensures that the state remains its referent point, for development as for security, thus "making a mockery of its stated commitment to individual security, local ownership and similar developmental orthodoxies" (Hills, 2006: 642). USAID's concern with the domestic politics, stability and democracy of recipient states has also inevitably led to accusations of meddling and interference. In 2014, shortly after civil society activists demonstrated in Nairobi to protest against alleged government corruption, lack of safety in public places, high unemployment and poverty, the Kenyan government claimed to have evidence that USAID had been trying to "destabilise" the country by financing anti-government demonstrations (BBC, 2014). In Cuba USAID has been accused of political subversion and essentially becoming a front for a carefully planned US intelligence operation (*The Guardian*, April 3rd 2014), using social media to wage cyber warfare against countries which do not abide by Washington's demands. As part of its "democracy promotion" programme, USAID was accused of using a Twitter-like service called "ZunZeo" to mobilise demonstrations that might trigger a "Cuban Spring". In addition to claims that it has sponsored and encouraged insurgencies in the name of democracy, USAID has also very much returned to its long-standing entanglement with countering them.

THE REVIVAL OF DEVELOPMENT-BASED COUNTER-INSURGENCY

At the end of the 1960s, JFK's "decade of development", some of those involved in US technical assistance programmes of modernisation were advocating an integrated "systems" framework for the execution of rural planning. Earl Kulp's (1970) *Rural Development Planning: Systems Analysis and Working Method* provided such an approach, tried by the Kenyan state amongst others (Chege, 1972), and alongside sections dedicated to planning in urban and rural spaces there is a chapter on basic counter-insurgency principles. Kulp, a management consultant with experience in US government rural development programmes in Africa and Asia and later a member of the US Supreme Court, noted that "rural development includes the paramilitary and civil aspects of counterinsurgency, the environment and the planning techniques of which are basically the same" (Kulp, 1970: 15). In the early 1970s and for the US, they were the same and in a range of contexts such as Vietnam, Palestine or Afghanistan, USAID has increasingly practiced development as a form of governmentality, its interventions becoming ever more concerned with the government and management of restless rural populations and the restoration of political order and stability (Attewell, 2017). Afghanistan in particular has long served as a crucible where various techniques for pacifying restless populations have been developed, tested and refined (McCoy, 2009). Over time, USAID has remained responsible for neutralising and dispelling perceived threats in their place of origin, becoming the humanitarian and developmental side to US interventionism. As a result, USAID has increasingly waged "liberal" warfare in battlespaces such as Iraq and Afghanistan but also been involved (alongside the DoD) in programmes like *Plan Colombia*, a counter-insurgency strategy devised in 1999 and framed as a crackdown on drugs.

In contrast to conventional warfare which seeks to control territory and destroy the opposing military, counter-insurgency is "population centric" in seeking to control society and targeting the "capillary" level of social relations (Parenti, 2011: 23). This "population centric" focus was one of the main recommendations of the expanded research programme on counter-insurgency funded by the RAND corporation in 1962, following the outbreak of communist-inspired nationalist insurgencies in the wake of decolonisation. RAND researchers focused on the issue of the insurgents' need for popular support and on the problems of modernisation, with counter-insurgency seen as providing the people with security from predations by states, insurgents and the negative consequences of development (Hosmer and Crane, 2006). As insurgency and civil war took root in Iraq and Afghanistan and as the US ran out of options to stem the spiralling violence, there have been attempts to revive the spirit of development-based counter-insurgency, long repressed in military circles following its failures in Vietnam (Belcher, 2012). Designed by its advocates to address the pacification and governance of populations following invasion and occupation, this involved revisiting and reworking the principles of counter-insurgency, drawing lessons from the Algerian War (1954–62) and the Malayan Emergency (1948–60) to distil techniques for pacification that could be readily employed in contemporary battlespaces.

As Greenburg (2018) notes through ethnographic observations of training activities on US military bases, military instructors specifically train troops to manage civilians by drawing on the colonial history of a variety of "small wars" in ways that erase differences between their specific historical geographies or that fail to recognise that in counter-insurgencies it is typically "impossible to determine where violence ends and power begins, and vice versa" (Belcher, 2018: 102). Further, Greenburg (2018) notes that Haiti's experience with disaster relief and recovery has also provided an important reference point here for how humanitarian response could serve as a particular theatre for rehearsing counter-insurgency tactics (Greenburg, 2018). As noted in chapters 3 and 4, within the "classical" counter-insurgency campaigns fought at the time of decolonisation, with its promise of progress and inclusion, development was already a valued strategic component (Duffield, 2010a) and it has again become central in this latest *rapprochement* with counter-insurgency doctrine. Through unprecedented collaborations involving officers, soldiers, policymakers, academics and journalists, Army Field Manual *FM 3–24 Counterinsurgency* was published in late 2006 (United States Army/Marine Corps, 2006), which subsequently became one of the most widely read and influential documents ever produced by the US military (Belcher, 2012).

FM 3-24 was heavily influenced by the "classical" school of Cold War-era counter-insurgency focused on defeating communist and anti-colonial insurgencies by strengthening weak governments that are seen by a critical mass of people in the host nation as illegitimate (Cancian, 2017). It frames insurgency as a contest between insurgents and governments over an undecided population, a contest whose outcome is principally determined by the relative capability of each side to govern people. The manual suggests that an undecided civilian population will support the side that they think can best provide services and advocates persuading civilians that the counter-insurgent army can best shield them from hardship. Classical counter-insurgency theory posits an insurgent challenge to a functioning (though often fragile) state, typically regarding insurgency as primarily rural and as something that occurs within one country or district, between an internal non-state actor and a single government (Kilcullen, 2006a). While the field manual does state that people might have grievances other than lack of state capacity, operations are overwhelmingly targeted at deficient states (Cancian, 2017) whilst "successful" counter-insurgency efforts "comprehensively address the host nation's core political–economic and socio-cultural problems" through a "synchronized application of military, paramilitary, political, economic, psychological, and civic actions": or, a productive form of "armed social work" (United States Army/Marine Corps, 2006; see also Kilcullen, 2006a; Roennfeldt, 2011).

For Kilcullen, the problem of "weaning jihadist combatants away from extremist sponsors, while simultaneously supporting modernisation, does somewhat resemble pacification in traditional counter-insurgency" (Kilcullen, 2006b: 113) and echoes colonial campaigns but also includes a number of new elements, including a shift from "revolutionary" insurgencies to "resistance" insurgencies where the intent to replace existing governments or create independent states is only partly evident and where, compared to the classical insurgencies of decolonisation, the economic relationship between insurgents and the population is very different (Kilcullen, 2006b: 119). The classic Maoist doctrine of insurgency likens the insurgent to a fish that swims in a sea constituted by the people with the insurgent reliant upon the people for food, shelter and information and in return, he or she identified with them and supported their struggles (Duffield, 2008: 157), but this dependent relationship has changed, even reversed, as insurgents draw on complex global networks of support. Through the development of adaptable trans-border shadow economies, remittances and diaspora networks, insurgents have become radically self-reliant, with many routinely having access to "more wealth, hardware and information than the impoverished populations amongst whom they are embedded" (Duffield, 2008: 158).

War can and should be understood as occurring not simply in the meeting of two adversaries on a battlefield but also through the humanitarian regimes that have become the means for governing the displaced, the refugee, the poor and the unwanted (Weizman, 2011). Humanitarian infrastructures, technologies and practices thus constitute key sites through which a relation of war is sustained and reproduced. Contemporary "humanitarianism", Weizman (2011) argues, has evolved into various technocratic collusions among those who work to aid the vulnerable and those who mete out state violence in the name of security. For Weizman, violence and humanitarianism are entangled at their level of constitution (see also Asad, 2015) and the collusion of these humanitarian technologies with military and political powers forms what he calls the "humanitarian present". Weizman (2011: 11) observes that international humanitarian law, for example, does not seek to end wars but rather "to 'regulate' and 'shape' the way militaries wage them" and by moderating the violence they perpetrate, Western militaries hope they "might be able to govern populations more efficiently". Consequently, US imperial endeavours tend to slip between "exemplary or performative forms of violence meant to intimidate and more 'humane' and developmental warfare intended to persuade" (Khalili, 2012: 4). As Bhungalia (2012) has argued in relation to Palestine, this has resulted in:

> a proliferation of sites and diversity of means through which US political and economic power is being articulated. Alongside its military and diplomatic interventions, the US is simultaneously extending its reach through a host of "development experts," humanitarian agents and "democracy promoters" charged with filtering, sorting and policing the Palestinian civilian population. While taking a new and perhaps more sophisticated form, these contemporary practices and strategies must not be dissociated from a longer history of counterinsurgency in Palestine. (Bhungalia, 2012)

Palestine has historically occupied a central position in global counter-insurgency and its experiences can be read contrapuntally with those of US counter-insurgency practices elsewhere (Khalili, 2012). Under the British Mandate (1917–48), Palestine served as a staging ground for the consolidation of British imperial policing and pacification strategies, and following mandate rule, as a testing ground for Israeli experiments in asymmetric warfare and demographic engineering, often with considerable US diplomatic and material support (Bhungalia, 2012; see Figure 6.3). In recent years counter-insurgency and pacification practices have been

FIGURE 6.3 Activist graffiti adorns the Israeli separation wall in the West Bank town of Bethlehem in July 2010. Photo by Ryan Rodrick Beiler/Shutterstock.

mobilised by USAID in contemporary occupied Palestine through humanitarian and development interventions and the networks of aid governance (Bhungalia, 2015). The foreign aid regime in the Palestinian territories has served to "mitigate the most deleterious effects of military occupation and dispossession" but at the same time has "further extended a regime of war and policing into ever-more intimate spaces of Palestinian everyday life" (Bhungalia, 2015: 2308). The various NGOs and development contractors through which USAID operates are thus "simultaneously tasked with dividing, surveilling and policing the Palestinian population on behalf of the US state" (Bhungalia, 2015: 2313; see Figure 6.4).

Contemporary US counter-insurgency doctrine aims both to establish "security" within a host country as an *a priori* condition for any political and economic reconstruction projects and to raise "cultural awareness" within the US military to better understand the root causes of an insurgency (Belcher, 2014). Such efforts have sought to show US forces engaged in a kinder, gentler, culturally sensitive occupation and thus a war Americans could feel good about fighting (Belcher, 2014; González, 2015). This has involved embedding social scientists in combat brigades in Iraq and Afghanistan through the controversial Human Terrain System (HTS) initiative

FIGURE 6.4 A 2012 poster by Hafez Omar from the Palestine Poster Project depicts the bloody aims of USAID. Courtesy of Hafez Omar.

which, during its eight years of existence (2006–14), cost taxpayers more than US$725 million, much of which went to two large defence contractors, BAE systems and CGI Federal (González, 2015). This was envisioned as an intelligence-gathering programme similar to the US Army's collection of ethnographic data on Vietnamese civilians for CORDS – a pacification programme set up by the South Vietnam and US governments in 1967 that then shared this ethnographic data with paramilitaries working for Operation Phoenix, a secret branch of CORDS (González, 2009). CORDS/Phoenix "census grievance" teams collected census and ethnographic data and interviewed people about their needs, complaints and sentiments toward the Viet Kong in much the same way the Human Terrain Teams were instructed to do in Iraq and Afghanistan (González, 2009: 28). HTS personnel conducted a range of activities including data collection, intelligence gathering and psychological operations and used outdated anthropological concepts, theories and methods, mostly from the 1930s and 1940s, a time when many anthropologists were employed by colonial governments to more effectively control indigenous populations. Less than a year after the first HTS was deployed to Afghanistan, the American Anthropological Association issued a statement expressing disapproval of the programme and an alternative "counter-counterinsurgency manual" (Network of Concerned Anthropologists, 2009) was produced whilst more than 1,000 anthropologists signed a petition pledging non-participation in counter-insurgency work. In recent years, there has been a gradual shift in the language, strategy and concepts used by the US military-intelligence community from the anthropology of human terrain to a growing interest in human geography and geospatial intelligence (Wainwright, 2016).

More recently, efforts to showcase the "humane" side of contemporary US counter-insurgency in Afghanistan and to secure areas like Helmand and Kandahar have involved various biopolitical reconfigurations involving varied development-related interventions (such as rebuilding destroyed homes, crop substitution and food security efforts) that target Afghan households, small farms and villages and the cultural institutions animating rural life (Belcher, 2018). This turn to the informal and intimate structures of Afghan life was a "rediscovery" of the scale on which counter-insurgencies have always been most violently felt: the colonised body and the home (Khalili, 2012; Belcher, 2018). This has also included the reconstruction of villages like Taroke Kalacha that seek to both confine and discipline Afghan bodies at the level of the household and to coerce local people to recognise the authority of local governing institutions and a foreign military. Village destruction and reconstruction of course has many precedents in US and British counter-insurgency operations, including the strategic hamlet programme in South Vietnam and the New Villages scheme in Malaysia (Scott, 2016; Sioh, 2010), but for

Belcher (2018) this phase of US-led operations (2010–12) marks a dramatic shift in the counter-insurgency strategy in southern Afghanistan. In such projects, fields and homes were rebuilt in a way that maximised military surveillance rather than the needs of the villagers themselves (Belcher, 2018: 103).

The counter-insurgency manual was updated and retitled in 2014 with even greater attention to culture and fewer references to Algeria and Malaya, instead attempting to recognise the complexity of Iraq and Afghanistan and that intra-state conflict takes many forms – revolutions, rebellions, coups d'état, insurgencies and civil wars (United States Army, 2014). The revised version also alludes to the billions of taxpayer dollars often squandered on futile reconstruction efforts. Instead, the manual insists that counter-insurgents not lavishly bestow aid projects on the local population, but rather address the *root causes* of insurgency since "deprivation that is not considered unjust is much less destabilizing than relative prosperity that is considered unjust" (United States Army, 2014: para 4-16). Additionally, the US doctrine of "stability operations" (FM 3-07) (United States Army, 2003, 2008) both builds on and exceeds previous counter-insurgency thinking, arguing that "the adversary is often disease, hunger or the consequence of disaster" (United States Army, 2003: 6-1), that fragile states unable or unwilling to provide for the most basic needs of their people represent "the greatest threat" to US national security (United States Army, 2008: vi) and that the weakness of these states "threatens the success of any development effort" (United States Army, 2008: para 1-44). Within the stability dimension of the US military's "full spectrum operations", civil affairs forces now pursue an "indirect" and "population-centred" approach and play an increasingly important role in the US military's preventative or peacetime engagements around the world, particularly in Africa.

THE US AND COUNTER-INSURGENCY IN AFRICA: DRAINING THE "SWAMP OF TERROR"

Historically, the US has always adopted a militarised foreign policy towards Africa (Majavu, 2014). In March 1960, during a year in which 16 European colonies in Africa became independent, the US Secretary of State, Christian Herter, told the US National Security Council that Africa had become "a battleground of the first order" (cited in Gleijeses, 2013: 6). One month later,

Vice-President Richard Nixon stated that Africa was "potentially the most explosive area in the world" (Gleijeses, 2013: 6) and the incoming Kennedy administration agreed. In recent years, Africa has once again been constructed in US foreign policy as an explosive battleground and its underdevelopment has again been framed as threatening, this time in terms of Islamic terrorist insurgencies in the context of the War on Terror. Since the bombing of US embassies in East Africa in 1998, which was followed by a US retaliatory strike against Sudan, Africa has increasingly been regarded by the US as the next front or battleground in the war on terrorism (Majavu, 2014). In the aftermath of 9/11 US officials scoured the globe for these "ungoverned spaces" with the potential, like Afghanistan, to foster anti-American extremists and US policy has since become increasingly attentive to emerging but not fully formed threats, many of which are seen as evolving in regions like Africa that are deficient in both "development" and "security" (MOD, 2009; United States Army, 2008) or home to weak and failed states. Top US foreign policy, military and intelligence leaders have constructed the idea of a Sahelian-Saharan front in the War on Terror, representing it as a "swamp of terror", indeed a "terrorist infestation" which "we need to drain", or as a "magnet for terrorists" (Keenan, 2009: 3), fabricating a fiction of the continent as a hotbed for terrorism where Islamist extremism has metastasised.

Contemporary US engagement in Africa is now dominated by security issues, particularly counterterrorism, counter-piracy and efforts to resolve internal conflicts. US air and drone strikes aimed at countering terrorist organisations in Africa have increased considerably under the Trump administration with multiple strikes against Al-Shabaab in Somalia and against Al-Qaeda in the Arabian Peninsula (AQAP) in Yemen (see Figure 6.5) along with pinpoint raids by small Special Forces teams in countries like Somalia and Libya. Many of these involve small-scale "secret wars" against Islamists, mainly linked to Al-Qaeda and often carried out under the aegis of the US Africa Command (AFRICOM). Over the last year alone, the number of US forces in Africa has increased by nearly 1500 to 7500 (not including special forces) and the US now has 34 status of forces agreements (or similar treaties) with African countries, 14 of which were signed or upgraded in the last decade, whilst US troops were deployed in 50 out of Africa's 54 states, many on clandestine missions (Booker and Rickman, 2018). Protests against US bases and troop deployments have

FIGURE 6.5 A man walks past graffiti, on a wall in Sanaa (Yemen), denouncing US drone strikes on November 13th, 2014. REUTERS/Khaled Abdullah/File Photo.

also become more common, however, taking place in Ghana, Niger, Cameroon, Liberia and several other countries (Booker and Rickman, 2018).

Some African countries (notably Somalia, Nigeria and Mali) were identified among other "weak states" vulnerable to jihadi influence spreading south from the Maghreb where locally formed militias with a medley of grievances have in recent years morphed into jihadi insurgencies that are viewed by the US as a threat to regional stability and the economic resurgence of states key to US interests such as Kenya and Nigeria. The radicalisation of Islamist factions, and ultimately the spread of Al-Shabaab militants, was however fuelled in part by the US-backed Ethiopian offensive in Somalia in 2006 but has also been enabled by the US-backed rebellion in Libya in 2011, which led to the overthrow of Muammar Gaddafi but also to arms from Gaddafi's regime circulating around North Africa and ending up in the hands of jihadi groups such as Boko Haram. Key to the US response to Islamic insurgency is AFRICOM, created in 2007 and one of six DoD "geographic combatant commands". AFRICOM's mission statement claims that, working with partners, it:

> disrupts and neutralizes transnational threats, protects U.S. personnel and facilities, prevents and mitigates conflict, and builds African partner defence capability and capacity in order to promote regional security, stability, and prosperity. (AFRICOM, 2017)

AFRICOM sees itself as advancing US national security interests primarily through joint military-to-military training and assistance programmes with dozens of African and European militaries, security cooperation activities and military operations. Military training programmes are specifically designed to increase Africa's counter-terrorism capabilities and prevent the creation of terrorist safe havens but also serve to protect US economic interests in the region and to secure future energy and resources (Lyman and Morrison, 2004). The US military has publicly insisted that its efforts in Africa are small scale, no more than a "light footprint" but by 2015 the number of missions, exercises, operations and other activities under AFRICOM's purview had more than doubled to include some 75 joint operations, 12 major joint exercises and 400 "security cooperation initiatives" (Frontera, 2016). Brigadier General Bolduc, the former commander of SOCAFRICA (Special Operations Command, Africa), claimed in 2016 that AFRICOM's activities were motivated by the fear that "Africa's challenges could create a threat that surpasses the threat that the United States currently faces from conflict in Afghanistan, Iraq, and Syria" (Bolduc cited in Turse, 2016). At the time Bolduc also noted that there were nearly 50 terrorist organisations and "illicit groups" operating on the African continent but identified only the Islamic State, Al-Shabaab, Boko Haram, AQIM, and the LRA by name or acronym, while mentioning the existence of another 43 groups (Turse, 2016) (see Figures 6.6 and 6.7). The Africa Center for Strategic Studies, a research institution under the DoD, now regularly publishes updated cartographic representations of the changing geographies of Africa's militant Islamist groups.

In Africa US troops are now conducting around 3500 exercises, programmes and engagements per year, an average of nearly 10 missions per day (Turse, 2017). Little substantive information has however been made public about what exactly these missions have involved and just whom US forces have trained. This is the age of "the everywhere war" (Gregory, 2011) as the traditional geographies of warfare have become increasingly blurred in an age defined not by clear battlefields and officially recognised combat zones (Moore and Walker, 2016), but by multidimensional

FIGURE 6.6 Nigerian soldiers hold up a Boko Haram flag that they had seized in the recently retaken town of Damasak, Nigeria, March 18th, 2015. REUTERS/Emmanuel Braun.

FIGURE 6.7 A Somali soldier holds a mortar gun at Sanguuni military base on June 13th, 2018, where an American special operations soldier was killed by a mortar attack five days previously, about 450 km south of Mogadishu, Somalia. More than 500 American troops have been working with the African Union Mission to Somalia (AMISOM) and Somali national security forces in counterterrorism operations and have conducted frequent raids and drone strikes on Al-Shabaab training camps throughout Somalia. Mohamed Abdiwahab/AFP/Getty Images.

and fluid "battlespaces" (Graham, 2009), shadowy campaigns against non-state actors in "borderlands" and undisclosed locations, and "war in countries we are not at war with" (Ryan, 2011). As Moore (2017) notes, this everywhere war is characterised by complex geopolitical and geoeconomic entanglements given the US military's outsourcing and increasing reliance on private companies that employ a global army of civilian labourers to provide logistical support for operations around the world. In Africa, private military and security contractors (PMSCs) increasingly perform a variety of services, from transporting and housing personnel, to shipping materials and food, to providing medical support and conducting surveillance. Nigeria, for example, brought in hundreds of South African mercenaries, with roots in Apartheid-era security forces, to assist in its offensive against Boko Haram (Varin, 2015).

There are various elements that make up the US military assemblage in Africa which collectively facilitate the movement of US soldiers and equipment across the continent (Moore and Walker, 2016). Networks of physical sites within this assemblage can be difficult to identify, however, and fluctuate as the military constantly shifts resources, personnel and equipment (Moore and Walker, 2016). The US military presence in Africa is also masked by the extensive use of covert special operations forces, secret facilities and PMSCs. Several FOLs (Forward Operating Locations) and drone bases have been established and discontinued in Ethiopia, Kenya and Mauritania due to changing political considerations, operational demands and resource limitations (Moore and Walker, 2016). There is also a growing network of cooperative security locations (CSLs) through which small numbers of US troops periodically rotate as well as contingency locations (CLs) (used only during ongoing missions) and a variety of logistics nodes across the continent and beyond.

The US is also currently building a new US$110 million military air base in Agadez, Niger, capable of deploying drones and giving the Pentagon another surveillance hub for launching intelligence, surveillance and reconnaissance missions against a plethora of terror groups in Western and North Africa. Niger in particular has positioned itself to be the key regional hub for US military operations whilst in Chad, the US Air Force has been flying drones and other aircraft from a French military base to search for hundreds of schoolgirls abducted by Boko Haram in Nigeria in April 2014. Ouagadougou in Burkina Faso (which has been the scene of jihadist attacks in recent years and sits between Al-Qaeda groups to the north and Boko Haram to its east in Nigeria) has also been the centre for a US programme code-named "Creek Sand" where US spy planes have carried out surveillance missions across Northern Mali and Mauritania, targeting fighters from Al-Qaeda and the Islamic Maghreb (AQIM) (Whitlock, 2012). The US has also established an alliance with the corrupt and authoritarian regime of Cameroonian President Paul Biya (see Figure 6.8) to establish a drone base at Garoua, enabling intelligence gathering on Boko Haram (Hammer, 2016). The enrolment of West African regimes into AFRICOM's counterterrorism network has however arguably "undermined governance and human rights and increased political instability across the region" (Moore and Walker, 2016: 705). For political and military elites in the Sahel, "binding themselves to AFRICOM's counterterrorism assemblage can be useful for better securing their own authority and privileges against potential challengers" (Moore and Walker, 2016: 705). This dynamic is not limited to Africa, of course, as the US has "repeatedly collaborated with murderous, antidemocratic regimes and ignored widespread evidence of human rights abuses" in countries that it relies upon for overseas bases of operation (Vine, 2015: 97).

There are also various bureaucratic and military practices that are instrumental in "forging alignment" between the US and African

FIGURE 6.8 US Marines and sailors work with Cameroon's *Fusiliers Marins* and *Compagnie des Palmeurs de Combat* to increase their capabilities to combat illicit activity and increase security in the waterways and borders of Cameroon. Photo by Cameroon's Fusiliers, courtesy of US Marine Corps.

states and militaries, and facilitating flows of money, weapons, knowledge, people and ideologies in the assemblage (Moore and Walker, 2016). These include status of force agreements, overflight permissions, multinational "military to military" training exercises such as Flintlock and African Endeavor, security cooperation programmes like Africa Contingency Operations Training and Assistance (ACOTA), and military assistance programmes such as the Section 1206 scheme, which allocates billions of dollars to the DoD to distribute to select countries to pay for equipment and training for counterterrorism purposes (Moore and Walker, 2016; see Figure 6.9). Key to many of these activities is Camp Lemonnier, a former French Foreign Legion base leased from Djibouti in East Africa which houses about 4,000 US military personnel and civilian contractors along with conventional forces specialising in training African militaries and several hundred special operations troops. It has become the "backbone" of covert missions across Africa and the Arabian Peninsula and one of the military's most important bases for drone missions in Somalia and Yemen (Turse, 2016).

US troops carry out a wide range of operations in Africa, including airstrikes targeting suspected militants, night raids aimed

FIGURE 6.9 A US Marine instructor teaches Mauritian Fusilier Marin soldiers how to fire an AK-47 automatic rifle during light infantry training on February 26th, 2015 in Nouadhibou, Mauritania. US Marines Photo/Alamy.

at kidnapping terror suspects, airlifts of French and African troops onto the battlefields of proxy wars, and evacuation operations in destabilised countries (Turse, 2014a). Above all, however, the US military conducts training missions, provides mentoring to allies and funds, equips and advises its local surrogates with whom it seeks to work on counterterrorism. The mantra has been "muscular soft power" (Sengupta, 2013), the process of preparing states to defend themselves while building up infrastructure and civic institutions. One example is the Trans-Saharan Counterterrorism Partnership (TSCTP), which operates civilian and military projects in Mali, Chad, Mauritania, Niger, Algeria, Morocco, Senegal, Nigeria and Tunisia in the name of promoting regional "stability". The US has also been aiding France's ongoing interventions in West and Central Africa in a burgeoning Franco-American alliance on the continent as both countries have steadily expanded their presence in West Africa in response to the presence of jihadist groups taking root in "weak" states, even setting up neighbouring drone hangars in Niamey, Niger's capital, to conduct reconnaissance flights over Mali.

The US military is also deeply involved in the field of reconstruction and development across the African continent, regarding humanitarian assistance as a form of "security cooperation", "civil affairs" programmes and "stability operations". Across Africa, the US military is now engaged in a panoply of aid projects with an eye towards winning the hearts and minds of Africans and so reducing the lure of extremist ideologies associated with groups such as Boko Haram and Al-Shabaab. Most contemporary US civil affairs activities on the continent are carried out in areas that US security strategists conceptualise as being "ungoverned" or "undergoverned" and thus exploitable by terrorists and other forces seen as contributing to instability (Bachmann, 2008; Bradbury and Kleinman, 2010). The concept of stability (or stabilisation) operations takes counter-insurgency thinking to another level: it extends and normalises military engagement in the fields of development and governance, even in peacetime contexts (Bachmann, 2014; Collinson, Elhawary and Muggah, 2010; Morrissey, 2015). The so-called civil-military operations, or CMOs, include "humanitarian assistance" projects like the construction or repair of schools, water wells and waste treatment systems, and "humanitarian and civic assistance" (HCA) efforts, like offering dental and veterinary care. Such efforts are engineered to improve US visibility, access and influence with foreign military and civilian counterparts in order to promote the security and foreign policy interests of the US (Turse, 2014b). These small-scale efforts are further divided into "community relations activities" like the distribution of sports equipment, and "low-cost activities" such as seminars on solar panel maintenance or English-language discussion groups (Turse, 2014b). US military personnel of the Combined Joint Task Force-Horn of Africa (CJTF-HoA) have, for example, carried out hundreds of small projects in East African countries since 2003 (including Ethiopia, Djibouti, Kenya and Uganda) which include constructing or refurbishing schools or health centres (see Figure 6.10), providing medical care wherein health experts offer short-term treatment for the population, and completing veterinary assignments or projects related to water access (Bachmann, 2017). In northern and north-eastern Kenya the US military teams have implemented approximately 200 projects over the last decade, with civil affairs projects almost exclusively concentrated in the regions close to the Somali border where the majority of the population is Muslim (Bachmann, 2017).

Compared to the scope and costs of other US military activities in Africa, including large military-to-military training and manoeuvres,

FIGURE 6.10 US Navy Lieutenant Cory Cole from the Maritime Civil Affairs team helps Kenyan students plant casuarina trees at Mjanaheri primary school in June 2012. Combined Joint Task Force-Horn of Africa (CJTF-HoA) planted 600 tree seedlings at the school as part of World Environment Day. PJF Military collection/Alamy.

secret operations and logistics investments (Bachmann, 2014; Turse, 2015), these civil affairs operations are "a relatively small but important element of the US military's post-counterinsurgency crisis prevention strategy and its emphasis on stability operations" (Bachmann, 2017: 11). So far, however, critical scholarship has paid little attention to this type of military practice, let alone the dynamics of giving and taking in affected communities (Bachmann, 2017). Such activities are however difficult to research, given that the little that is known about US civil affairs operations in Africa often comes from within the military establishment itself or from scholars close to it (Farrell and Lee, 2015). Official news releases often include "feel-good" stories about what the projects have achieved but many have been deeply problematic. An investigation by the DoD's Inspector General found failures in the planning, execution, tracking and documentation of such projects with military officials failing both to identify how their projects even supported AFRICOM's wider objectives on the continent and to ensure that local populations were equipped to keep the small-scale projects running or sustainable (Turse, 2014b).

CONCLUSIONS: (RE)MILITARISING DEVELOPMENT

USAID narrates its work as being about "supporting US national interests" and "making Americans safer at home and abroad" (USAID, 2017) and alongside diplomacy and defence efforts, it claims to be tackling the things that foster "violent extremism, instability, transnational crime and other security threats" by addressing such challenges as "extreme poverty, food insecurity, pandemic disease, conflict, violence, and poor governance" (USAID, 2017). So much of what USAID does in the name of aid, assistance and development, past and present, has been geopolitically oriented and motivated – focusing on governance, political stability and democracy and on securing states by defeating terrorism and countering insurgency. In many ways, USAID's current configuration illustrates both the increasing securitisation of development and the power of the "security–development nexus" but also the ways in which development is increasingly framed as a technology of counter-insurgency. Feeding on a geopolitics of fear and constructing narratives of "threat", failure and crisis in the borderlands, the nexus constitutes a growing field of development and security actors, aid agencies and professional networks, that call forth the conditions of need and insecurity that they then seek to intervene upon. The aid given by these actors is not simply about nation states or one-way coherent flows of money, however, but involves complex networked elements or assemblages, constituted by flows of capital, knowledge, influence, practices, material objects and people (Roberts, 2014). A key part of this is the growing use of private military and security contractors and the powerful contracting assemblage centred on the Beltway region in the greater Washington, DC area (Roberts, 2014) which means that selected for-profit US firms benefit considerably from the sums US agencies have spent in the name of securing developing areas.

Alongside USAID, the DoD and State Department have become increasingly important in the framing and delivery of US foreign assistance and, as a result, US military operations encroach ever more on domains traditionally associated with development and diplomacy whilst military actors and objectives have become increasingly influential in both the formulation and delivery of aid programmes. In the process of reconstruction as war, "non-traditional" military practices around development (e.g. civil affairs programmes in East Africa) have proliferated in an attempt to sustain new practices of civil–military integration revived and advanced in the counter-insurgencies of Iraq and Afghanistan. Such interventions have however been beset with tensions and controversy and are a bundle of contradictions in pursuing strategic-military objectives as part of stabilisation operations whilst simultaneously aiming to address immediate local socio-economic needs. More research is needed to understand how these development projects work in practice, how the US military mobilises and affects the actors involved in different ways and draws local populations into encounters with foreign militaries and their strategic agendas (Bachmann, 2017). In Iraq and Afghanistan in particular, the disbursement of US foreign assistance has been plagued by allegations of fraud, corruption and abuse with interventions characterised by poor planning, shoddy construction, mechanical failures and inadequate oversight whilst billions of dollars have been squandered on counter-insurgency failures with "stories of ruined roads and busted buildings, shoddy schoolhouses and wasteful water parks, all in the name of winning hearts and minds" (Turse, 2014b).

Underdevelopment has long been scripted as "dangerous" by USAID but the idea of poverty as a recruiting ground for menacing strategic threats has been monotonously rediscovered by

USAID (and a wide range of other donor agencies) in and through the global war on terrorism (Duffield, 2010a). Building on its Cold War experience of assisting counter-insurgency efforts in places like Vietnam, USAID interventions are increasingly concerned with the government and management of restless rural populations and the restoration of political order and stability (Attewell, 2017). The historical focus of US assistance on large-scale iconic development projects, conceived in the course of Cold War modernisation, has now been replaced by the pursuit of development as a form of governmentality, with USAID increasingly focused on neutralising and dispelling perceived threats in their place of origin, particularly in battlespaces such as Iraq and Afghanistan and increasingly in Africa, depicted as the new "swamp of terror". In this sense it is important to consider the biopolitics of counter-insurgency and the ways in which such efforts seek to reconfigure biopolitical landscapes, to discipline bodies or to coerce people to recognise the authority of local governing institutions and a foreign military (Belcher, 2018). USAID claims that "[s]trong American leadership will promote development and provide life-saving assistance" (USAID, 2017) but during the Cold War US foreign assistance played a key role in supporting authoritarian regimes and in creating conflict and instability. Today the US remains the largest arms exporter globally (accounting for 33% of global arms exports), offering a wide array of weaponry and services and making arms deliveries to at least 100 countries (SIPRI, 2017). Although Washington sold to a global market, states in the Middle East (also some of the largest historical recipients of US aid) received nearly half (47%) of all US arms transfers between 2012 and 2016 (SIPRI, 2017). Enabling countries like Israel to consolidate and secure their statehood through huge arms transfers and long-standing military training programmes makes a real difference to regional, national and local balances of power and the result has not always been greater security and stability.

USAID's work is also very much about geoeconomics, about making recipient states "safe" for the spread of capitalism and enabling and accelerating their neoliberalisation, urging societies to develop in accordance with appropriately neoliberal state institutions. In this sense, as Roberts, Secor and Sparke (2003) have argued in conceptualising "neoliberal geopolitics", it is important to think about the inter-articulation between neoliberalism and the violence of American military force, particularly given that the recent conflicts in Afghanistan and Iraq can be seen in part as a form of neoliberal "roll-out". It is the ways in which US geopolitical interests intersect with geoeconomic strategies through a "security–economy nexus" (Coleman, 2005) that are important here. USAID has thus developed from and advanced both geopolitical and geoeconomic objectives and ideals while at the same time reshaping the balance of political and economic forces and institutional actors in the political structures of recipient states along with the US state itself and the wider international system (Simpson, 2008). In this sense it is important to remember that practices of aid are co-constituted by international donors and recipient states (Hyndman, 2009: 883). To get a better understanding of how this works in practice, however, more research is needed on the messy realities of American aid, on how USAID projects and practices articulate in specific sites and contexts and around the ways in which populations targeted for development negotiate, contest and/or co-opt American aid interventions (Bhungalia, 2016: 89).

Celebratory and redemptive narrations of the role of American aid in spaces of insecurity like Afghanistan have been common in some popular geopolitical texts like *Three Cups of Tea* (TCT), which tells the story of the life and work of American mountaineer-turned-humanitarian Greg Mortenson, and his efforts to counter terrorism in Northern Pakistan and Afghanistan through the creation of schools. Popular with the American public as a story of humanitarian development, TCT embodies a depoliticised and dehistoricised representation of Northern Pakistan and includes essentialising narratives of terror, rural "ignorance", backwardness, underdevelopment, "extremism" and global terrorism at the same time as it promotes a liberal interventionism and reproduces the West-affirming terms of development discourse (Ali, 2010). TCT became implicated in a participatory militarism in which an ethnographically sensitive military (along the lines of CORDS and HTS) strives to "listen" and "build relationships" to "serve people" in order to occupy more effectively and for longer. In this sense, the story reflects the wider collusion between the technologies of humanitarianism and military power that characterises the "humanitarian present" (Weizman, 2011). Mortenson has since faced accusations that the book contains multiple fabrications and also that he mismanaged donations to the Central Asia Institute, a non-profit community education organisation that he co-founded (Goldberg, 2015).

In contemporary US foreign policy Africa's "ungoverned spaces" and "failed states" have increasingly been scripted as a battleground in the war on terror, leading to an expanding, near-continent-wide campaign utilising the core tenets of counter-insurgency strategy (Turse, 2014b). US foreign policy concerning Africa has however often struggled to differentiate between conflicts and disputes that are localised in their remit and those that present a threat to the US, but has also seen the US strike agreements with autocratic regimes with dubious human rights records (e.g. Cameroon) in pursuit of its anti-terror agenda. Half a century ago, the US was concerned about "dangerous, pro-Communist" African radicals and there remain concerns that US policy towards Africa may mirror its "anti-communist" support for autocratic regimes during the Cold War, with "anti-terrorist" support for comparable regimes in the post-Cold War era (Adebajo, 2008b: 233–234; Rupiya and Southall, 2009).

Central to US interventions on the continent has been the geopolitical assemblage of AFRICOM and an expanding network of facilities such as Camp Lemonnier, drone bases, cooperative security and contingency locations, along with a variety of logistics nodes across the continent and beyond aimed not just at advancing US security and foreign policy objectives but also at protecting US economic interests in the region (e.g. around access to hydrocarbon resources in West Africa). US officials often point out, however, that AFRICOM's activities are still "dwarfed" by America's non-military assistance to Africa including the President's Emergency Plan for AIDS Relief (PEPFAR), and Power Africa, Obama's programme to boost the electrification of the continent. US drone strikes in Africa (e.g. against Al-Shabaab militants) have become commonplace but have bred widespread resentment due to the civilian casualties they often incur (e.g. in places like Yemen and Somalia). In 2017 the Trump administration lifted the Presidential Policy Guidance rules put in place by President Barack Obama to govern counterterrorism strikes away from conventional war zones and with a view to preventing civilian casualties. By designating regions as areas of "active hostilities" where the Guidance rules do not apply, Trump lifted restrictions on counterterrorism operations in Africa in April 2017 and the ensuing deaths of several US special forces troops soon after drew attention to US counter-insurgency efforts on the continent. In October 2017, Nigerien forces and American Green Berets were ambushed by members of the Islamic State in the Greater Sahel (ISGS) during an intelligence-gathering mission along the border with Mali, with four US Green Berets killed in the fighting and two others wounded (Fall and Koura, 2017).

Trump engaged in a politicised and widely publicised confrontation with the widow of one of those killed over the words he used in a "consoling" phone call, but the events underlined how little the American public (and much of the political establishment) knew of the missions conducted by the US State Department and DoD in Africa.

Some commentators and observers have, sometimes rather crudely, drawn out some of the contrasts between contemporary US and Chinese policy towards Africa with the former seen as ideological and the latter as pragmatic (Copley, 2014). Turse (2014a), for example, discusses the case of Mali where US Special Operations forces have been teaching infantry tactics to Malian troops whilst the Chinese have invested heavily in developing national infrastructure:

> For the Chinese, Africa is *El Dorado*, a land of opportunity for one million migrants. For America, it's a collection of "ungoverned spaces", "austere locations" and failing states increasingly dominated by local terror groups poised to become global threats, a danger zone to be militarily managed through special operations and proxy armies.

Some of these points of contrast have, however, perhaps been overstated. There have been a number of incidents involving the kidnapping or killing of Chinese labourers in places like Ethiopia, Mali, Nigeria and Sudan that have raised awareness about the political risks around Chinese investment in Africa and the profile of security issues in China–Africa relations. In fact, the deeper Beijing ventures into the African continent the more it is beginning to stumble upon various security challenges (Holslag, 2009). Further, China's own history with significant political interventions around economic development since the inception of the CCP under Mao's rule has resulted in a strong belief in the necessity of economic growth to maintaining internal order whilst China's "views on how to contribute to international peace and security are very much influenced by a development-security nexus" (Benabdallah, 2016: 21). China is also no stranger to post-war reconstruction through its recent engagements with several African countries, including Sierra Leone, Liberia, Sudan, Angola and the DRC. China similarly runs capacity-building and vocational training programmes with African high-ranking army officials, police forces, peacekeepers and private security personnel who participate in Beijing-sponsored annual training exercises and military-to-military exchanges as an essential part of China's Africa security strategy. China has also transferred arms and provided training to the Nigerian and Cameroonian militaries for their operations against Boko Haram and supported the AU mission against Al-Shabaab in Somalia. Further, China has its own domestic experiences of countering insurgency to draw upon and has participated in a range of counter-insurgency exercises, including jungle warfare training (Wayne, 2008). There is also the likelihood that China will support insurgents – and counter-insurgents – as part of its foreign policy (Odgaard and Nielsen, 2014).

China's military spending has increased almost every year since 1989 from US$21 billion in 1988 to US$215 billion in 2015 (SIPRI, 2017) and indirectly China also plays a key role in shaping the contemporary geographies of peace and security in Africa in that Chinese military equipment is now being used by more than two-thirds of African countries (IISS, 2017). There are also increasing obligations for China as a self-styled "responsible world power" to play a greater role in peace and security on the continent. In July 2017 China opened a new support and logistics base in Djibouti intended to assist with China's growing number of peacekeeping and humanitarian missions and as part of China's wider OBOR initiative, with around 1,000 personnel to eventually be stationed there. There have also been suggestions that Beijing has plans to build an overseas naval supply and recuperation facility in the Seychelles to assist its resource exploration and anti-piracy efforts, raising considerable alarm in India which has significant interests in the country and itself had plans to build a military base there (although the agreement was cancelled by the Seychelles in June 2018). Beijing's new naval port, if completed, will be located in close proximity to the Pentagon's strategic military establishment, Diego Garcia, owned by the British and located in the central Indian Ocean. It is to this question of the expanding role of China (and other so-called "rising powers" or "emerging economies") in the global South that we now turn in chapter 7.

Chapter 7

The Rise of the South

INTRODUCTION: THE REVIVAL OF SOUTH–SOUTH DEVELOPMENT COOPERATION

> The South has risen at an unprecedented speed and scale... collectively bolstering world economic growth, lifting other developing economies, reducing poverty and increasing wealth on a grand scale... Global economic and political structures are in flux... the principles that have driven post-Second World War institutions and guided policy makers need recalibration, if not a reset, to accommodate the growing diversity... and to sustain development progress... (UNDP, 2013: 1–2)

IN recent years several UN agencies have suggested that the South is on the rise. A 2004 UNCTAD secretariat report, for example, declared that "a new geography of trade" was "reshaping the global economic landscape" with the South "gradually moving from the periphery of global trade to the centre" (UNCTAD, 2004: 1). In 2013 the UNDP even entitled its annual Human Development Report (HDR) The Rise of the South (UNDP, 2013), noting that the value of trade between developing countries (South–South trade) had overtaken that of developing country exports to the global North in 2012, a trend that was "radically reshaping" the twenty-first-century world. The share of global GDP generated in the South has grown significantly along with the demand for mineral and energy resources needed to fuel the domestic growth of emerging hegemons like China and India. The global geographies of aid and investment are shifting too as (re-)emerging economies become major sources of outward FDI flows to other parts of the global South. Asia alone is now home to some 197 of the Fortune 500 companies for 2017, more than any other continent. Emerging economies have also increasingly begun to develop an internationalist profile and to assert themselves as humanitarian, peacekeeping, peacebuilding and "policekeeping" actors on the world stage. As a result, the nature of both aid and international development cooperation is slowly being "reset" and "recalibrated". This has further raised the spectre of South–South development cooperation (SSDC) articulated at Bandung in 1955 and spurred on by the dependency debates of the 1960s and by calls for a New International Economic Order (NIEO) in the 1970s. SSDC was institutionalised by the UN with the opening of a special office within the UNDP in 1974 (UNOSSC) which, in the true spirit of modernisation, framed cooperation as a largely technical exercise.

Countries of the South that have experienced significant economic growth in recent decades not only espouse the cause of "developing countries" (which they claim to be champions of) but are also keen to emphasise that they themselves belong to this group and that they are articulating progressive visions of horizontal, less hierarchical and more equal interaction and cooperation that are mutually beneficial or "win-win". Many of these claims and the narratives of solidarity that they rely on are reminiscent of Third Worldist coalitions of the past (Narlikar, 2013: 603), even leading some to suggest that "the world is witnessing a refurbishment of China's old Three World Theory" (Chan, 2016) or that China's rise brings the possibility of a "new Bandung" (Arrighi, 2007). A stringent positive definition of SSDC is hard to find, however; the concept remains broad and indeterminate although historically "self-reliance" and "self-help" have been regarded as central elements (Chaturvedi, 2013). Further, SSDC suggests a "natural" congruity between states across the Global South that are in fact very different (McEwan and Mawdsley, 2012). SSDC practices should thus not "be understood as an unproblematic unitary force, but as constituted by complex and often contradictory national prerogatives and interests" (Gray and Gills, 2016: 564).

This chapter explores the recent revival of SSDC and examines some of the different conceptions of aid and development assistance being formulated by (re-)emerging donors from the South. It focuses on how both the framing and delivery of this assistance are shaped by geopolitical agendas, discourses and imaginations but also on Africa's growing significance in international relations and the ways in which four (re-)emerging economies (China, India, Brazil and South Korea) have vociferously courted the continent. Since many of these (re-)emerging donors have opted out of the international regime put in place by Western governments after World War II to track overseas development finance activities, it can be difficult to get reliable data on the precise levels of assistance being provided. Many of these donors have also blurred and loosened the definition of what constitutes "aid". What is becoming clear, however, is that globally ODA from OECD countries is, in relative terms, of declining importance and attraction and represents a shrinking proportion of transnational transfers (Shaw, 2014), although this changing picture is subject to significant regional variation. The old order of global development, predicated upon a vertical donor–recipient relationship, is in decay and key institutions like the OECDs Development Assistance Committee (DAC) are facing a "crisis of legitimacy" that sees criticism from recipients, non-state actors and emerging donors, thus "challenging its very nature as the pre-eminent donor forum" (Fejerskov, Lundsgaarde, and Cold-Ravnkilde, 2016: 14). A number of new sources of finance have further undermined the power of the IFIs including transfers from faith-based organisations and new private foundations (e.g. Gates, Clinton, Mo Ibrahim) (Moran, 2014), along with remittances from diasporas, carbon taxes/trading, climate change funds, Sovereign Wealth Funds, controls on money laundering and of course aid and private investment flows from (re-)emerging donors themselves (Shaxson, 2012; Shaw, 2014). Such trends have even led some observers to proclaim the "end of development" (Brooks, 2017).

The "Asian financial crisis" of 1997–98 was also billed by some observers as ostensibly a harbinger of the "end of late development" (and a reassertion of the West's political and economic dominance over the global South) but ten years later the 2008 global financial crisis, having originated in the West itself, arguably turned that narrative on its head, undermining the credibility of the IFIs to prescribe the pathway to successful economic growth (Gray and Murphy, 2013). The 2008 economic crisis was also seen by some to be part of a "global rebalancing process" whereby the established global North–South axis is being superseded by an East–South turn, one that "holds significant emancipatory potential" (J N Pieterse, 2011: 22). The crisis also opened up space for emerging economies to play an increasingly active role in the reform of global economic and political governance, to the extent that a "regime change" was regarded as a distinct possibility (J N Pieterse, 2011: 22). For some, this was the beginning of a transition from a unipolar US hegemony to one of "emancipatory multipolarity" (J N Pieterse, 2011) in which the countries of the South now have a position at the head table, or even a broader underlying "global centre shift" or "hegemonic transition" (Gills, 2011). Often this depicts hegemonic conflict as centred in tensions between a "declining" US state and a "rising" Chinese state (Arrighi, 2007; Anderson, 2017). The US-led capitalist world order of the post-war era is clearly experiencing significant shifts in the centres of politico-economic power following the rise of the South, but the decline of US hegemony is not the same as the end of US power and many traditional international donors like the US and the European Union continue to hold significant influence. More "pessimistic" views have also been expressed that the rise of the South has been overstated and is firmly located within Western global hegemony, offering only a limited, "within-system" challenge (Stephen, 2014).

This chapter examines the challenges to global governance collectively posed by the (re-)emerging Southern donors and questions the extent to which their rise has produced "new" or "emancipatory" modalities of "development" or has had transformative impacts on the world order. As Stuenkel (2016) notes, many readings of the current international order are limited in that they depict a "post-Western world" from a narrow-minded Western-centric standpoint where the West is conceived as the sole actor entitled to shape the norms by which the international system is disciplined. This is often framed in terms of inter-state rivalry, through a myopic agenda that typically revolves around the question of China's "challenge" to the West as a zero-sum competition (Hirono and Suzuki, 2014: 458). Some analyses see only an erosion of global governance's most liberal principles or a "wrenching [of] global relations into flux" (Shaw, Cooper and Chin 2009: 27). Others reject the idea that the imminent post-unipolar world will be necessarily chaotic and unstable and argue that non-Western actors do not so much seek to undermine Western institutions or to question the foundations of Western liberal order, as to forge parallel institutions (that emulate the West) and a parallel global order (Stuenkel, 2016). Either way a fragmentation of global governance appears inevitable and for some may even be creative in that it reflects the broader forces of change in the global political economy (Acharya, 2016).

There has also been much discussion of the idea of "socialising" countries like China and India into the current international development system by involving them in current structures and making them "responsible stakeholders" leading to greater norm convergence, such that they come to value the norms that underpin the system and eventually seek to uphold and potentially enforce them (Narlikar, 2013). By binding new or emerging powers into the current architecture of international development, so the argument goes, their challenges to pre-existing conceptions and practices can be managed and contained, suggesting an underlying agenda to "domesticate" rivals (Abdenur and Da Fonseca, 2013). Also significant here are the labels used to describe these countries as "emerging" markets, economies or powers, as "new" or "non-traditional" development donors or simply the "rising powers" (Cooper and Flemes, 2013; Gray and Murphy, 2013; Six, 2009). Many of these labels are characterised by a historical amnesia that fails to recognise that countries like China, India and Brazil have been active across the global South for many decades and that their status as donors or their economic "rise" is not new and has important historical antecedents. Grouping these actors under a unitary analytical category also considerably downplays some of the substantial differences between them and fails to recognise that their individual discourses have not yet converged on a common narrative around SSDC. They also often overlook the significance of and sensitivity around labels like "rising powers" within these countries. China, for example, has been keen to counter that the "co-development" it seeks to promote overseas is friendly and mutually beneficial or "win-win", although this often obscures the many parallels with Western approaches (Driessen, 2015).

One of the most influential terms is the BRICs acronym first coined in 2001 by Jim O'Neill, then at Goldman Sachs, which regarded Brazil, Russia, India and China as future motors of global accumulation, but this rests upon and reworks previous "emerging markets" discourses going back to the 1980s and represents a further shattering of prior meta-geographical demarcations (categories such as "Developed" and "Third World") (Sidaway, 2007, 2013). The conceptual genealogy of the term "BRICS" can be traced back to the investment discourses, financial industry narratives, statistical tools and common classificatory regimes (which differentiate economic performance, risk and return expectations) used by investment banks like Goldman Sachs (Wansleben, 2013). The BRICs concept re-narrated a specific cluster of large emerging economies as safe, long-term financial asset providers and has since come into widespread use as a symbol of the apparently epochal shift in global economic power away from the developed G7 economies towards the "developing world", along with a wider realignment of world economic and ultimately political power (Power, 2015). This has also been followed by a plethora of acronyms created by rival banks, investment firms and investment funds in search of opportunities and pastures new including the "Next Eleven" (N-11), the CIVETs and EAGLES (Emerging and Growth-Leading Economies), the "7% Club", MIST and MINT.

This chapter uses the term (re-)emerging economies and argues that rather than singling out influential players like China (or the individual policies that led to their rapid economic growth) it is necessary to adopt a relational view of their (re-)emergence vis-à-vis each other (and the US as the pre-existing global hegemon) and sets out to contextualise their growth within the wider neoliberal capitalist order. China's new global role, for example, cannot be reduced solely to an analysis of state policies post-1978, but rather "has largely grown out of the changing nature of transnational capitalist production imperatives" (Hart-Landsberg and Burkett, 2005: 36). In many ways, it is too easy to single out countries like China as an exception "without addressing the structural and institutional forces that are driving not only China, but also other emerging powers, to look with covetous eyes at Africa's natural resources and markets" (Luk, 2008: 13). To an extent it is also necessary to look beyond the BRICS in order to capture the increasingly significant roles played by other (re-) emerging actors such as Turkey (Baran, 2008; Apaydin, 2012; Shinn, 2015) or Israel (Yacobi, 2015). Arab donors such as Saudi Arabia, Kuwait and the UAE also have a long-standing history of development cooperation in the Middle East and North Africa which gained momentum after the petroleum booms of the 1970s (Mawdsley, 2013). In 2011, with a stock of investments amounting to US$19 billion, Malaysia was in fact Africa's most important Asian investor (Steinecke, 2016), ahead of China and India in terms of the size of its FDI, although this had dropped back to US $12 billion by 2017 (UNCTAD, 2017). There is also Japan, one of the most well-established Asian donors, whose integrated approach in Asia, using tied concessional loans supplied together with technical assistance and grant aid with the aim of promoting FDI from Japanese companies and enhancing trade, has been highly influential in SSDC (Morikawa, 2005; Raposo, 2014a, 2014b; Yamada, 2015). The competition for strategic influence in the South between these (re-)emerging donors (with many fearing China's rise in particular) is an important driving force pushing them to further and widen their international engagements.

The chapter focuses in particular on what the "rise of the South" has meant for Africa and argues that this needs to be situated and understood within the context of wider transformations in the global political economy, since this "rise" is linked to and collectively shaped by broader transnational capitalist dynamics (Ayers, 2013: 250). What the role of China and growing economies such as India tells us about Africa and international relations is thus:

> not limited to Africa as a case-study or passive entity in which changing configurations of power continue to play out or as a region wholly bound by structural social and economic forces... Today, the rise of China suggests a shifting terrain of international relations in which Africa is at the core, with the potential opportunity to make aggressive use of the space created by the presence of China to exert greater agency in the international system. (Harman and Brown, 2013: 19)

Although Africa has become much less peripheral in this shifting terrain of international relations, Asia's centrality also needs to be acknowledged here, particularly in the context of OBOR where China's focus on its nearest neighbours has intensified. Given the centrality of energy and natural resources in the global outreach strategies that (re-)emerging economies have developed in search of new markets, there has been talk of a "new scramble" for Africa (Carmody, 2011) and suggestions that a new round of dispossession is underway, entrenching highly uneven dynamics of accumulation and leading to the production or reproduction of (often fortified) enclaves – enclosed areas of capital-intensive extraction, predominantly in mining and commercial agriculture – alongside high-surplus labour environments marked by displacement, destitution and "disposability" (Ayers, 2013; McNally, 2011). These "new enclosures" are seen to herald the intensification of primitive accumulation, deepening Africa's integration into global extractive circuits and once again "locking" the continent into trading in and dependency upon primary commodities (Fulquet and Pelfini, 2015). Each of the (re-)emerging donors has pursued "resource diplomacy" – a form of economic statecraft where resource relationships are strategically manipulated by states to obtain access to and control of resources. Indeed, resources have been so significant in the relations these (re-)emerging economies have forged with other countries of the South that they have been framed as a security issue understood as critical to sustaining their own rapid growth and economic performance. In a throwback to the age of modernisation there has also been a clear preoccupation with the development of infrastructure, particularly roads, railways, dams and other energy projects. The chapter considers the twin imperatives of sourcing resources and opening markets that have been key to the geopolitical and geoeconomic strategies pursued by (re-)emerging economies in Africa and is divided into four parts which consider the cases of Brazil, China, South Korea and India, respectively.

BRAZIL AS A "CONDUIT FOR PAN-SOUTHERN ACTION"

Historically Brazil's domestic and foreign policies have been entangled with development thinking in a variety of ways, including through the "insurgent" development theory of the *dependentistas* (e.g. Milton Santos, Fernando Cardoso and Enzo Faletto) (Slater, 2008). In a 2004 speech to the UN General Assembly, Brazilian President Lula da Silva quoted Frantz Fanon ("if you so desire, take it: the freedom to starve to death") in reference to the common legacies bestowed by colonialism upon "those silenced by inequality, hunger and hopelessness" (Lula da Silva, 2004). Positioning Brazil firmly as a developing and emerging economy, Lula's speech conveyed a sense that Southern countries could and should make their own decisions and take charge of their own fate but also noted the asymmetric international structures "impeding" development. Ideas of Brazil as a leader and builder of Southern partnerships and coalitions, as a mobiliser and coordinator of pan-Southern voices, as a "conduit" or "interlocutor" for pan-Southern interaction, and a "bridge" or "balancer" between global North and South, have been present in Brazilian foreign policy for many years (Dauvergne and Farias, 2012). This has enabled Brazil to "advance its own interests behind a pan-southern value-creating, integrative fascia" (Burges, 2013: 593) as its long-standing ambition to be viewed and accepted as a major world power and desire to gain greater influence within existing institutions have led to the construction of "a benign, conciliatory, consensus-creating persona for Brazilian diplomacy" (Burges, 2013: 577). Brazil's moral leadership and the use of "opinion-shaping instruments" have, however, led to significant collective action on some shared challenges such as global health (Kickbusch, 2011).

Under Lula, Brazil's foreign policy priorities took a firm Third Worldist orientation reminiscent of the independent foreign policy tradition established before the introduction of an authoritarian government in 1964 (Almeida, 2007) which explicitly and fundamentally incorporated cooperation around "development" into Brazil's foreign policy (Costa Vaz and Inoue, 2007). Prior Brazilian attempts at driving national development forward through expanded links with Africa or even Latin America had little in the way of concrete results (Dávila, 2010) but Lula's administration (2003–11) pushed for a more representative global governance architecture (Kahler, 2013) and spoke of building a "new trade geography of the South" (Amorim, 2017). Lula's foreign policy built on previous attempts in the 1980s to diversify Brazil's trade linkages that used the South American trade bloc Mercosur as a platform for reducing the country's dependence on North American and European markets (Dávila, 2010) and was also an extension of some of the global positioning, reframing and coalition-building strategies used during the Cardoso administration (1995–2003), when stabilising Mercosur and South America became critical elements of Brazil's reinsertion into the international community (Ban, 2013).

In quite prosaic terms Brazil has often talked up its Southern credentials with reference to cultural diversity and having expertise and technologies that fit the needs of "developing countries", due to greater proximity (vis-à-vis Northern donors) in terms of economic and institutional development, culture and language (in the case of some African countries) and even agro-climatic conditions (Cabral et al., 2013).

To avoid subordination to the agendas of dominant countries Brazil has pursued a foreign policy strategy of "global power diffusion" (Christensen, 2013), forging alliances with large peripheral countries and some of the leading states of the South, regarded as key allies in building these new economic and political geographies (Bernal-Meza, 2010) but also in facilitating Brazil's rise and the realisation of its foreign policy aims such as permanent membership of the UN Security Council (UNSC). In the past, Brazil has sought to transform the global order by working against US unilateralism (Burges, 2013), attempting to push the US (and Canada) out of the management of South American and wider inter-American affairs. The idea that Brazil should at the very least be a sub-regional leader has long been part of the conceptual superstructure of Brazilian foreign policy (Dávila, 2010), with South America consistently regarded as a platform for Brazil's geopolitical ambitions in the regional and global arenas and for its competitive insertion in the global economy (Christensen, 2012). Brazil's recent turn to Africa is also not new and can be traced back to diplomatic attempts in the 1960s and 1970s to establish economic, political and cultural relations with the continent (de Freitas Barbosa, Narciso and Biancalana, 2009) when, after siding with Portugal against colonies demanding independence, Brazil changed its position and began recognising newly independent African states. The presence of Brazil's biggest companies in Africa also dates to the 1970s when the Brazilian military regime supported the internationalisation of domestic companies as a means to secure resources and foster development (Ban, 2013).

The onset of the global economic crisis in 2008 prompted Brazil to intensify its search for new partnerships in the South. At the centre of its attempts to do so is the Brazilian Cooperation Agency (Agência Brasileira de Cooperação, ABC), created in 1987 inside the Ministry of External Relations (MRE) and responsible for coordinating the negotiation, implementation and evaluation of Brazilian technical cooperation projects (both bilateral and multilateral), although it occupies a relatively low-grade position in the government hierarchy (Cabral et al., 2013). In 2003 a new working department was created within the MRE specifically with a view to fostering SSDC with Africa, leading to several new cooperation framework agreements between Brazil and African regional organizations like the African Union (2007) and the Southern Africa Development Community (SADC) (2010). Shortly after the Lula government took office in 2003 the MRE began to organise the Brazil-Africa Forum on Politics, Cooperation and Trade, as a symbolic event to herald a new era of Brazil–Africa relations (Stolte, 2015). There is also the Africa-South America Summit (ASA), a tri-annual bi-continental diplomatic conference which began in 2006 along with preferential trade agreements between Mercosur and the Southern African Customs Union (SACU) agreed in 2009.

Under Lula's administration presidential visits to Africa reached record levels and the number of Brazilian embassies there more than doubled, with the MRE facing significant staffing challenges as a result (Dávila, 2010). Although not a member of the NAM, Brazil has participated as an observer and has had additional interaction with African states via the G77, where it has played an active role. The Brazilian state has frequently drawn on these historical intersections to construct a narrative of SSDC that stresses Brazil's shared history and solidarity with African partners (Saraiva, 2012), emphasising the "special" nature of its relations with the continent, a rhetoric that is heavily based on the African heritage within Brazilian culture and on Brazil's historical debt to Africa for perpetuating slavery. The Brazilian public's perceptions of the continent however continue to be shaped by representations in the media and the education system that mix idealised "*Mama África*" narratives of Afro-Brazilian history with negative stereotypical representations of the continent (Oliva, 2009).

Brazil constructs itself as the purveyor of ideas to push events forward (e.g. around agricultural or urban development or in relation to clean energy in the South) rather than as a provider of significant volumes of tangible resources (Dávila, 2010: 587) and claims that its cooperation approach is guided by the principle of "solidarity diplomacy", bringing together elements of altruism (supporting those in need) and reciprocity (forging mutually beneficial partnerships) in a horizontal relationship between Southern peers (Cabral et al., 2013). According to ABC, what distinguishes Brazilian cooperation efforts are their non-interference and "demand-driven" nature, their acknowledgement of the value of local experience and the absence of conditions and associations with commercial interests (ABC, 2011: 3). Brazil rejects being labelled as a "donor" (a term it associates with the perceived vertical nature of North–South cooperation) and has avoided the use of labels such as "failed", "fragile" and "weak states", considering them to be stigmatising (Abdenur and Marcondes De Souza Neto, 2014). More generally, Brazil often emphasises that it can offer its own "tested" solutions to development problems whilst its technical and scientific cooperation initiatives have long deployed a discourse of "mutual learning", promoting exchanges of knowledge gained from "successful" social and economic development experiences (e.g. in health and agriculture) and playing a "key role in promoting capacity development in developing countries" (United Nations General Assembly, 2009: 3). Brazil's *Bolsa Familia* (Family Allowance) cash transfer programme for tackling poverty and its *Fome Zero* (Zero Hunger) programme to eradicate hunger and extreme poverty, both launched by Lula, have been seen as opportunities for knowledge sharing (particularly with Africa), with the latter being backed by a cooperation agreement with the FAO.

Indeed, technical cooperation agreements are the most visible modality of the country's cooperation portfolio and are increasingly used as "a central instrument" of Brazil's foreign policy (Dauvergne and Farias, 2012) and the one most explicitly used as a tool of diplomatic affairs (Cabral et al., 2013). Brazil rarely provides direct concessional loans, emphasizing instead scientific–technical cooperation and technological transfers (Fulquet and Pelfini, 2015). Typically, this consists of the transfer and adaptation of expertise, skills and technology mainly through training courses, workshops, consultancies, exchange programmes and, occasionally, the donation of equipment (Cabral and Shankland, 2013: 6). Africa has been the second-largest recipient of technical cooperation after Latin America and the Caribbean (COBRADI, 2013) with agriculture the main focus of ABC's efforts there. Available ABC funding for cooperation projects in Africa fell 25% between 2012 and 2015, however, due to cuts brought on by the global recession (Mello, 2015). In Africa ABC has worked through the Agricultural Research Institute, Embrapa, established in 1973 to promote technological development and support the development of the *Cerrado*, the vast tropical savannah of over 200 million hectares spreading across the central regions of Brazil whose significant agricultural potential it helped to unlock. Brazilian imaginaries of agricultural development have been important in its relations with Africa and are dominated by an inflated optimism about the power of technological modernisation that sometimes borders on techno-utopianism (Cabral et al., 2013). Brazil's use of "biofuels diplomacy", aimed at expanding the world market for ethanol products and internationalising Brazilian biofuels companies, has also allowed it to use its technical expertise (in clean energy) to successfully position itself as a development-oriented global power (Dauvergne and Farias, 2012; Power et al., 2016).

Beyond ABC, Brazilian development cooperation with Africa involves over 170 institutions among federal government organs (IPEA and ABC, 2013) as well as other Brazilian private institutions (NGOs, foundations and corporations) acting in a wide array of areas including education, health, urban and rural development, agriculture, energy and the environment (Fulquet, 2015: 90). Responsibilities for other modalities of cooperation, such as debt relief, concessional lending and emergency relief, are spread across several institutions including the MRE and the Ministries of Finance, Development, Industry and Commerce, along with the International Trade Chamber (CAMEX) and the External Credit

Assessment Committee (Cabral, 2011). There is also the Brazilian Development Bank (BNDES) created in 1952, the federal government's main financing institution which supports external trade and internationalization and is a key ally and resource for major Brazilian corporations such as Vale.

Brazil's "South–South" agenda creates significant opportunities for its internationalising businesses, regarded as industrial champions and the main sources of Brazil's export prowess with commodities such as soya, steel, cotton and oil, often supported by preferential credit from the state (Amann and Baer, 2003; Baer, 2008). Brazil's biggest international companies specialise in areas of civil construction, energy, mining, engineering and other natural resource-based activities (e.g. Odebrecht, Andrade Gutierrez, Petrobras, Vale) and the Brazilian state has strong interests in these areas, retaining control over sectors deemed "strategic" for development (e.g. banks, oil, electricity and aerospace). The oil giant Petrobras, for example, is state-owned whilst Vale, one of the world's biggest mining companies, is partly controlled through a state-owned bank (BNDES) and government-related pension funds (Aldrighi and Postali, 2010) despite the privatisation of Vale in 1997. Brazil's robust and well-articulated civil society has forged some strong and productive global links with movements in Africa that are actively monitoring the behaviour of these domestic actors from Brazil in Africa. Further, Brazilian non-state actors, from the MST to the agroecology movement, have also been active in formulating alternative framings of development and are beginning to mobilise, questioning dominant development cooperation models within Brazil whilst reaching out to build alliances with civil society groups in Africa (Cabral et al., 2013).

In the ten years after Lula's administration came to power in 2003 Brazil signed nine new defence cooperation agreements with African states and Brazilian military officials have been involved in a number of training and capacity-building exercises across the continent (Seabra, 2014). Brazil has also recently been playing a growing role in peacekeeping initiatives in Africa whilst "peacebuilding" is one area where Brazil has sought to exercise global leadership, championing the UN Peacebuilding Commission and chairing it in 2015 (Call and Abdenur, 2017). In many ways these initiatives have been beset with contradictions, however (Harig and Kenkel, 2017), in that whilst the Brazilian government has been keen to advance peace and security in Africa it has played an enabling role for Brazil's defence industry to considerably expand and diversify military exports to African states (Muggah, 2017).

Under Dilma Rousseff (2011–16), Brazil's foreign policy lost the impetus and centrality it had during the previous eight years and there is evidence that Brazil is now breaking with its SSDC focus (Vieira and Menezes, 2016). Since President Temer took office in August 2016, Brazil's international engagements have undoubtedly tilted back towards the sphere of Western interests, with its foreign policy in closer alignment with the US. Temer's administration has prioritised bilateral negotiations in a move away from Brazil's investment in multilateralism within the WTO and has deliberately dismantled many of the policies and political legacies of President Lula, including key foreign policy objectives such as building close links with left-leaning regional players like Venezuela and Bolivia and reform of the UNSC, one of the most important demands of Brazilian diplomacy for at least three decades.

CHINA AS EMERGING GLOBAL DEVELOPMENT HEGEMON

When Mao died in 1976 China was on the verge of ruin and still coming to terms with the collective trauma of the Great Leap Forward and the Cultural Revolution, combined with the additional threat of impending aggression from Communist Vietnam. Since that time China's economy has been transformed by the privatisation and de-nationalisation of Chinese economic enterprises, by the opening of its markets to foreign capital and through the partnerships its state cadres forged with transnational investors and local entrepreneurs (Nonini, 2008). GDP growth rates in China averaged at 9.71% between 1978 and 2015 (UNDP and CDB, 2017: 4) and China's economic rise has led to some discussion amongst Western scholars as to whether China is set to become the next global hegemon with imperialist ambitions (Agnew, 2012; Luttwak, 2012). It has also produced a revival of "Sinomania", reminiscent of the Cold War constructions of "Red China" as a focus of fear and anxiety and "a totalitarian nightmare more sinister even than Russia" (Anderson, 2010: 3).

Japan's ODA to China, which began in 1979, played an important role in China's rapid modernisation alongside China's domestic policies and its opening up to the global economy. For much of the post-World War II period, Japan's aid programme was utilised to bolster market access for Japanese firms, expand exports and gain access to key natural resources in developing countries (Potter, 2002, 2009; Reilly, 2013). By the end of 1978, Japan had signed 74 contracts with China to finance large-scale turnkey projects such as bridges, dams, highways, airports and port facilities, all financed with repayment in oil and all intended to attract further private investment. Japan subsequently funded hydroelectric and thermal power plants, urban water supply, telecommunications, highways and fertiliser plants, sectors that formed the backbone of China's modernisation (Brautigam, 2009: 51). Interestingly, in the early 1990s Japan slowly began to shift away from this approach of explicitly linking its development assistance to resource extraction just as China (and several other (re-)emerging donors) adapted and adopted it (Reilly, 2013), with China's historical experience as an "aid" recipient itself clearly shaping its contemporary foreign assistance modalities.

China's "going out" strategy, launched at the turn of the century to promote Chinese investments abroad, has encouraged a huge surge in outward-bound FDI. In 2000 China launched the Forum on China–Africa Cooperation (FOCAC), a diplomatic instrument organised around regular Sino-African ministerial conferences and designed to consolidate its engagement in Africa, to resist American hegemony and to encourage African countries to follow Chinese pathways to development whilst containing and limiting recognition of Taiwanese sovereignty (Taylor, 2004). At that time China–Africa trade was US$10 billion but by 2014 it had risen more than 20-fold to US$220 billion, although it has fallen back since due to lower commodity prices. China also became the second-largest global source of outward FDI (behind the US) for the first time in 2016, representing a total of US$183 billion (UNCTAD, 2017).

China's (re)turn to the South, which has often been misunderstood (Chan, 2016), regularly invokes histories of SSDC – indeed the principles that guide Chinese foreign assistance today draw on historical discourses of harmonious intra-Asian relations and notions of non-interference, mutual benefit and peaceful coexistence agreed upon by Jawaharlal Nehru and Zhou Enlai at Afro-Asian solidarity summits in the 1950s, partly in response to rising border tensions between the two countries (Strauss, 2009). Chinese state actors and SOEs have also been quite adept at tailoring their strategies to the histories and geographies of recipient states to ensure resource, market and investment access, working with rather than against the grain of African state–society formations in what Carmody and Taylor (2010) refer to as "flexipower", or "flexigemony". Historically, China–Africa cooperation has often involved symbolic "friendship" or "prestige" construction projects such as the building of national sports stadia, government ministries and a variety of monuments to the glory of China's state partners in Africa (Power, Mohan and Tan-Mullins, 2012). China has also financed and constructed the new African Union HQ in Addis Ababa (see Figure 7.1) as a significant statement of its commitment to the continent although in January 2018 AU officials accused China of hacking its computer systems every night for five years and downloading confidential data (*The Guardian*, January 30th, 2018). The allegations were denied by China, but the events seemed to underline the risk African nations take in allowing Chinese technology companies such prominent roles in developing their telecoms systems and infrastructures.

Some US$94.4 billion worth of loans were extended to African states and SOEs by the Chinese government, banks and contractors between 2000 and 2015 (CARI, 2017). In much of the debate about Chinese investment flows to Africa, however, the size and scale of Chinese lending are often exaggerated and there has sometimes been a failure to disaggregate the different types of official financial flows involved or to distinguish between official foreign assistance and more commercially oriented sources and types of state financing. Aiming at supporting national exporters in

FIGURE 7.1 Chinese Premier Li Keqiang delivers a speech at the African Union Conference Centre in Addis Ababa, Ethiopia on May 5th, 2014. Xinhua/Alamy.

their competition for overseas sales, Chinese export credits are subsidised and modestly concessional loans to third countries (export buyer's credits) or Chinese firms (export seller's credits) (Brautigam, 2010). There are also natural resource-backed lines of credit (e.g. from the Export-Import Bank or Exim) and so-called "mixed credits", a kind of "package financing" involving combinations of both buyer's and seller's credits and concessional loans. Other funds available to Chinese enterprises and entrepreneurs investing in Africa include the China-Africa Development Fund, which had invested a total of US$4.5 billion by the end of 2017 in 91 projects across 36 countries (People's Daily, 2018) by sourcing private equity from the China Development Bank (CDB) (Xinhua, 2017). Indeed, banks and financial institutions have been at the heart of enabling the state's plan to encourage the "go-out" of Chinese capital with agencies like Exim Bank, the CDB and the Bank of China playing a key role in the facilitation of Chinese foreign cooperation and investment flows. Exim describes foreign assistance as a "vanguard" supporting Chinese exports and investments while contributing to sectors such as transportation, telecommunications and energy (Export-Import Bank of China, 2006: 21) and its preferential loans have financed over 90% of China's infrastructure projects, although the Ministry of Commerce (MOFCOM) has begun providing some investment and trade credit financing. Chinese lending in Africa also has very specific geographies: Angola received US$21.2 billion in cumulative loans between 2000 and 2014, roughly a fifth of all Chinese loans, whilst just five countries accounted for over 50% of all loans (Angola, Ethiopia, Sudan, Kenya, DRC) (Hwang, Brautigam and Eom, 2016).

Chinese lending volumes are often measured against those of the World Bank but comparing China's developmental experience with the Washington Consensus is inherently political (Ferchen, 2013) and Chinese loans (which are far more commercially oriented than those available from the IFIs) are not being rolled out as part of a strategic attempt to challenge and altogether replace the IMF's public debt norms (Malm, 2016). Perhaps a more pertinent question to ask here concerns the extent to which the neoliberal adjustment programmes rolled out by the IFIs in Africa have *paved the way* for the recent surge in Chinese investment on the continent. Although there is a state-led "go out" policy (initiated in 1999 to encourage Chinese investments abroad) and whilst some SOEs are managed by the State-owned Assets Supervision and Administration Commission of the State

Council (SASAC), rather than being part of a strategic "master plan" Chinese assistance is arguably primarily driven from below by various state-owned, private and hybrid companies, linked predominantly to sub-national governments, seeking business opportunities by lobbying Chinese state agencies to initiate aid-funded infrastructure and construction projects for them to undertake (Hameiri and Jones, 2016) (see Figures 7.2 and 7.3). This is a much more fluid and disaggregated system than is often acknowledged then, with key players jockeying to maximise their impact on decision-making (Hameiri and Jones, 2016).

Some of the largest sectors of cooperation being financed are transportation, energy and mining. There are close to 350 Chinese-funded and Chinese-built overseas hydropower projects either completed, under construction or at the Memorandum of Understanding (MOU) stage, most of them in Southeast Asia (38%) and Africa (26%) (Urban, Siciliano and Nordensvard, 2018; see Figure 7.4). Most of these are large dams with a capacity over 50MW at a time when other nations and organisations, particularly those from the OECD, have withdrawn from the dam-building industry. Consequently, Chinese policy banks are becoming the largest sources of energy finance for governments across the world, providing US $225.8 billion globally in energy financing since 2000 (China Energy Finance database, 2018), whilst Chinese renewable energy

FIGURE 7.2 Chinese and African workers on the construction site of a station and railway at Libala, Angola in 2011. Dieter Telemans/Panos.

FIGURE 7.3 A Chinese supervisor working for China GEO-Engineering Corporation gives instructions to Zambian workers who dig trenches for water pipes on March 23rd, 2007 in Kabwe, Zambia. Photo by Per-Anders Pettersson/Getty Images.

FIGURE 7.4 Chinese workers stand in front of the Merowe dam in Sudan. The main construction work was undertaken by a Chinese joint venture company established between China International Water & Electric Corp, and China National Water Resources and Hydropower Engineering Corporation. Photo by Luis de las Alas-Pumariño.

companies are playing an increasingly significant role in Africa's low-carbon transitions (Shen and Power, 2017). China's heavy involvement in natural resource extraction in Africa has attracted more attention and controversy, however, with the country often criticised for taking advantage of the vulnerable dependency of African economies on increased foreign investment to spur their growth. China–Africa engagement cannot be reduced to a mere quest for resources, however, and has to be understood in the context of its wider diplomatic strategic pursuits and global foreign policy objectives, its efforts to sustain its own domestic economic and human development, to ensure Taiwan's reunification and to counter secession drives by minority areas within China (e.g. Tibet) (Besada, 2013: 83).

There has been much talk about the unique and exceptional nature of China's approach to foreign cooperation. Within China debates about a "China Model" have even spawned a mini publishing boom (Ferchen, 2013). Here, the focus is often on China's "pragmatic", "ideologically neutral" and "disinterested" government, depicted as having successfully replaced the political and ideological intensity of the Mao years with a simple, "do-what-works" focus on economic development (see Zhao and Wu, 2010; Yao, 2010a, 2010b). China has been held up as a "model" of economic development and poverty alleviation that other countries from the South can learn from in escaping the prescriptions of Anglo-American neoliberalism (Stiglitz, 2002) or as a means of "defying the conditionalities of the Bretton Woods institutions" (Manji and Marks, 2007: 136). There is no single, unitary and coherent "Chinese model", however (Friedman, 2009), and such narratives considerably downplay the global interconnectedness and transnational dynamics shaping China's economic development strategies both domestically and overseas (Power, 2015). As it has grown, China's economy has increasingly become intertwined with the fate of many other economies and its economic development is co-produced by capital accumulation strategies both on the "inside" (in the domestic restructuring of SOEs and macroeconomic strategies) as well as the "outside" (in the global search for new markets and corresponding foreign demand for financial capital) (Lim, 2010). Calling this a "Chinese" "model" is also problematic given the extent to which China has looked to and drawn from East Asian examples of (developmental) state practice (Power and Mohan, 2010), or has followed the tracks of Japanese and Korean mercantilism.

China is often depicted in much of the debate around Sino-African relations as a "unitary rational actor" (Constantin, 2006: 17), as implied by the "China Inc." euphemism that depicts the country as a kind of single monolithic corporation. China is however a hybrid assortment of institutions, practices and disciplines of power that have juxtaposed older elements of Maoist governance (e.g. central planning and an ideology of socialist paternalism toward "peasants" and "workers") with elements of market liberalisation in a kind of "slow-tempo improvisation" (Nonini, 2008: 156) aimed at developing China's forces of production, preserving the position and legitimacy of the Chinese Communist Party (CCP) and consolidating the foundations of accumulation for China's "cadre-capitalist" class (So, 2005). Michel, Beuret and Woods (2009: 108), for example, very simplistically liken Beijing's role to the Godfather:

> Borrow from the Chinese and you are drawn into the bosom of its – highly profitable – family. Beijing is the Godfather, engaged in everything from textiles to infrastructure to uranium and oil. His bids are all interlinked and his motivation is constant.

There is no single Chinese "state capitalism", however, but rather a loosely coordinated, sub-national network of "regional capitalisms", and although there are "Chinese ways" of capitalism these are plural rather than singular and represent a range of globally networked but distinctively Chinese "sub-models" of capitalism or regional styles of capitalist development that together call into question the privileging of the national scale (Zhang and Peck, 2016). In China "state" and "business" take on many forms according to the way Chinese provinces and businesses are organised, reflecting diverse political and business cultures and forms of "state capitalism". As Gu et al. (2016) have argued in relation to China–Africa cooperation in the field of agriculture, much state–business interaction is also informal, unplanned, negotiated, decentralised, uncoordinated and run through highly diversified routes, including business associations, migrant networks and a complex range of both central and provincial-level companies and enterprises. China has also been changing quickly and its sheer geographical and population size defies simple descriptions (Ferchen, 2013).

The lines that separate inside and outside, domestic and foreign, state and non-state and public and private in contemporary

China are blurry and it is difficult to identify where Chinese state interests end and corporate interests begin (Breslin, 2013c). Under President Xi Jinping, for example, the CCP has increasingly moved to create party cells within businesses, long a feature of SOEs but also now a part of corporate life at private companies and foreign joint ventures (Hornby, 2017). Sinologists have emphasised the rise of a "fragmented authoritarianism" (Lieberthal, 1992), within an increasingly "deconstructed" state (Goodman and Segal, 1994), whose internal architecture is "more like the European Community of the 1970s than the USA today" (Breslin, 2013b: 70). Such disaggregation of statehood (the divestment of power and control to semi- and fully private actors, and the devolution of authority and resources to sub-national agencies) has generated internal differences over external relations, not a single "national" position (Goodman and Segal, 1994). These fragmented and decentralised state apparatuses and quasi-market actors are increasingly pursuing their own independent interests and agendas overseas (Duchâtel, Gowan and Rapnouil, 2016), generating conflict-ridden, incoherent policy output, often mistakenly interpreted as "grand strategy" (Hameiri and Jones, 2016). The political economy of China's foreign cooperation is thus not determined by "centralized geo-graphing alone" (Gonzalez-Vicente, 2011: 2).

In stark contrast to the Maoist period (Sutter, 2008), the foreign policy duopoly of the foreign and defence ministries has evaporated and the conventional perception of Chinese foreign policy being decided by a centralised leadership in Beijing is outdated (Lanteigne, 2009) as a variety of actors with an interest and influence in China's foreign policy-making, at multiple territorial scales, struggle to exert power and authority (Jessop, 2009). Although China's foreign policy-making still remains highly centralised, processes of information processing, deliberation and decision making, along with the management of foreign relations, are no longer controlled by a handful of individual leaders (Zhao, 2008). With the growing role of economic actors involved in China–Africa relations, agencies like MOFCOM or the Ministry of Foreign Affairs (MOFA) have found it difficult to assert themselves and the primacy of national interests (Gill and Reilly, 2007). In terms of foreign policy, the list of key actors now includes (amongst others) China's National Reform and Development Commission (NDRC), the Ministry of Public Security, the People's Liberation Army (PLA) and banks like Exim and the CDB but also a wide variety of Chinese firms from small and medium-scale enterprises to SOEs. Empowered to independently control their international economic relations, provinces (many "twinned" with particular African countries due to historical pairing arrangements) and special administrative regions (SARs) (e.g. Macau) also represent quasi-autonomous foreign policy actors, concluding agreements with local and national states in the South.

The interaction between Chinese national and corporate interests and the partnership between the government and China's national oil companies are key to understanding the rapid progress of China's energy diplomacy (Zhang, 2015). The China National Petroleum Corporation (CNPC), for example, is the most significant Chinese actor in Sudan and South Sudan and has indirectly steered China's foreign policy there:

> Without its presence, China's involvement in crisis diplomacy in both countries would be minimal. Instead, in the face of conflict in the Sudans in recent years, MOFA discarded its stringent adherence to the non-interference principle by engaging in conflict resolution and urging the warring sides not to target Chinese oil investments and personnel. (Patey, 2016: 3)

Chinese SOEs operating in Africa like CNPC are often portrayed as the "handmaiden" of the Party-state, working to sustain the existing authoritarian political system and/or acting as a "Trojan horse" to help China compete globally (Yi-Chong, 2014: 823). The different commercial interests of competing SOEs can however result in the pursuit of market-driven relations with host governments and firms that run counter to the foreign policy objectives of the Chinese state (Liou, 2009). There are also approximately 100,000 sub-national SOEs compared to just 117 (albeit often very large) SOEs controlled by the central government (Szamosszegi and Kyle, 2011). Further, Chinese national oil companies went overseas long before the central government adopted the "go out" policy and many have a long history of working in Africa, going back to the 1960s. Chinese SOEs are also highly autonomous, market-driven entities selling commodities on global markets, not simply "delivering" them to China (Hameiri and Jones, 2016). Large SOEs may dominate in terms of the amount invested overseas, but it is private (rather than local) firms that are numerically superior in China's investments in Africa, representing 85% or so of all companies operating on the continent, although the significance of truly independent private companies is often under-reported (Gu, 2009). Contrary to the assumption of a strong, directive, collusive state, in the African context many Chinese enterprises do not have much knowledge or understanding of the state's policies relating to "going global" and investing in Africa (Gu et al., 2016) and follow commercial rather than policy objectives in deciding to establish themselves overseas (Gu, 2009).

Questions of military cooperation and security have often been neglected in the China–Africa literature (Alden et al., 2017). Although China was initially reluctant to play a security role in Africa, the protection of Chinese overseas interests has increasingly been declared a foreign policy priority. Since 2004, China has conducted multiple non-combatant evacuation operations, the largest of which involved 35,860 Chinese employees from Libya during the 2011 civil war in a vast, interconnected effort involving various state actors but also Chinese SOEs such as the CNPC, China Rail Construction and the shipping magnate COSCO (Duchâtel, Gowan and Rapnouil, 2016). The PLA has gained valuable operational experience in these encounters and through its growing role in attending to humanitarian crises (e.g. around Ebola). A 2016 Pentagon report to Congress on the Chinese military characterises Beijing's objectives in UN deployments as "improving China's international image, obtaining operational experience for the PLA, and providing opportunities for gathering intelligence" (United States Department of Defense, 2016: 21).

In 2012 the China–Africa Cooperative Partnership for Peace and Security was formed, bringing defence into the FOCAC process (Chun, 2017), whilst in June 2018 Beijing held the inaugural China-Africa Defense and Security Forum attended by representatives of some 49 African states. Beijing's closest defence alliances are arguably with Zimbabwe and Angola (Firsing and Williams, 2013) but Algeria, Egypt, Ghana, Nigeria, South Africa, Sudan, Uganda and Zambia also have high levels of military cooperation with China (Alessi and Xu, 2015). Of these ten countries, six are either suppliers of oil, gas and other critical resources or have substantial Chinese commercial investments (Conteh-Morgan and Weeks, 2016), although China's strategy for military and defence cooperation is complex and cannot be reduced to a simple story of access to or protection of natural resources. Beijing's engagement in conflict resolution and deepening involvement in regional and international peacekeeping efforts do mark an important shift towards a more flexible and pragmatic understanding of its traditional support for non-interference (see Figure 7.5). This has manifested in greater

Chinese involvement in UN peacekeeping operations and the first combat troops being deployed in northern Mali (2013) and South Sudan (2014). China's involvement in peace talks and peacekeeping activities in Sudan and South Sudan has been a "laboratory" for developing China's diplomatic involvement in other security crises (Duchâtel, Gowan and Rapnouil, 2016).

Paradoxically, alongside its growing commitment to African security China is also the foremost exporter of arms to Africa with Chinese military equipment now being used by more than two-thirds of African countries (IISS, 2017). China supplied Khartoum with a large stockpile of weapons that ended up in Darfur and sent a shipment of arms to Zimbabwe in 2008 during a period of heavy civil unrest following disputed elections (Spiegel and Le Billon, 2009). The Zimbabwean regime was also supplied with instruments of opposition control such as radio-jamming equipment to disrupt opposition party broadcasts and riot control equipment to suppress protests and demonstrations (Conteh-Morgan and Weeks, 2016). Supported by favourable financing, more than 130 Chinese companies are actively involved in selling their wares in the South as China has become a major global manufacturer and exporter of

FIGURE 7.5 Chinese military personnel associated with the former United Nations Organization Mission in the Democratic Republic of the Congo (now MONUSCO) work on a road rehabilitation project to allow greater access to a power plant in the east of the country. UN Photo/Marie Frechon, April 2008.

low-tech, affordable firearms and related ammunition along with a growing range of other equipment used in law enforcement (e.g. handcuffs, riot control gear, restraint chairs, direct contact electric stun weapons and spiked batons) (Amnesty International, 2014).

Whether in relation to military cooperation, arms sales, infrastructure and technology transfer or humanitarian intervention, there are important questions about African agency in mediating China–Africa encounters (Mohan and Lampert, 2012; Corkin, 2014). This however risks unwittingly reinforcing "internalist" explanations of African underdevelopment, implying that African elites are in the "driving seat", or are "copilots" in relations with China and able to fundamentally reshape the nature of their state-societies when the power of elites has often been confined to incremental bargaining rather than structural change and can be highly circumscribed (Carmody and Kragelund, 2016). Further, "African agency" is not a simple opposition to "Chinese power", since the two concepts are inter-constitutive, underlining the value of assemblage thinking which sees power residing not in actors but in networks (Carmody and Kragelund, 2016). The varied and often contradictory impacts that Chinese actors and networks have on the political economy of African states and, more globally, on the "extraversion dynamics" (Bayart, 2000) that characterise them, require further research, however. As Chinese entities develop transnational interests, they extend their "governance frontier" beyond China's territory, promoting state transformation elsewhere and so understanding how other states experience China's "rise" (e.g. in terms of power redistribution, the reconfiguration of their economies, increases in conflict etc.) requires further attention. Existing IR approaches arguably overlook these dynamics because their focus is overwhelmingly *systemic* (Hameiri and Jones, 2015). More research is also needed on questions of "race", racism and racialisation in China–Africa relations (e.g. the "race"/labour conjuncture) (Sautman and Hairong, 2016). Anthropological perspectives on micro-level social and economic dynamics (Driessen, 2015; Siu and McGovern, 2017) offer a more processual approach to China–Africa relations and are essential to understanding the everyday sociocultural and micro-political exchanges involved but also in complicating Cold War-derived understandings of both China and Africa and helping to make sense of the mutual construction of seemingly unconnected social formations across continental divides (Siu and McGovern, 2017; see Figure 7.6).

FIGURE 7.6 A Chinese businesswoman talks on her mobile phone outside a shop at the Chinese market in Luanda, Angola on March 31st, 2007. Photo by Per-Anders Pettersson/Getty Images.

SOUTH KOREA: EXPORTING A STORY OF DEVELOPMENTAL "SUCCESS"

South Korea's economic transformation from a war-destroyed and largely agricultural country with a per capita income of US$67 in 1953 to membership of the OECD in 1996 is often hailed as a remarkable "rags to riches" success story, one that is now being offered as a "model" for other states of the global South (Kalinowski and Cho, 2012). Founded on a "development first, democracy later" philosophy, this story is said to have particular appeal to many authoritarian and hybrid regimes in Africa (Darracq and Neville, 2014: 18). South Korea's economic success during the high-growth developmental decades rested upon political repression, however, and the historical legacy of the Park Chung-hee regime (1963–79) has been a thorny ideological issue domestically (Kim, 2015). As a front state in the Cold War, South Korea secured considerable amounts of US aid, with Korean companies gaining access to export markets in North America (Stubbs, 1999, 2005). The dynamic growth and industrial transformation of South Korea's economy, especially its construction firms, are attributable not only to the actions of the Korean developmental state but to the effects of a Cold War geopolitical economy that made access to technological and engineering learning opportunities available to South Korean contractors on unusually favourable terms (Glassman and Choi, 2014). As such "it is useless to partition the economic performance of states like South Korea from geopolitics and transnational class issues" (Glassman and Choi, 2014: 1177). Crucial South Korean industrial conglomerates, *chaebol,* were extensively involved in offshore procurement contracting for the US military, from the Vietnam War era forward, effectively becoming players within the US military–industrial complex (Glassman and Choi, 2014).

The G20 summit in Seoul in 2010 was regarded as belated international recognition of the country's success story. The formulation of the "Seoul Development Consensus" on how to tackle global poverty and volatile markets through the establishment of financial stability nets along with the "Seoul action plan" were seen as a huge success for Korea as an emerging player and "issue leader" in the field of development cooperation, as was the OECD High Level Forum on Aid Effectiveness held in Busan in 2011 (Kalinowski and Cho, 2012). In this sense the South Korean state has sought to make a distinctive contribution to global thinking on aid and development (specifically around the aid effectiveness agenda) and further the nation's "intellectual leadership" as a "knowledge champion" based on its widely admired development experience (Kim, 2015). South Korea has also constructed a "benign" (or at least "neutral") international identity for itself as a "moral nation", partly as a means to secure a greater stake in global norm-building processes (Kim, 2015).

As a relative latecomer in the race to secure resources abroad, South Korea has feared being left behind by more powerful economic risers such as China (Taylor, 2006; Darracq and Neville, 2014) and has in recent years increasingly used high-level diplomacy to establish strategic partnerships with resource-rich countries and to advance its own claims to be a more prominent global power. Like Brazil and China, South Korea has also turned to the South in search of energy security and new markets for its companies. Faced with saturated domestic markets, South Korean firms have been supported by state agencies offering generous credit programmes, investment funds and tax benefits in support of overseas resource development projects. South Korean oil companies however face stiff competition from traditional Western giants including BP, ENI, Chevron and Total, as well as increasingly prominent state-owned Asian behemoths such as Sinopec and CNPC (Darracq and Neville, 2014: 13).

In recent years the intensity of debate in South Korea over aid policy issues has grown, as has the number of actors involved in aid policy-making (Kim, 2015). ODA (around 40% of which is provided in the form of loans) is disbursed by the Korean International Cooperation Agency (KOICA), supervised by the Ministry of Foreign Affairs (MOFA). A key reference point has been the *Saemaul Undong* or New Community Movement instituted by President Park Chung-hee's regime (1963–79) which was instrumental in transforming South Korea's fortunes in the 1970s and regarded rural development and modernisation as key to breaking the poverty cycle. KOICA now has *Saemaul Undong* replica projects in some 21 African countries (KOICA, 2017). South Korea also operates a "Knowledge Sharing Program" (KSP) which aims to "export" Korea's development experience and expertise to developing countries, with the selection of recipients closely linked to the list of countries being targeted by its resource diplomacy efforts and foreign policy objectives (Kim, 2015). There have been concerns however amongst some civil society representatives that the KSP is "donor-centred" and involves an "overdose of patriotism" (Kim, 2015: 78). Since joining the DAC in January 2010, South Korea's policy of tying much of its aid has come under greater scrutiny and it has faced increasing pressure to conform to DAC standards that favour grants over loans and the provision of ODA to countries with very low levels of development (Darracq and Neville, 2014).

Although Asia has been the primary location of Korean outward FDI, Korea's ODA to Africa in the total budget rose from 2.7% in 2002 to 23.8% in 2014 and by 2016, of the US$1.54 billion South Korea spent on direct bilateral aid, just under a third (US $414 million) went to Africa (OECD, 2017). This was spurred by the Roh Moo-hyun government (2003–08) which launched a significant programme of re-engagement with Africa in the shape of KIAD (Korea's Initiative for Africa's Development), one of the stated objectives of which is to boost South Korea's presence in the international community. Bilateral trade between South Korea and African countries jumped fourfold between 2000 and 2011, rising from US$5.7 billion to US$22.2 billion, with exports to Africa increasing fivefold in the same period (Darracq and Neville, 2014). Technical expertise (e.g. in agriculture) is shared through the Korea-Africa Food and Agriculture Cooperation Initiative (KAFACI) which provides training and education for African agriculturalists. There is also the Korea-Africa Forum (KAF), a ministerial forum launched in 2006, and the Korea-Africa Industry Cooperation

Forum (KOAFIC), started in 2008, which has become a platform for private sector participation in Africa. The Korean state has also been actively lobbying African counterparts on behalf of South Korean companies in African markets to secure favourable government tenders, backed by a rapidly expanding South Korean diplomatic network on the continent. Chinese experiences of providing development bank loans to finance infrastructure projects reimbursed by natural resources have been influential in South Korea, even though Korea has frequently castigated China's actions in Africa, accusing Beijing of natural resource plunder and overlooking issues of development (Darracq and Neville, 2014). South Korea also has its own Peace Corps-style organisation, *World Friends Korea*, with around 700 volunteers active in 30 African countries, aimed at "helping people around the world while enhancing Korea's brand value" (World Friends Korea, 2017). Some 168 South Korean NGOs were also engaged in international aid service and advocacy activities as of early 2012, many of which derive from religious associations (Kim, 2015: 74).

Seoul's strategy (past and present) in Africa should also be viewed "through the prism of inter-Korean relations", since events in the Korean peninsula were largely responsible for Seoul's initial diplomatic embrace of Africa and "remain a key factor in its operations in the region" (Darracq and Neville, 2014: 8). Seoul established its first diplomatic relations in Africa in 1961, beginning with Benin (then Dahomey) and was concerned at the speed with which Pyongyang had forged diplomatic relations with newly independent African states, fearing that this would ultimately lead to UN recognition of the Democratic People's Republic of Korea (DPRK), at the expense of the South. Further, the DPRK's entry into the NAM in 1976 gave Pyongyang an international platform in lieu of UN recognition before 1991 to engage with African states, to acquire wider recognition and to promote an "anti-Seoul" campaign (Darracq and Neville, 2014: 8; see Figure 7.7). Pyongyang today operates military cooperation "on a large scale" with several African countries to evade sanctions imposed for its nuclear and ballistic missile programmes (UNSC, 2017). This includes Angola (where it has trained members of the presidential guard in martial arts) and Uganda, a regional security ally for the US (where it has been training Ugandan air force pilots and technicians), along with Benin, the DRC, Egypt, Mozambique, Namibia, Nigeria, Libya, Seychelles and Zimbabwe (ISS, 2016). The UN Security Council has alleged that North Korea "uses its construction companies that are active in Africa to build arms-related, military and security

FIGURE 7.7 The African Renaissance Monument in Dakar, Senegal, completed in 2010 and built by Mansudae Overseas Projects from North Korea. Zoonar GmbH/Alamy.

facilities" (UNSC, 2017: 27) and indeed North Korean-built arms factories have appeared in the DRC, Ethiopia, Namibia, Madagascar and Uganda. Along with competition from China then, South Korea's desire to counter and contain the international influence and diplomacy of the North has been a key driver of its growing desire to engage with SSDC, particularly in Africa.

INDIA–AFRICA DEVELOPMENT COOPERATION

Historically, India has played a key role in the making of the "Third World" and the idea of development. During the Cold War, Western diplomats were "put off by India's flexible nonalignment, which for a time was a pretext for a close relationship with the Soviet Union" (Cohen, 2001: 66). Jawaharlal Nehru, one of the leading figures of Third Worldism, had great admiration for the Soviet Union's self-transformation into a global power. Under Nehru (1947–64), India perceived Africa as a single regional bloc and adopted a uniform policy on the continent of colonial and imperial opposition and non-alignment (Dubey, 2010a, 2010b). Nehru's notion of non-interference, articulated in various Afro-

Asian solidarity and non-alignment fora and one of the five principles of the Panchsheel Treaty signed by India and China in 1954 (which came to the fore at the Bandung conference), was influential in African political thought, however (Mazrui, 1980), and continues to be influential in Indian foreign policy (Narlikar, 2013). Soske (2017) places India, the Indian diaspora and the Indian Ocean city of Durban at the centre of the development of an inclusive philosophy of nationalism by the African National Congress (ANC) in South Africa. Gandhi, who worked as an expatriate lawyer in South Africa to protect the civil rights of the Indian community (Desai and Vahed, 2015), was also an inspiration to many in Africa, particularly his doctrine of non-violence (de Waal and Ibreck, 2013).

Even though India does not explicitly rely on the language of Third Worldism to justify its engagement with the South today, a strong moralistic framing of global distributive justice does permeate its demands for 'policy space' and 'development'" (Narlikar, 2013: 605). India is however not necessarily perceived as a "champion" of the South, despite the historically important role it has played in various "incubators" of South–South initiatives, from Bandung and the NAM to the G-20, the UNDP, UNCTAD and IBSA (the India–Brazil–South Africa Dialogue Forum). India's approach has also arguably not been radical enough concerning the transformation of existing international economic or political structures, nor has it always articulated a concrete agenda to promote Southern interests, or the interests of non-members (Alden and Vieira, 2005). Traditionally, India has offered assistance to other countries through training and capacity-building, mainly through the Indian Technical and Economic Cooperation scheme (ITEC) founded in 1964 (Price, 2011). Beyond the state, however, it is important to recognise that there are many non-state and sub-state actors involved in shaping and defining India–Africa development cooperation. Indian NGOs and civil society actors have been very active across the continent in a wide variety of sectors and for many years.

Jim O'Neill (2011: 73) once wrote of the resentment of Western practices he had encountered amongst Indian elites, "development among them". Similarly, Indian officials have often expressed a preference for practical projects, rather than the "more theoretical positioning" usually associated with prescriptive development policy (Price, 2011: 27). Rather than using the language of "development assistance" or "aid", India's growing presence as an exporter, donor, investor and financier in the global South is often couched in the language of SSDC, partly as a way of getting past domestic sensitivities around giving foreign aid when India itself has high levels of poverty and inequality (Price, 2011). India is now one of the most significant (re-)emerging donors, but it was also the world's eighth-largest recipient of ODA as recently as 2008 and fourth overall from 1995 to 2009 (IRIN, 2011). Responsibility for the administration of ODA programmes is spread over several ministries and, as with China, India's assistance is still predominantly bilateral. In recent years India has started to play a more involved role in a wide spectrum of multilateral and regional institutions and become a large contributor to UN peacekeeping missions. India is known to be good at "multilateral diplomacy" (Malone, 2011: 270) and some rationalists have argued that as India rises, its growing integration in the world economy will lead to a convergence of its interests with other players, with greater stakes in the system producing a sense of ownership and willingness to invest in it (Narlikar, 2013).

Indian officials have been at pains to emphasise the longstanding continuities in India's commitment to and engagement with Africa. South Asians share a long, intensive and diverse history with Arab and Swahili traders in the Indian Ocean region, but it was not until the mid-nineteenth century that they began to settle in East Africa (Oonk, 2017). Created in part by colonial geographies of migration, India now has a very large diaspora and some very strong cultural connections on the continent (Yengde, 2015; Dubey and Biswas, 2016). India's post-colonial alliances with Africa were initially forged during the 1950s and 1960s in the context of debates about non-alignment and decolonisation (Dubey, 1990; Beri, 2003). In the early stages of India's independence, at a time of widespread domestic poverty, it extended lines of credit to various African infrastructural projects worth US$3 billion (Dubey, 2010). The Export-Import Bank of India (set up in 1982 for the purpose of financing, facilitating and promoting India's foreign trade) has since extended billions of dollars in credit lines to African countries.

India's FDI outflows stood at US$6 million in 1990 but peaked at US$21.1 billion in 2008 before falling back to US $5.12 billion in 2016 (UNCTAD, 2017) and in that period Africa has become increasingly important to India's economy and foreign policy. India's "Focus Africa" programme, launched in 2002 to significantly enhance Indian trade with the continent, has grown from 7 to 24 African member states. Total trade between India and Africa was just under US$1billion in 1991 and increased almost fivefold between 2005 and 2015, reaching US$52 billion in 2015–16 (*The Hindu*, 2017). The volume of India's aid to Africa also grew by 57 times between 2001 and 2017 (Vivek, 2017). Additionally, since 2008 India's Duty-Free Tariff Preference scheme has extended preferential market access to 48 LDCs (34 of which are in Africa) through removing import duties on 98.2% of all tariff lines. Although Africa has risen in importance in India's global aid strategy, the bulk of Indian assistance continues to be directed at Asian recipients, particularly Bangladesh, Bhutan, Nepal (which all share borders with both India and China), Myanmar and the Lao PDR. The bulk of India's aid to neighbouring countries is in sectors that hold mutual economic-strategic interest, such as transport, energy and democracy (Bhogal, 2016).

Given Indian policymakers' uncertainty about what is and is not successful in terms of domestic policy, there seems little likelihood of an Indian state-led strategy to export a particular "model" of development on a large scale to other countries (Price, 2011). Rather than concentrating on state-led development assistance, the Indian government has instead set itself up as an "enabler" for its private sector, particularly state-owned firms (e.g. in oil and infrastructure). Indeed, some 75% of India's aid is tied to the provision of goods and services from Indian suppliers (Tharoor, 2016), creating considerable market opportunities for Indian companies, products and services. Indian companies have been active in Africa for many years – the multinational conglomerate Tata, for example, opened a subsidiary in Zambia in 1977 and now has 11 different country offices in Africa with interests in automobiles, steel and engineering, chemicals, information technology, hospitality, food, beverages, energy and mining. There are also hundreds of thousands of small and medium-sized Indian enterprises active on the continent (Alden and Verma, 2016).

Energy and the extractive industries, particularly oil, coal and uranium, have been at the forefront of India's Africa strategy (Chakrabarty, 2017), with the continent seen as a possible source of raw materials and energy to fuel India's industrial growth and ensure its energy security. India's oil and coal imports from Africa and the significance of the Indian Ocean to India's economic development and security are regarded as particularly immense (Vines and Campos, 2010). Renewable energy technologies have also become increasingly important (Power et al., 2016), as demonstrated by the International Solar Alliance which India helped to establish in 2015 and the key role Indian firms are now playing in the roll-out of solar energy technologies.

There is considerable anxiety in India about China's growing influence on the continent and a growing desire to mimic the close economic and commercial ties forged by Beijing in recent years. As with many other (re-)emerging donors, India has also staged its fair share of elaborate regional summits underlining the virtues of SSDC. In April 2018 India announced plans to open 18 new embassies in Africa by 2021 (in addition to the 29 embassies it has already established) in order to further strengthen economic cooperation and links to the Indian diaspora in Africa (Africa News, 2018). Like China and Brazil, India also now has several defence cooperation agreements with African countries such as Kenya, Madagascar, Mozambique, South Africa, Tanzania and Uganda (along with a growing volume of arms exports to Africa) (Ranjan, 2016).

CONCLUSIONS: THE "EMANCIPATORY" POTENTIAL OF (RE-)EMERGING DONORS?

> Western modernity is no longer uncritically viewed as the future of developing countries. (Humphrey, 2007: 16)

Central to the self-representation of the West in Africa has been the claim that its role is essentially altruistic or beneficent, in ways that are dominated by an enduring notion of trusteeship (Mawdsley, 2008). In recent years, this altruism and beneficence has often been constructed through a contrast with the purportedly opportunistic, exploitative and deleterious role of the emerging powers, particularly China (Ayers, 2013), so often portrayed as the "villainous other". The rise of a non-Western power is thus depicted as somehow an "unknown", an unprecedented and potentially dangerous development. The solutions prescribed by Western donors (e.g. trilateral development cooperation) are based upon an implicit assumption that these donors need to be "socialised" into multilateral fora and the "liberal" way of "doing development" (Gallagher, 2011). There are multiple echoes of Cold War discourses of the "Third World" here: the "Sinomania", for example, surrounding China's rise or the idea of defenceless and powerless fledgling "Third World" nations being "sucked" into the orbit of these "dangerous" others who must be brought into a liberal political and economic order through the universalisation of a more "liberal way of development". There are also the ways in which these donors themselves draw on Cold War discourses of

SSDC or recall Cold War histories of cooperation along with the revival of triangulation, where during the Cold War, developing countries pursued relations with multiple states and played would-be suitors off against each other. The dynamics of development diplomacy have shifted, however, from the age of Cold War polarity where groups of countries stuck together in blocs and "rejected" other blocs towards "a world of numerous overlapping, often issue-specific and quite probably fluid alliances and groupings" (Breslin, 2013a: 628).

The rise of (re-)emerging donors like China, Brazil, India and South Korea has opened up opportunities for other countries of the South to increase their influence in the world, not just through the assistance they offer in a quest to increase their spheres of influence and the "followership" they seek in the South (Vickers, 2013) but also in the way that they have attempted to make the global order more inclusive, equitable and multilateral. Brazil and India, for example, have become much more central to WTO negotiations through the G20 (see Figure 7.8) whilst the World Bank and DAC have been pressed into reform, potentially recalibrating influence towards emerging powers (Verschaeve and

FIGURE 7.8 UN Secretary-General Ban Ki-moon, Chinese President Xi Jinping and US President Barack Obama stand together at the G20 summit in Hangzhou, China on September 3rd, 2016. Photo by Wanghanan/Shutterstock.

Orbie, 2015). Although it has introduced new actors and added new discursive frames (alongside its multiple Third Worldist resonances), SSDC must be understood in its historical, political and economic context, and in relation to the particular, diverse and often contradictory interests of states and capital. Forms of SSDC are rooted in specific state–society configurations (Gray and Gills, 2016) and within different varieties of capitalism.

What many (re-)emerging donors have in common, however, is the way they conceive of their development assistance as a form of "soft power", making heavy use of discourses of SSDC to advance their desire for a stronger bargaining position within the global political and economic order (Gray and Murphy, 2013), to boost their trade and to create markets and opportunities for their companies. As Abdenur (2014: 92) has noted, "[f]rom a geopolitical point of view, the BRICS helps China to counter US hegemony without direct confrontation" by boosting China's multilateralism and enhancing its reach, reinforcing calls for change (e.g. in relation to the UNSC or IFIs) without coming across as overly aggressive. More generally, SSDC is an important part of how these states actively manage their international reputation through diplomacy and improve their image in diplomatic forums like FOCAC. International power projection is, however, relational and requires close attention to the geopolitical position of states within the global state system, as well as to the response that development cooperation projects elicit from "receiving" states and the increasing capacity of donor states to directly and indirectly alter geopolitical and developmental spaces outside their borders, contrary to the many proclamations they make around "non-interference".

The (re-)emerging donors have formulated new aid modalities which pose a challenge to traditional, established modes of development cooperation (Six, 2009) and models of donor–recipient relations, introducing competitive pressures and weakening the bargaining position of Western donors by offering alternatives to aid-receiving countries (Woods, 2008). In several cases, there is a strong preference for working bilaterally whilst responsibility for the delivery of assistance is spread over several ministries. China has also actively used forms of "hybrid" financing tools, blurring the lines between aid, trade and FDI but India, Brazil, South Korea and others are also now increasingly making use of similar approaches. Paradoxically, these new donors represent models of economic success, yet they have been until very recently, or are still, recipients of international aid, which again belies the notion of a unified epistemic donor community. While some parts of China boast development indicators comparable with those of Sweden and Singapore, in the recent past there have been others for which the analogy of Sudan or Honduras is more appropriate (cf. Heileg, 2006). Even though officially just over 43 million people still live on less than 2,300 Yuan (US$350) per year, the poverty line set by the government (Chow, 2018), China remains "an archipelago of capitalist urban formations within a sea of rural underdevelopment" (Zhang and Peck, 2016: 64). Similarly, profound socio-economic realities in Brazil and India echo the economic chasm of the wider North–South divide and consequently there remain very real domestic political sensitivities around giving overseas aid.

The (re-)emerging donors have broadened the range of sources of development finance available and have been instrumental in establishing regional development banks such as BancoSur or the Asian Infrastructure Investment Bank (AIIB) which have begun to erode the primacy of the IMF and World Bank as lenders in Asia and Latin America. There is also the BRICS' desire to establish a parallel mechanism to the World Bank in the shape of the "New Development Bank" that lends to infrastructure projects both within the BRICS and across the global South. Consequently, there have been suggestions of an emerging hegemony led by China or the BRICS generally, by means of regional banks and investment funds or through transnational infrastructural corridors such as China's "One Belt, One Road". Indeed, infrastructure has been at the heart of various forms of SSDC and this is having a range of social, political and economic impacts across the South. This chapter has also made the case for a relational view of the (re-)emergence of these donors (and the different assemblages of SSDC that they construct) vis-à-vis both each other and existing global actors that continue to hold significant influence in Africa such as the EU and US.

Whilst frequently heralded as "new", "alternative" and "pathbreaking", SSDC builds upon pre-existing forms of international development and neoliberal policy frameworks and adds new variants of statecraft to facilitate capital accumulation and new forms of market socialisation (Amanor, 2013). Indeed, the (re-)emerging economies' reassertion of the role of the state in development is "one of the most consequential events of the global economy" (Ban and Blyth, 2013: 250). Despite the global trend towards privatisation, many of these states continue to protect and nurture their own industries and seek to make them competitive within a global economy. They have, for example, reserved mining and energy sectors for SOEs and maintain significant regulatory control of these industries through interventionist policy tools. The statist nature of their capitalisms is also reflected in "fuzzier" distinctions between public and private sectors of the economy. These are not the unitary states of much IR theorising and globalisation has fundamentally changed them, reshaping their interrelations with other states in the process (Hameiri and Jones, 2016). That these states are becoming increasingly fragmented, decentralised and internationalised is noted by some IPE and global governance scholars but has often been neglected in IR (Hameiri and Jones, 2016). Partly as a result of the enduring Eurocentrism in IR the rise of non-Western powers has thus remained "undertheorized, resulting in an impoverished vision of a world order where Western hegemony is no longer guaranteed" (Hirono and Suzuki, 2014: 445). It is also necessary, however, to go beyond the privileging of nation-states as the fundamental units of analysis in many IR/IPE debates since such accounts of geopolitical competition as distributive conflicts often neglect the wide range of other actors involved in SSDC.

The focus on energy and natural resources (and the extractivist economies and enclavic developments that they give rise to) is also seen by many as likely to only further entrench the uneven dynamics of accumulation, extraversion and dependent relations that have historically characterised the African continent. Some have also questioned the transformative and supposedly "emancipatory" potential of these donors due to their neglect of civil society relations in SSDC (Banks and Hulme, 2014) or key issues like social justice and poverty eradication (Palat, 2008). Whilst it is often claimed that these donors offer an escape from the conditionalities favoured by established Western counterparts, in practice projects *are* often tied to the use of labour and input procurements. Thus, it has been argued that the BRICS countries pose only a "within-system" challenge because of the extent to which they rely on neoliberal modes of capitalist development, but also because of their dependence on and level of engagement with existing institutional structures (Stephen, 2014).

In a curious case of the empire taking back the theory of dependency, in March 2018 Rex Tillerson (then US Secretary of State) accused China of "encouraging dependency" in its approach to Africa, arguing that its use of "opaque contracts, predatory loan practices and corrupt deals that mire nations in debt and undercut their sovereignty" stood in direct contrast to that of the US which "incentivises good governance" (Tillerson cited in VOA, 2018). Similarly, a report to the US State Department in May 2018 warned of China's "debtbook diplomacy" where it extends huge sums in loans to countries that cannot afford to repay and then strategically leverages the debt (Parker and Chefitz, 2018). China has also regularly been depicted as a "strategic competitor" to the US and a "revisionist power" in recent US National Security and Defence Strategies and there have been growing tensions between the US and China in Africa, often focused on the close proximity of both countries' military bases in Djibouti (see Figures 7.9 and 7.10).

The BRICS group is sometimes seen as an inner cabinet, a star chamber, a kind of "Security Council" of the G20 (Chan, 2016) but, as with IBSA, differences between its members remain substantial. It does not share a common geopolitical outlook and this is clearly evident in tensions between the anti-US (China and Russia) and more Western-friendly (India and Brazil) alignments of its members, as well as in deep Sino-Russian bilateral security disputes (Wilson, 2015). China's (re)turn to the South, as a form of "planetary post-colonialism" (Sidaway, Woon and Jacobs, 2014), is also arguably in a league of its own compared to the other BRICS and its economy, exports and official foreign-exchange reserves are considerably bigger (Rothkopf, 2009). The BRICS often claim to be deepening a South–

FIGURE 7.9 A Chinese warship docks in the port of Djibouti in the Gulf of Aden on February 6th, 2016. Photo by Vladimir Melnik/ Shutterstock.

FIGURE 7.10 Chinese People's Liberation Army personnel attend the opening ceremony of China's new military base in Djibouti on August 1st, 2017. STR/AFP/Getty Images.

South dialogue on development yet key SSDC fora such as IBSA involve only informal consultations and small-scale technical partnership agreements or discussions and so the depth of this "dialogue" remains very shallow. Resources (and a common interest in resource diplomacy) have been a key platform for cooperation among BRICS and forms of resource cooperation, including energy security measures, commodity price stabilisation and development-focused financial agreements, have been called for in every BRICS Summit Declaration since they began in 2009 (Wilson, 2015). The summit culture of groupings like the BRICS or IBSA also mimics that of the G8 and UN and struggles to get beyond "the platitudinous and empty promises that all the established and rising powers seem to make in various forums" (Narlikar, 2013: 601). Few observers believe, then:

> that well-choreographed encounters, handpicked initiatives, or lofty plans signify that diverse and potentially antagonistic states are either willing or able to translate their combined economic prowess into collective geopolitical clout. (Brütsch and Papa, 2013: 300–301)

The (re-)emerging donors are increasingly being drawn into transnational structures of production, denationalising their economies and it is important therefore to contextualise their growth within the wider neoliberal capitalist order and to explore the internal relations, dynamics and processes of the capitalist global political economy in order to properly locate this intensification of geopolitical competition and the specific dynamics of accumulation that their growing presence is creating in the South. Again, this requires attention to the ways in which geopolitical interests intersect with geoeconomic strategies (e.g. in relation to the different forms of resource diplomacy that are being deployed). Attention to South–South value chains and production networks is also very valuable here (Horner, 2016; Horner and Nadvi, 2018). A variety of sub-state and non-state actors from (re-)emerging economies have become key foreign policy actors in their own right and play an important role in driving SSDC. The varieties of capitalism that have emerged within these economies are designed "to take advantage of the neo-liberal global economic system by deeply integrating with it while keeping state control intact" (McNally, 2013: 35). Their adoption of a "neoliberal" strategy of development has been selective and subject to translation, filtered through specific institutional, cultural and production contexts and through actors who do not simply "cut-and-paste" new economic policies developed in foreign "labs" (Ban, 2013). The extent to which "new" or "alternative" paradigms for development cooperation can really be created within this neoliberal framework is questionable since so often the "development" being promoted here acts primarily to facilitate capital accumulation.

Chapter 8

Conclusions: Development and (Counter-)Insurgency

THE EXCESS OF DEVELOPMENT

> If development can be seen as a formula for sharing the world with others, in its present configuration many seem destined to die before their time, while others are able to live beyond their means. (Duffield, 2010a: 57)

> I don't think we said in the *Dictionary* (nor today) that developers are dead; they continue their destructive enterprise. What is dead is its promise. (Esteva and Escobar, 2017: 2570)

> [T]he winds of war are blowing in our world and an outdated model of development continues to produce human, societal and environmental decline. (Pope Francis, Christmas Message, December 2017)

As a "formula for sharing the world with others" (Duffield, 2010a: 57), it has been claimed that development has "evaporated" (Esteva, 1992: 22) and now lies in ruins on the intellectual landscape as an "outdated monument to an immodest era" (Sachs, 1999: 1–2) or "a mined, unexplorable land" (Esteva, 1992: 22). The shadows cast by development are said to obscure our vision (Sachs, 1992, 1999), emphasising the need to fracture its gaze so as to render possible the dissemination of other knowledges. In part, the problem is that, as Gudynas (2011: 441) observes, "development" is a "zombie category", not really alive anymore but not quite dead either. Yet there is also the paradox that development "can be declared defunct and yet in the next step promoted as the only way forward [which] is deeply embedded in modern culture" (Gudynas, 2011: 441). As Sachs (2017: 2575) notes, the immodest era of expansive modernity that characterised development at its zenith in the 1960s is now over and this can be seen in the formulation of the recently agreed Sustainable Development Goals (SDGs) which refocus development away from "progress" to "survival" and make do without the ambitious plans for sky-high growth that were once promised, now seeking instead only to "secure a minimum for a dignified life universally". In some quarters the SDGs have been heralded as promoting a broader, more holistic and universal conception that promises to "leave no-one behind" and that understands that "all countries and regions of the world are interlocked in a mutual process of development" (Gills, 2017: 156). In focusing on poverty and inequality in every country, rather than just on extreme poverty in developing nations, some argue that international development is "no longer framed as an aid-financed gift to the poor" (Hulme, 2015) but rather as an activity that all countries can participate in defining. Just as it is declared defunct then the SDGs promote development as the best way forward.

Many words in the development lexicon evoke futures possible, others carry traces of worlds past but in many ways "development" has become a sort of performative word: saying by doing (Rist, 2010: 20). A certain degree of simulative potential has always been inherent in the declarations of the UN (Sachs, 2017: 2575) whereby quantitative data serve to enable comparison in time and space, constructing deficits along the time axis between groups and nations (akin to those in Rostow's stages of development) in a "deficit-creation dynamic" that has given the idea of development a purpose to exist for the last 70 years. The Human Development Index (HDI), like the archetypal measure of development, GDP, is one such deficit index, categorising countries according to a hierarchy. More generally, development practitioners often use an Orwellian "doublethink" that obscures, disguises, distorts or even reverses the meanings attached to development in order to portray failures of development policy as successes (Brooks, 2017).

The SDGs, with the scales and indices of its 17 core goals and 169 sub-goals, clearly follow in the tradition of this "deficit creation dynamic" with sustainability becoming (much like development itself) a kind of shibboleth – devoid of meaning but full of moral righteousness. Ironically development requires endless consumption and much of what is important to individuals – clean air, safe streets, steady jobs etc. – typically lies outside its field of vision as recognised in key measures like GDP (along with the wider distribution of wealth within this "growth") (Pilling, 2018). Development experiences non-Western peoples as somehow incomplete or lacking in the essentials for a proper existence (Mehta, 1999) whether it be a lack of education, an absence of capacity or the inability to use environmental resources sustainably, absences that mean life cannot be lived properly and that call forth development interventions and institutions as a response to this inadequacy through a "will to improve" (Li, 2007). Historically, development has promoted solutions in the form of externally provided support and the moral or educative trusteeship of experts (Mitchell, 2002), aiming to bring incomplete or underdeveloped life to its full potential (Duffield, 2008, 2010a). In this way, development must enrol its subjects as always "becoming", in order to continuously legitimise the superiority of its ideology. The ways in which development practitioners narrate and represent these deficits and disparities also constitute key space-producing practices that have important geopolitical implications (Strüver, 2007).

Despite the universalising impulse of the SDGs, development agencies still render global South and global North as if they imply a rigid dichotomy between geographic areas and historically established power structures (Wolvers et al., 2015). This book has argued that it is necessary to critically engage with Southern theory and to locate the South at the intersection of entangled political geographies (Sparke, 2007) but also to consider its historical role alongside other influential metageographies like "the tropics" and the "Third World". As Comaroff and Comaroff (2012: 47) suggest, "the south cannot be defined, *a priori*, in substantive terms. The label bespeaks a *relation*, not a thing in or for itself" (emphasis in original). The global South is not a stable ontological category symbolising subalterneity (Roy, 2014a) and has been deployed here in this more relational sense as a "concept metaphor" that interrupts and disturbs some of the many conceits of neoliberal globalisation that assume a single hegemonic form of capitalism flattening difference wherever it travels. Such an approach brings into view the different reinventions of development that are taking place but also the multiplicity and heterogeneity of capitalism's futures, acknowledging the different models and varieties of capital accumulation that are emerging within the South (along with the many different forms of resistance and contestation this has engendered).

Development has, since the earliest days of decolonisation, promised to slay the dragon of backwardness and underdevelopment but the regularly promised annulment of global poverty that this has rested upon has proven elusive. More familiar has been a recurrent and indignant rediscovery of the persistence of poverty and the idea that it is a recruiting ground for the strategic threats menacing the liberal order, which serves a variety of (geo)political functions and helps to validate a liberal way of development (Duffield, 2001). In this sense, conceptualising development as governmentality and examining the way in which development creates governable objects and spaces is particularly valuable today given the "biopolitical turn within aid policy" (Duffield, 2010a: 55) where development and underdevelopment are increasingly conceived *biopolitically*, in terms of how life is to be supported and maintained, within what limits and level of need people are required and expected to live or around what minimal provisions are needed for survival (in the case of the SDGs). Rather than building physical things or redistributing material resources, development has become more concerned with changing how people think or act (Duffield, 2001: 312), regulating conduct in the name of advancing stability and security.

The book has also argued that development can productively be conceptualised as a heterogenous assemblage or *dispositif*, or as a complex ensemble of institutions, discourses, resource flows, programmes, projects and practices. The different combinations of elements within this assemblage have particular geopolitical and geoeconomic effects that require and reward further critical scrutiny. Some of the early post-development scholarship conceived of the development *dispositif* or apparatus as a "machine-like" kind of entity (Ferguson, 1999) that reproduces itself by virtue of the unintended, unplanned, yet systematic side effects it brings about through the institutions, agencies and ideologies that structure development thinking and practice. The common assumption in many critiques of development is that its projects or processes are instigated from outside, whether through the introduction of new values, new production and consumption structures, or new ways of relating to nature, the social and the body, but how those desires are internalised and articulated is a key issue here that needs to be addressed in moving post-development scholarship forward.

Sachs, in the preface to the new edition of his *Development Dictionary*, acknowledges that while development "was an invention of the West" it was "not just an imposition on the rest" and he now recognises that the global South has become "the staunchest defender of development" as well as the extent to which the idea of development has "been charged with hopes for redress and self-affirmation" (Sachs, 2010: viii). To an extent, therefore, the idea and promise of development retains for many of the world's population "the potency it has acquired since the Enlightenment and industrial revolution" (Andreasson, 2017: 2644) and it is likely that the "ideational influence and normative power of development by means of economic growth and modernisation will remain an attractive proposal" (2644). Clearly then "we cannot *not* desire development" (Wainwright, 2008: 1, emphasis in original). Rather than seeing post-development and psychoanalysis as incompatible and diametrically opposed, however, as Kapoor (2017) does, it has been argued here that these two approaches can very productively be combined. Following Deleuze, the development apparatus can and should also be understood as a "desiring machine" (De Vries, 2007: 25), as a social body constituted by assembling heterogeneous desires (Deleuze and Guattari, 2002 [1980]). Indeed, as De Vries (2007) suggests, this also helps to understand why development so often fails or falls short since it relies on the production of desires which ultimately it cannot fulfil. There is, in other words, a certain "excess" in the concept of development that is central to its very functioning.

POST-DEVELOPMENT, STATE POWER AND INSURGENCY

> The long agony of development as a myth is clearly ending. . .. In my view, development is no longer a myth, a taboo, a promise or a threat. It is an obsession, an addiction, a pathological mania that some people suffer, in their minds, their emotions or their behaviour ... and also a tool of domination and control. . .. (Esteva and Escobar, 2017: 2567)

> The era of expansive modernity is over. The more this insight sinks in globally, the more the talk of development and thus also of Post-development will fade. . .. Consequently, it is about time that someone declared the end of the Post-development era, some 25 years after we declared the end of the Development era. (Sachs, 2017: 2584)

Many of its critics fail to differentiate between the heterogeneous positions subsumed under the heading of "post-development" and have arguably not fully grasped their political implications (Ziai, 2004: 1058). Recalling Polanyi's notion of capitalism's "double movement", whereby market forces unleash social havoc that in turn generates demands for social justice, Hart (2001: 650) claims that post-development critiques can be understood as "expressions of the opposing forces contained within capitalism" (see also Kiely, 1999). For Hart, post-development's so-called radicalism is integral to the development of capitalism. A similar notion that post-development has somehow "surrendered" to capitalism is also articulated by Kapoor (2017). Although widely critiqued and condemned, with even some of its original proponents now saying that it is "about time someone declared the end", post-development approaches are particularly useful in understanding the nexus between geopolitics and development. First and foremost, post-development scholarship has provided insightful and productive reflections on the very meaning of "development" and the origins of the concept, including the historical context of colonialism and in tracing the discursive construction of a "Third World" during the Cold War. Post-development also usefully insists upon a re-politicisation of development and poverty and for these issues not to be reduced to "technical problems" (Ferguson, 1990; Nustad, 2007), along with its recognition that development studies are interwoven with relations of power and Eurocentrism (Ziai, 2017). Post-development is also highly relevant in making sense of the hegemony of neoliberalism but also of the complex spatialities that characterise contemporary geographies of development and the reconfigurations of postcolonial sovereignty that they involve. Further, post-development envisions the possibility of a political community that can be explored beyond the state system (Nakano, 2007) and thus can be useful in reimagining the state and in understanding the creation of spaces for autonomy (Escobar, 2012). In this sense, post-development comes closest to fracturing the development gaze and to enabling other ways of imagining futures:

> For an initiative to be considered post-development it should contribute to the dismantling of the physical and discursive hegemony of development so that new locally grounded futures may be imagined and pursued. This includes freeing bodies, minds and community processes from the pursuit of development and opening up new socio-political spaces in which local imaginaries can be enacted and empowered. Crucially, in the context of foreign aid, locally based communities should have control over the actions and initiatives of external actors operating in their locality. (McGregor, 2007: 161)

This emphasis on the local and on communities having local control over development actions and initiatives has been seized upon by some critics who refer to it as "wobbly romanticism" ("only the rich get lonely, only the poor live hospitably and harmoniously") and "implausible politics" ("we can all live like the Mahatma, or would want to") (Corbridge, 1998: 139). As Sachs (2017: 2574) has recently acknowledged in reflecting on the early post-development scholarship, "we were opposed to the idea of Development, in chronopolitical, geopolitical and civilizational terms" and occasionally that opposition was sketched in quite bold contrast through references to a single or homogeneous "development project" even though there neither was nor is such a monolithic or singular construction, even at the zenith of modernisation in the 1960s and early 1970s (Simon, 2007).

Some commentators have suggested that as long as post-development fails to offer a constructive political programme for dealing with poverty, it will be stuck in ineffective agnosticism (Ziai, 2017: 2719) and thus remain dispensable in Development Studies (Pieterse, 1998). Post-development has however exerted considerable influence on Development Studies (and other disciplines) and a number of its central features (the intensive reflection on the study, meaning and origins of development and its imbrication with relations of power and Eurocentrism) have often been accepted and implicitly endorsed, leading to a curious mixture of rejection and integration as its radical anti-establishment position is explicitly disavowed in some critiques, but then implicitly some of its tenets are endorsed without giving credit where it is due (Ziai, 2017). Post-development does however require a more sophisticated employment of Foucault's analysis and must understand power in less absolute and unitary terms (Nilsen, 2016: 273), allowing more room for the agency of actors at all levels of insertion in the development *dispositif.* Although power is constantly exercised through discourses of development and the North has long attempted to make and mould the South in its own Eurocentric image, development is also constantly challenged and consequently reshaped. This requires a greater focus on the internal dynamics of development's "regimes of truth" and the political communities and knowledges that they generate but also closer attention to the heterogeneity of development and its actors and a clearer sense of how policy ideas, models and frameworks travel, or how they are received, reworked or rejected as they move across networks and assemblages at various spatial scales.

During decolonisation, development was reconfigured as an inter-state relation of governance and moved from colonial bureaucracy into a series of institutions of external expertise arranged "to help and support the newly discovered underdeveloped state" (Duffield, 2008: 148). Where "an obsession, an addiction, a pathological mania" (Esteva and Escobar, 2017: 2567) did quickly emerge was around the idea of creating "steady" states in the periphery, safe for the expansion of capitalism. Cold War modernisation mythologies implied that in order to "catch up" or to capture the secret of the success of wealthier states, countries of the Third World should internalise some of their features (Arrighi, 1991), but developmentalism was plunged into crisis by the fact that many states lacked the means of fulfilling the expectations, accommodating the demands or satisfying the desires that the lodestar of development had created (Chatterjee, 2004). This book has argued that the state is best approached not as a fixed and static entity and not a thing, a system or subject but as an assemblage of powers and techniques, an ensemble of discourses, rules and practices, as something to be discovered, as contingent and evolving, always emergent and becoming. States emerge as an imagined collective actor partly through the narration of statehood and the idea of development is key to that process. States seeking to control and consolidate both frontier spaces and populations domestically or to "civilise the margins" in the name of vanquishing rural "backwardness" have long deployed the idea of development, but territoriality is also produced from below in "counter-mappings" to those of the state and it is these competing claims to territory and place that can reveal a great deal about how geopolitics intersects with development. Territory has taken on particular significance across Latin America in the last three decades via a range of social movements that have mobilised and "re-invented" it as knowledge and practice (Porto-Gonçalves, 2012; Halvorsen, 2018; Zibechi, 2012).

Infrastructures can also provide a useful point of entry here, since they help to bring states into being, to show their strength and power and to narrate and perform their presence and role. Across the South, infrastructures have been enrolled in processes of nation-building, modernisation and development and in geopolitical imaginaries of territory, nationhood and sovereignty and they are also often named as a key object of struggle by social movements. State power is in part the capacity of infrastructures to order, arrange and make legible and they are thus key to "seeing like a state" (Scott, 1999). Given the partiality and fragmentation of infrastructures and access to them in the South, examining the uneven extension of biopolitics through the governmental logics of development can be very instructive in terms of what it reveals about wider state–society relations (Bakker, 2013). It is also productive to examine the resource–state nexus (Bridge, 2014), particularly given the rise of (re-)emerging economies like China, India and Brazil where state-led developmentalisms (in which resources and extractive industries figure prominently) have returned to centre stage.

The idea of *exporting* development models and success stories has a long history, beginning with the "Sinatra doctrine" of modernisation thinking (Frank, 1997) and coming to the fore in debates about the East Asian "developmental state" (itself the product of intersections between development and security in the Cold War). Development practitioners know all too well, however, that formal models are slippery in application, finding "fraught accommodation with the political economy of place, history, production and territorial government" (Craig and Porter, 2006: 120). Within every development agenda there lie "invisible assumptions" about people and places, assumptions which can be found to "lie at the root of the litany of unintended outcomes in developmental work" (E Pieterse, 2011: 2). Creating desires that cannot always be fulfilled, the excess of development sows the seeds of its own destruction. The reformist drive of colonial developmentalism, for example, intended to avoid anti-colonial resistance, generated the very conditions for its own demise as state-controlled development projects created expectations on which colonial authorities could not deliver (Cooper and Stoler, 1989). Indeed, subaltern political possibilities are "always present and inherent in the historical material processes of development" (Paudel, 2016: 1047) and the political geographies they create provide valuable insights into the nexus between geopolitics and development. Development is in many ways a dramatic and complex struggle over the shape of futures (Pieterse, 2001: 1) and so a focus on the different spaces of insurgency that development creates is instructive, particularly as the geographies of developmental empowerment and subaltern rebellion have often become entangled and overlap (Paudel, 2016).

Attention to the geographies and histories of social movements can also contribute to ongoing efforts to decolonise the political geographic imaginaries of critical development studies (Silvey and Rankin, 2010). There has been much talk about decolonising Political Geography and IR knowledges, but decolonisation is not a metaphor (Tuck and Yang, 2012) and is about much more than vocabularies and imaginaries: it is also necessary to decolonise structures, institutions and praxis (Esson et al., 2017). Indeed, the same could be said of development – it is not only development as a discourse or idea that requires decolonisation but also the very structures, institutions and practices that maintain it. The approach that has come closest to appreciating this most fully is post-development, although it has often ignored or overlooked important questions of how race and racism have variously been folded into projects of development. The "border thinking" (Mignolo, 2000) of scholars working on the concepts of *Buen Vivir* and decoloniality has a stronger and perhaps more promising concern for the intersectionality of race and indigeneity along with a valuable anti-racist standpoint, raising the possibility of non-Eurocentric modes of thinking and of countering modernist narratives (Escobar, 2007b: 180).

Dominant theories of social movement studies are typically modelled on the experiences of the global North yet the social movements they reference typically tend to concentrate on matters involving society rather than the state, while those in the South "tend to take a stand against the state, challenging the existing system of control inside and outside of each nation state" (Shigetomi, 2009: 6). In part, social movements arise because of the disconnect and disjuncture between citizens and the state and are "trying to narrow the distance between themselves and the ever more institutionally remote national state" (Davis, 1999: 609). In this sense, as Simone (2016: 85) has noted, it is important to focus on the meanings ascribed to notions of development that emanate from the repositories of the marginalised "without debilitating judgement" being placed upon them. Questions of place are key here in understanding this "anti-geopolitics", or "geopolitics from below" (Routledge, 2017) and its role in reshaping development, since "if we are to understand the social movements in the South – the conditions in which they have emerged, their strategies and their reach – we must look at the particular socio-political, institutional and economic contexts in which they are rooted" (Polet, 2007: 1). It is also productive in mapping the contested geographies and geopolitics of development to examine the tensions and contradictions in development between hegemonic (state) "discourses of control" and oppositional "discourses of entitlement" (as articulated by social movements) (Nilsen, 2016: 272).

There are now thousands of alternatives to and beyond development, a position of "One No and Many Yeses", to reference the Zapatistas' notion of creating a world in which many worlds can be embraced (Esteva and Escobar, 2017). Social movements have skilfully linked localised struggles to the dynamics of global power structures and mobilised to achieve progressive changes across spatial scales but are also playing a key role in democratising development and promoting more participatory and deliberative forms of political decision-making, although popular struggles have often quickly been diminished and co-opted by states. Non-state-centred terrains have historically been important spaces of resistance and destabilisation of the development project (cf. Escobar, 2008) but social movements do not always seek an outright rejection of development in its entirety, often seeking instead to negotiate and change its meaning and direction (Nilsen, 2016). Resistance and insurgency are also not simply assertions of otherness that reject development outright but are tangled up with desires for equality, dignity and redress and are practices of meaning- and claims-making hinging on oppositional appropriations.

RE-CENTRING AFRICA AND DEVELOPMENT IN POLITICAL GEOGRAPHY AND IR

> Global development can in its most general sense be understood as an improvement in the quality of international relations. (Hettne, 2010: 50)

As Hettne (2010) correctly observes, development is in many ways about seeking to improve the quality of international relations. Subaltern geographies of development have, however, often been neglected in both IR and Political Geography given their established focus on the core of the world system, as have the multiple forms of subaltern geopolitics associated with social movements in the South, but this book has called for a greater engagement with the complex and rich experiences and scholarship of different places and for engaging with other geopolitical traditions, knowledges and ways of thinking about the political. What is also required is a greater degree of engagement with the various historical entanglements between tropicality, orientalism and geopolitics, with the Cold War "area studies complex", with the histories, spaces and sites of diplomacy around modernisation (or SSDC) and with the military origins of development thinking. The book has also argued for the need to better understand the spaces, networks and circuits for international and transnational scholarly work through which knowledges of development and geopolitics flow and has sought to "provincialise" and "post-colonialise" Political Geography through a focus on Africa, partly as a way of challenging the view of the continent as essentially inadequate, a place of systemic failure in terms of its ability to engage with and partake in the modern world (Andreasson, 2017). The decolonial approaches pursued by scholars like Arturo Escobar and Walter Mignolo attempt to go further by switching away from this post-colonial provincialising of Western claims and instead seeking to rethink the world *from* Africa, *from* Latin America, *from* indigenous places and *from* the marginalised academia of the global South (Grosfoguel, 2010; Radcliffe, 2017).

In many ways, "development has always existed in relation to a state of exception" (Duffield, 2007a: viii) and usually it is insecure, collapsed, weak and fragile states (particularly in Africa) that provide the "other" against which model, steady states in the periphery are imagined and constructed. As Duffield (2001, 2007a) shows, the anxieties caused by the disavowal of the object produce all sorts of images of the Third World as a phantasmic obscene space of excess and abjection, representing everything that the West is not. Through their predominant focus on Western states, IR and Political Geography have also perhaps unwittingly played a role in the normalisation of particular states as the benchmark for analysis and in the creation of certain assumptions and teleological arguments in which states in the South can be depicted as "deviant", "weak" or "failing", with no "real" sovereignty. The legacies of their imperial and orientalist histories have meant that both IR and Political Geography have been slow in showing how many of the fundamental assumptions that fortify discourses of statehood, state failure and good governance have an orientalist makeup and colonial trajectories. From the different colonial governmentalities and developmentalisms and the resistances they engendered to the post-colonial challenges of taking over and transforming colonial bureaucracies, colonial constructions and spatialisations of the state have had significant and enduring consequences. They are an example of what Ann Stoler (2016) terms imperial "durations": the hardened, still-present traces, constraints and confinements of empire that continue to impact in the contemporary moment.

In many ways Africa has had something of an "ambivalent, tension-filled relationship" to the discipline of IR (Abrahamsen, 2017: 125) and scholars of Africa routinely accuse the discipline of sins of omission and misapprehension given the preoccupation with great power politics and the states seen as making the most difference. On both sides, however, there are now growing efforts at dialogue and mutual learning, with ambitions to "bring Africa in from the margins", to demonstrate the "lessons" IR can learn from Africa and to include more Southern voices. Yet, as Abrahamsen (2017: 130–131) argues, Africa does not enter the discipline of IR as a neutral object of study but is instead already overdetermined and embedded in diverse struggles. The function of then adding a series of African cases or illustrations to IR (of "add Africa and stir") does not fundamentally change or challenge theories and deeper assumptions of what constitutes "the international". In this sense:

> An African IR would be just another provincial IR, a substitute or evil twin of a Western IR, and would do little to facilitate the development of theoretical concepts and frameworks that allow us to theorize the international or the global – wherever it may be located. (Abrahamsen, 2017: 127)

In many ways post-development "has had little to say about Africa" (Matthews, 2004: 374) and all too often Africa is seen as only acted upon, but if combined with assemblage thinking it is possible to study Africa simultaneously as a place in the world and of the world, i.e. in a manner that appreciates its specificity and its globality (Abrahamsen, 2017) and that recognises the interdependence of both the unique local social relations and the global spatio-historical context in which African places are situated (Massey, 1993). As opposed to regarding them as ontologically given, this perspective understands politics and society in any location as assembled from "a multiplicity of actors, actants, knowledges, norms, values, and technologies, some local, some global, some public, and some private" (Abrahamsen, 2017: 133). The "international", as IR's object of study, is thus potentially found in any location and can be traced in its specificity from the ground up in assemblages that "inhabit national settings but are stretched across sovereign boundaries" (Abrahamsen, 2017: 133). By studying Africa from the ground up, as it is being constantly assembled by a multiplicity of local and global forces, the continent's politics and societies can be captured as both unique and global or as a "window on the contemporary world and its articulation in particular settings" (Abrahamsen, 2017: 127). What is also helpful in studying Africa from the "ground up" is a more "grounded" approach to the place-specific "micro-geopolitics" (Dwyer, 2014) and local histories and actors that shape how development is received, experienced and encountered in particular landscapes and by different populations, along with greater attention to its "everyday geographies" (Rigg, 2007).

MODERNISATION AND COLD WAR GEOPOLITICS

Much of the institutional architecture of development was formed in the Cold War period and was consequently closely shaped by foreign policy priorities and objectives. The Cold War was fought not just with military hardware but with competing models of the modern and different ideas of the correct pathways to development, with loans and grants and with technicians and planners. For Walt Rostow, underdevelopment was a danger to "infant" nations and foreign aid was nothing less than a "weapon" in the US strategic arsenal that should be used as a foreign policy tool to advance US national security objectives. Conversely, foreign aid programmes from the socialist, Second World became a way of fomenting revolution and resistance to US imperialism in the Third World. Consequently, the idea of modernisation was shot through with geopolitical discourses, practices and imaginaries. My intention here has been to illustrate that modernisation was a global, transnational project, shaped by a variety of places, projects and individuals, not just following from the West to the Rest or a US export. At the height of the Cold War battle for supremacy and ideological influence, a variety of competing models of modernisation were on offer but it was not always a straightforward choice between them – recipients often selectively appropriated elements of the approaches taken by different donors, sometimes combining elements of state planning, cooperatives and state farms, with US models of rural development. In this way it is important to recognise that Third World peoples and political leaders were able to negotiate the multiple modernities and political geographies of the Cold War in various ways.

The foreign aid programmes set up in pursuit of modernisation had a variety of geopolitical impacts on recipient states, becoming an important part of domestic security arrangements and a tool that recipient governments used to stabilise social and political relations at home (Hyndman, 2009) but also enabling various regimes to consolidate their grip on power through, for example, the modernisation of domestic military and police forces. Cold War foreign aid was in many ways all about *conversion* –

transforming recipient states into a model of the modern fashioned in the donor's own image, persuading them of the pitfalls of communism or capitalism and transforming their economies, infrastructures and state institutions in the process. The pursuit of modernisation through foreign aid was also all about *convergence* as it was believed that industrialisation and the transfer of technology and modern values would pull all recipient nations towards a common point (modern, industrial societies), thus creating steady statehood and political order in the periphery.

Across the Third World, roads, railways, airports, schools, hospitals, factories, textile mills, state farms and river basin development projects were constructed that were heavily determined by geopolitical considerations. From Afghanistan to Angola, what began to emerge were a number of Cold War "tournaments of modernisation" (Cullather, 2002: 530) where different models of modern development were enacted and showcased by donors. Aid quickly became a tool to reward allies and prevent threats emerging and in the pursuit of wider geopolitical aims of dominating, pacifying, protecting, strengthening or transforming recipient countries. In many ways, the result was a massive and widespread militarisation of the Third World. What followed was a bloody arc of US-backed military coups that stretches from Iran in 1953 (via Guatemala, Congo and Indonesia) to Chile in 1973, as the US sought to constrain transgressions of the narrow limits of capitalist nation-building (see Silver and Slater, 1999). Along with the US, China, the USSR and Cuba also intervened in a range of Third World conflicts but also often lacked sufficient understanding of the recipient contexts or of the complex nature of the struggles they had entered.

The Cold War around modernisation and development was also fought through popular geopolitics and propaganda wars for hearts and minds where donors sought to narrate the "progress" their assistance was enabling through popular geopolitical texts including films, radio programmes, comics and posters and which created significant affective communities, as in the idea of "Red Africa" created in Soviet propaganda (Nash, 2016). In particular, US attempts to modernise the Third World were all about defining America and Americans and in this sense foreign aid also provides a valuable "window into the national political soul" (Taffet, 2007: 4) and reveals a great deal about the values and world view of donors. US theory and practice on modernisation were also framed as much by events domestically as internationally. The establishment of domestic welfare states in the 1930s and 1940s in particular paved the way for the foreign aid regime (along with the Marshall Plan) and the willingness to consider governmental programmes of assistance to people overseas. Modernisation theorists and practitioners became obsessed with that which had worked elsewhere (Andreasson, 2010) and transferring these "lessons" and models to the periphery, to the point that what would work in recipient countries almost became of secondary importance. In drawing out such "lessons" and exemplars, Cold War donors often drew upon domestic experiences of modernisation and development (such as the Tennessee Valley Authority or Virgin Lands campaign). Projections of the spheres of influence that each power was building in the Third World were however never as unified, constant or unilateral as mapmakers suggested and the reality was a far more complex and dynamic patchwork of shifting allegiances. Interest peaked and troughed at different historical moments and was always subject to a wide range of national, regional and global geopolitical dynamics. Projects were to an extent supply-led but were also very much demand-driven, but frequently struggled with issues of implementation and were often resisted and reworked by recipients and intended beneficiaries whilst their reception "on the ground" was far more complex than anticipated by those who dreamt them up.

DEVELOPMENT AND PACIFICATION

It is important to remember that the idea of development took shape in the context of a military power struggle among emerging states in a turbulent and industrialising Europe (Cowen and Shenton, 1996) and from its very beginnings development intended "to compensate for the negative propensities of capitalism through the reconstruction of social order" (111). With its in-built sense of design, "development" has thus always represented forms of mobilisation associated with order and security (Duffield, 2002). With its promise of redemption through making full and wholesome what would otherwise remain incomplete, development has a long history then as a strategic response to various threats, a role that is still not widely appreciated since as a practical technology of security, development exists in the here and now and its benefits are always cast as a future yet to be realised (Duffield, 2008). Recalling the competing development projects put forward by colonial states and anti-colonial movements in the death throes of empire, Cullather (2009: 509) argues that:

> At the theoretical level, development was a generic, universal vaccine against communism in postcolonial settings, but at the level of implementation it looks more like a kind of combat, an exchange of economic and administrative thrusts and counterstrokes aimed at specific political goals.

This notion of development as a "kind of combat", involving "thrusts and counterstrokes" aimed at fulfilling political objectives, goes to the very core of the geopolitics–development nexus.

The sense that development could be used as a kind of antidote or "vaccine" to counter the "contagion" of insurgency and disorder in the Third World was very much a product of the age of decolonisation and the Cold War. During this time development became ever more closely associated with pacification, with the government and management of restless rural populations and with the restoration of political order and stability (Attewell, 2017), typically attempted (e.g. by agencies like USAID) through a series of villagisation, resettlement and infrastructure projects. Such schemes illustrate the presence of an emerging governmental rationality where development increasingly came to be fashioned as a set of biopolitical compensatory and ameliorative technologies of security (Duffield, 2006b), seeking to accomplish stable rule by creating certain sorts of governable subjects and objects (Watts, 2003). The word pacification originates from the Latin *pacificare* meaning "to make peace" but as a locus of disciplinary (bio)power the primary concern of development has often been about *violently* fixing potentially insurgent populations within an extended archipelago of highly securitised spaces (Duffield, 2007a, 2008, 2010a).

The Third World has long been constructed as a space of insurgency but was also an important source of insurgent theory and alternative geopolitical imaginations. In particular, questions of "race" (and other axes of difference) played a key role in the mobilisation of those who sought to rework the meanings of development (Prashad, 2007). The theme of poverty reduction also came to be closely linked to widespread anxieties about racialised violence in American cities and the wars of insurgency in the global South (Roy and Crane, 2015). The Third World was never a place, however, but was an ideological project, conceived in various locations and refracted through distinctly different routes and through various socio-cultural contexts. An exaggerated aura of permanent insecurity and chaos was ascribed to the "Third World" (and more recently the global South), stoking a "geopolitics of fear" that this chaos and insecurity will "leak" out into

Western homelands or that its peoples will "invade" or "infect" donor countries in the North if adequate aid transfers are not provided. In constructing and producing the *threat* of a crisis, various enforcement measures to eliminate them, to "liberate" others and to guide them to the trappings of liberal democracy and capitalist development, are sanctioned and enabled. This has been a key feature of the "security–development nexus" which itself can be understood as a *dispositif* or "constellation of institutions, practices, and beliefs that create conditions of possibility within a particular field" (Slater, 2008: 248). Here, development has found a reinvigorated and renewed purpose in the context of post-Cold War violence and insurgency, increasingly being transformed into conflict resolution and a technology of counter-insurgency (Duffield, 2007a, 2008). The idea of development as a form of containment has been rehabilitated and instead of anti-colonial resistance or communist contagion, the focus is on containing terrorist insurgencies and the mobility of underdeveloped life given the global outlawing of spontaneous or undocumented migration which has seen a shift in the focus of security from states to the people and movements based within them. The "shotgun marriage" between development and security (Luckham, 2009) has further blurred the contours of the shapeless, amorphous and amoeba-like entity of development but crucially it also constitutes a field of development and security actors, aid agencies and professional networks that call forth the conditions of need and insecurity they seek to provide solutions for and then intervene upon.

Rather than upholding a threatened state or rebuilding a failing one, recent interventions in Iraq and Afghanistan have sought to break and then remake the state and thus comprise a different model of militarised power, that of "reconstruction as war" (Kirsch and Flint, 2011), which constitutes a fertile arena for pursuing a critical analysis of geopolitics but also the intertwining of the geopolitical and geoeconomic in contemporary development. Military setbacks in Iraq and Afghanistan have also reignited a long-dormant interest in development-based counter-insurgency (Duffield, 2010a) as the US has sought to stem the spiralling violence and this has involved a rediscovery of colonial wars of counter-insurgency such as those of Malaya and Algeria. The transfer of counter-insurgency techniques and knowledges between these very different spatial and temporal contexts deserves much further critical attention, however. Palestine, for example, has historically served as an important stage for the consolidation of British imperial policing and pacification strategies (and later as a testing ground for Israeli experiments in asymmetric warfare and demographic engineering) whilst the counter-insurgency techniques honed by the British Army in Africa were subsequently used in countries like Northern Ireland and Cyprus (Hughes, 2013). US imperial violence in Vietnam, where "population-centred" approaches foregrounded development as key to counter-insurgency and pacification, has also been a key reference point. This "armed social work" again involves extensive efforts to reconfigure biopolitical landscapes, only this time it is supposedly more humane, culturally sensitive counter-insurgency. Indeed, US imperial endeavours slip between exemplary or performative forms of violence meant to intimidate and more "humane" and developmental warfare intended to persuade (Khalili, 2012: 4). In particular what stands out here is the increasing role of military actors in development, which characterises the "humanitarian present" (Weizman, 2011) where various technocratic collusions occur among those working to aid the vulnerable and those who mete out state violence in the name of security. This militarisation of development sees humanitarian language increasingly recruited to justify military operations but also a growing role for military actors in the design and delivery of foreign assistance, where, for example, reconstruction is embedded within combat brigades, becoming explicit tactics for countering an insurgency.

In recent years the US has increasingly focused attention and its foreign aid programmes on Africa's "ungoverned spaces" and "failed states". Consequently, Africa's fragility has become highly securitised – its problems seen as threatening to US national security, and its insecurities placed in the context of the War on Terror. The US has always adopted a militarised foreign policy towards Africa (Majavu, 2014) and in constructing the continent as a "swamp of terror" contemporary US engagement in Africa has become increasingly obsessed with security issues, particularly counterterrorism, counter-piracy and efforts to resolve internal conflicts. In particular, the idea of "stability operations" builds on and exceeds counter-insurgency thinking but also further normalises military engagement in the fields of development and governance. US civil affairs forces now increasingly pursue an "indirect" and "population-centred" approach using humanitarian assistance (from medical care to infrastructure projects) as a form of "security cooperation" through a panoply of aid projects, aimed at winning the hearts and minds of Africans and so reducing the lure of extremist ideologies associated with groups such as Boko Haram and Al-Shabaab whilst improving US visibility, access and influence with foreign military and civilian counterparts.

The US military publicly insists its efforts in Africa are small scale, no more than a "light footprint", but there is no doubt that the various elements that make up the US military assemblage in Africa have expanded considerably with an increasing network of surveillance and drone bases tracking and targeting jihadist groups taking root in "weak" states. Such strategies have important consequences for African states as some of the continent's political and military elites have sought to bind themselves to AFRICOM's counterterrorism assemblage as a means of better securing their own authority (Moore and Walker, 2016). There are however no clear battlefields and officially recognised combat zones in this age of "everywhere war", only multidimensional and fluid "battlespaces" (Graham, 2009), and shadowy campaigns against non-state actors in "borderlands", whilst the nature of insurgencies (and the relations they have with the impoverished populations amongst whom they are embedded) is also changing.

In this context USAID has become a part of the national-security structure along with its ever more ambitious ventures into the domestic security environments of recipient countries, becoming a kind of "quasi-security" agency concerned with crafting particular kinds of states and promoting political order and stability in the interests of US national security. The DoD is also increasingly preoccupied with addressing the roots of instability and extremism in "weak" and "failing" states and preventing their collapse into conflict by building counterterrorism capabilities, which operate alongside USAID's historical efforts to enhance the capabilities of civilian police and paramilitary forces in the name of long-term "institution building". Foreign aid is however not a story featuring only nation states as the actors or one-way coherent flows of money and may be more effectively conceptualised in terms of a set of complex networks, networked elements or assemblages, constituted by flows of capital, knowledge, influence, practices, material objects and people (Roberts, 2014). Key network or assemblage elements in US foreign assistance include multiple government agencies, contractors, industry organisations and lobbyists, as well as the actual contracts, sub-contracts, grants and procurements. Thinking of foreign aid in terms of the "security–economy nexus" provides a better understanding of the various ways in which US geopolitical interests intersect with geoeconomic strategies, or of how geopolitical forms are "recalibrated by market logics" (Smith and Cowen, 2009: 24). These long-standing geopolitical priorities related to the promotion of security, democracy and liberal peace have increasingly overlapped and intersected with the neoliberal agenda around aid effectiveness (Duffield, 2008), but also with wider projects of neoliberalisation that involve making countries "safe" for the passage of US capitalism. In this sense, USAID (as with many other Western agencies involved in aid disbursement) is increasingly embroiled in the machinations of the global neoliberal roll-out (Essex, 2013).

SSDC AND THE CHANGING DYNAMICS OF DEVELOPMENT DIPLOMACY

In addition to Africa's growing prominence in international relations, the continent has also been vociferously courted by a variety of (re-)emerging economies. In many ways, the self-representation of the West in Africa continues to be dominated by an enduring notion of trusteeship, but Western development interventions are today increasingly constructed as altruistic and beneficent in contrast to the supposedly opportunistic, exploitative and deleterious role of the (re-)emerging powers, particularly China (Ayers, 2013), so often portrayed as the "villainous other" in ways reminiscent of Cold War Sinomania. Here, the rise of a non-Western power like China is depicted as somehow an "unknown", an unprecedented and therefore potentially dangerous development. This has been further fuelled by an escalating trade war between the US and China, with the Trump administration seeking to justify the imposition of import tariffs on Chinese goods on the grounds of national security. There has also been an assumption that these donors need to be "socialised" into multilateral fora and the "liberal" way of "doing development" (Gallagher, 2011) and that as they rise, growing integration in the world economy will lead to a convergence of interests with other players since greater stakes in the system will produce a sense of ownership and a stronger willingness to invest in it.

In part these represent attempts to *contain* the challenge of the "rising powers" but there are also multiple echoes of the Cold War here. This includes various depictions of China's rise as threatening and alarming and the idea of a defenceless and powerless "Third World" being "sucked" into the orbit of "dangerous" others who must be brought into a liberal political and economic order through the universalisation of a more "liberal way of development". There has also been a revival of Cold War discourses of South–South cooperation, constructing visions of horizontal, less hierarchical and more equal interaction and development cooperation that are mutually beneficial or "win-win". The "triangulation" of foreign aid during the Cold War, where recipient countries pursued relations with multiple states and played would-be suitors off against each other, has also made something of a return in many African states. The dynamics of development diplomacy have shifted since the age of Cold War polarity, however, and rather than fixed bloc-type alliances, there has been a rise of multiple, overlapping and fluid alliances, issue-specific groupings and constellations of power (Breslin, 2017).

Soon after the Asian financial crisis of 1997–98 some observers proclaimed the end of late development and suggested it would lead to a reassertion of the West's political and economic dominance over the South, yet the origins of the contemporary global financial crisis lie in the West itself and this has clearly undermined the credibility of Western models and institutions (such as the IFIs) to prescribe the pathway to successful economic growth (Gray and Murphy, 2013). The old order of global development, predicated upon a vertical donor–recipient relationship, is clearly in decay, with key actors like the DAC facing a crisis of legitimacy. The consequent fragmentation of global governance appears inevitable and for some observers may even be creative (Acharya, 2016), opening up space for emerging economies to play an increasingly active role in the reform of global economic and political governance.

Many of the claims made in the name of SSDC and the narratives of solidarity that they rely on are reminiscent of Third Worldist coalitions and diplomatic traditions of the past with their references to harmonious intra-Asian relations and notions of non-interference, mutual benefit and peaceful coexistence. Indeed, SSDC must be understood in its historical context, and as rooted in specific state–society configurations and different varieties of capitalism. The elaborate summits staged by India or Japan intended to mimic those of China, to boost international presence and to woo African partners by underlining commitment to the virtues of SSDC, often reference Bandung and other Third Worldist summits and such sites and spaces of diplomacy deserve further critical scrutiny. Much of the debate about these donors is characterised by a historical amnesia however and there remain very substantial differences between them whilst their individual discourses are yet to converge around common narratives. Frequently heralded as "new", different and "path-breaking", SSDC builds upon pre-existing forms of international development and neoliberal policy frameworks and adds new variants of statecraft to facilitate capital accumulation and market socialisation. The extent to which "new" or alternative paradigms for development cooperation can really be created within this neoliberal framework is questionable. Many of these donors have also neglected civil society or key issues like social justice and poverty eradication, which also provides grounds upon which to question their supposedly "transformative" or "emancipatory" potential.

The rise of the South has increasingly been heralded as reshaping the global economic landscape and credited with creating "new" international economic geographies, or in the words of former Brazilian President Lula, a "new trade geography of the South". Certainly, (re-)emerging economies have introduced competitive pressures into the existing world of development assistance, weakening the bargaining position of Western donors by widening the options available to aid-receiving countries, although this has varied both spatially and temporally but also in part due to commodity cycles (Carmody, 2011). They have also broadened the range of sources of development finance on offer (through regional development banks that are beginning to erode the primacy of the IFIs as lenders in Asia and Latin America) and challenged the traditional configuration and discourse of "donor–recipient" relations. These "new" aid modalities have blurred and loosened the definition of what constitutes "aid" and are often characterised by a diffusion of responsibility for ODA programmes across several state agencies. Indeed, one of the characteristic features of SSDC is the huge range of actors involved: Brazilian development cooperation with Africa, for example, involves over 170 institutions including federal government organs and private institutions (NGOs, foundations and corporations). It is also necessary however to go beyond the privileging of nation-states as the fundamental units of analysis here in order to recognise the growing number of transnational connections being forged through SSDC by a range of non-state actors (e.g. civil society groups).

In some ways there is a sense that these donors are providing something more practical and less prescriptive in contrast with Western aid and it also appears that ODA from OECD countries is of declining importance and attraction, although this changing picture is subject to significant regional variation. It is often claimed (re-)emerging donors offer an escape from the conditionalities favoured by established Western counterparts, although in practice projects are often tied to the use of labour and input procurements. Further, the summit culture of groupings like the BRICS or IBSA mimics that of the G8 in a variety of ways but specifically in struggling to get beyond the platitudinous and the empty rhetoric and promises. To an extent then the rising significance of these actors is located within Western global hegemony in offering only a limited, "within-system" challenge given their level of engagement with and dependence upon existing institutional structures (Stephen, 2014) and the continuing predilection for neoliberal modes of capitalist development. It is also misleading to suggest that these actors have essentially completely superseded existing actors and hegemons – the decline of US hegemony is not the same as the end of US power and many traditional international donors like the US and the EU continue to hold significant influence in the framing and practice of global development in regions like Africa.

Many of the (re-)emerging economies portray themselves as both "developing country" and "champion of developing nations" and deploy a discourse of "mutual learning", promoting exchanges of knowledge gained from "successful" social and economic development experiences and "tested" solutions but also seeking to "export" them. Paradoxically, despite their much-heralded economic success, some of these donors have been until recently, or are still, recipients of international aid themselves and several experience strong domestic sensitivities around the giving of aid internationally in the context of high levels of poverty and inequality at home. Several of these (re-)emerging donors claim that their own colonial history, cultural diversity, expertise or technologies provides a better "fit" for the needs of developing countries, due to greater proximity (vis-à-vis Northern donors), the absence of conditions and associations with commercial interests, their "demand-driven" nature and the way they acknowledge the value of local experience. Brazil and India depict themselves as the purveyors of ideas to push events forward rather than providers of significant volumes of tangible resources, emphasising instead scientific–technical cooperation and technological transfers of expertise and acting as an enabler for the private sector rather than concentrating on state-led development assistance. Approaches to SSDC have also been closely shaped by the domestic experiences and imaginaries of development and modernisation in these (re-)emerging economies. The transformation of Brazil's *cerrado*, for example, has provided a model and inspiration for Brazilian agricultural cooperation in Africa (Cabral et al.,

2013) and China's historical experience as an "aid" recipient itself has clearly shaped its contemporary foreign assistance modalities, for example in the provision of development bank loans to finance infrastructure projects reimbursed by natural resources (an approach shaped by its relations with Japan and now widely used by others). Many (re-)emerging donors have been creative with the semantics of development, being careful to avoid stigmatising labels such as "failed", "fragile" and "weak states" and keen to depict themselves as leaders and builders of partnerships and coalitions across the South, or as interlocutors and mediators acting as a "bridge" or "balancer" between North and South.

It is important, however, to examine how both the framing and delivery of this assistance are shaped by geopolitical agendas, discourses and imaginations. SSDC is a key component of how the (re-)emerging powers seek to project an image of themselves and their power and this requires close attention to the geopolitical position of states within the global state system. The projection of an international identity as a "moral nation" and development "success" (e.g. South Korea) is also a means to secure a greater stake in the global norm-building process. In constructing a "benign" (or at least "neutral"), conciliatory, consensus-creating political persona for diplomatic purposes and in claiming to pursue global leadership, these countries are in part seeking global *followership* (Vickers, 2013; Breslin, 2017) and are using the diplomacy of development to advance this. Many (re-)emerging donors explicitly conceive of their development assistance as a form of "soft power" and so their use of SSDC discourse can in part be understood as rhetoric with which they have sought to negotiate a stronger bargaining position and to create new markets for their companies, which again requires attention to ways in which geopolitical interests intersect with geoeconomic strategies. Further, their international development cooperation efforts are also very much shaped by domestic geopolitics. South Korea's desire to counter and contain the international influence and diplomacy of the DPRK has been a key driver of its growing desire to engage with SSDC, particularly in Africa. Similarly, China's relations with Taiwan and its desire to counter the international recognition of Taiwanese sovereignty, along with the Sino-Soviet split, were historically very significant in China's embrace of Africa.

There is a degree of symbiosis in the relations forged between (re-)emerging donors and the partners they seek to work with across the South. Changes in the global economic order and the international state system over the last 20 years have created new opportunities and pressures for countries (like Malaysia, South Korea and Turkey, for example) to become more involved in the global politics of development and their emergence is at least to some extent a result of these transformed economic and political relations. The rise of countries like China, India and Brazil then has opened up opportunities for other countries of the South to increase their influence in the world, both through the assistance being offered by these donors but also in the way that they have been seeking to make the global order more inclusive, equitable and multilateral. The book has argued that it is necessary therefore to adopt a relational view of their (re-)emergence vis-à-vis each other (and existing global hegemons) and to contextualise their growth within the wider neoliberal capitalist order. Fear of being left behind by other more powerful economic risers such as China has certainly been a key driver of SSDC initiatives in, for example, India, Japan and South Korea. In order to properly locate this intensification of geopolitical competition and the specific dynamics of accumulation that their presence is creating in Africa and other parts of the South, it is necessary to situate this in the context of the internal relations, dynamics and processes of the capitalist global political economy.

The (re-)emerging donors have increasingly become part of transnational production networks and value chains and have denationalised their economies and societies but their adoption of a "neoliberal" strategy of development was not simply a "cut-and-paste" of economic policies developed in foreign "labs"; rather, it was selective and subject to translation, filtered through specific institutional and production contexts, cultures and actors. The specific varieties of capitalism that have emerged within these economies are often characterised by deep integration with the neoliberal global economic system while keeping state control intact. Indeed, a reassertion of the role of the state in development is a key feature of many of these (re-)emerging economies, where public/private distinctions in the economy are often "fuzzy" and as states continue to protect and nurture their own industries, to make them competitive within a global economy and actively seek to create market opportunities for them. State agencies from (re-) emerging economies have offered generous credit programmes, investment funds and tax benefits, backed by rapidly expanding diplomatic, trade and business networks, in support of their national corporations seeking to invest overseas and often faced with increasingly saturated domestic markets.

(Re-)emerging economies have also increasingly begun to develop an internationalist profile and to assert themselves as humanitarian, peacekeeping, peacebuilding and "policekeeping" actors on the world stage. Most now operate forms of military training and capacity-building exercises and are also starting to play a growing role in peacekeeping initiatives in Africa. Alongside their claims to be advancing peace and security, however, it is important to remember that the domestic defence industries from these countries play a huge role in arms transfers to Africa and in the creation of instability. Questions of military cooperation and security are often neglected in China–Africa studies (Alden et al., 2017) but China is now a major global manufacturer of low-tech, affordable firearms and related ammunition along with a growing range of other equipment used in law enforcement, exporting them to more than two-thirds of African countries (IISS, 2017). Further, military forces from China and India, for example, have gained significant field experience, expertise and intelligence through their engagement in African theatres of conflict. China's involvement with peacekeeping and peace talks in places like Sudan and South Sudan has, for example, served as an important test site or "laboratory" for China's diplomatic involvement in security crises (Duchâtel, Gowan and Rapnouil, 2016). Beijing's engagement in conflict resolution and deepening involvement in regional and international peacekeeping efforts mark a shift towards a more flexible and pragmatic understanding of its traditional support for non-interference, manifest in greater Chinese involvement in UN peacekeeping and the first combat troops being deployed in Africa along with China's increasing involvement in counter-insurgency.

Partly as a result of the enduring Eurocentrism in IR, the rise of non-Western powers has remained under-theorised and there has been a privileging of nation-states as the fundamental units of analysis in many IR/IPE debates, which depicts geopolitical competition as a series of distributive conflicts between rival states. The (re-)emerging economies are not the unitary states of much IR theorising and have fundamentally changed, however, reshaping their interrelations with other states in the process. Several (re-)emerging donors have witnessed a general shift towards "regulatory statehood" whereby central executives have withdrawn from "command and control" activities to merely set broad targets for diverse national, sub-national and private bodies (Dubash and Morgan, 2013). Indeed, the transformation of statehood under globalisation is a crucial dynamic shaping the (re-)emergence and conduct of these donors (Hameiri and Jones, 2016) as their states become increasingly fragmented, decentralised and internationalised.

China's (re)turn to the South in particular has often been misunderstood in this sense (Chan, 2016). China is not a unitary rational actor, nor is there a single Chinese "state capitalism", but many. The disaggregation of statehood has generated a multiplicity of different actors involved in foreign policy along with internal differences over external relations. Far from there being a single "national" position or "grand strategy", fragmented and decentralised state apparatuses and quasi-market actors pursue their own independent interests and agendas, generating conflict-ridden, incoherent policy output. China–Africa interaction is often informal, unplanned, negotiated, decentralised, uncoordinated and run through highly diversified routes (business associations, migrant networks and diasporas) and a complex range of both central and provincial-level actors from small and medium-scale enterprises to SOEs (Gu et al., 2016).

Contrary to the image of an all-powerful monolithic state centrally and coherently coordinating all of China's foreign assistance from Beijing, its strings all pulled by the CCP as "Godfather", the reality is far more complex in terms of the diversity of actors involved with assistance, often driven primarily from below by various state-owned, private and hybrid companies, linked predominantly to sub-national governments, seeking business opportunities by lobbying Chinese state agencies to initiate aid-funded infrastructure and construction projects for them to undertake overseas (Hameiri and Jones, 2016).

THE SHIFTING SPATIALITIES OF CONTEMPORARY DEVELOPMENT

As Chinese entities (e.g. SOEs and state agencies) develop transnational interests, they extend their "governance frontier" beyond China's territory, effecting state transformation elsewhere (Hameiri and Jones, 2016) and this question of how other states experience China's "rise" requires much further research, as does the increasing capacity of China and other (re-)emerging powers to directly and indirectly alter geopolitical and developmental spaces outside their borders, contrary to the many proclamations around "non-interference". Of particular value in doing this are anthropological perspectives on the micro-level social and economic dynamics around SSDC which offer a more processual perspective (Siu and McGovern, 2017) and help to make sense of the everyday sociocultural and micro-political exchanges and interactions as they are lived and experienced in a variety of places and contexts. China's engagement has often been depicted as simply a "scramble" for natural resources but whilst foregrounding the significance of the extractive dimensions of its "cooperation" this downplays the wider geopolitical context of its diplomatic strategic pursuits and global foreign policy objectives, its effort to solidify its position as a global power, to sustain its economic and human development, to ensure Taiwan's reunification or to counter internal secession. There has also frequently been a failure to disaggregate the different types of official financial flows involved in China–Africa cooperation whilst the size and scale of Chinese lending (which also has quite specific geographies) have often been exaggerated. Although China is by far and away the most powerful of the (re-)emerging economies in terms of the scale and scope of its SSDC efforts, it is of course not the only show in town and the heavy focus on China–Africa cooperation has to an extent obscured the growing significance on the continent of other (re-)emerging donors like Saudi Arabia, Kuwait and the UAE, Turkey, Israel or Malaysia. Further, despite the attention focused on the competition between these donors in Africa, China and many other (re-)emerging economies typically focus much of their development cooperation efforts on neighbouring countries within their own immediate regional contexts.

Many commentators have argued that China–Africa interactions have encouraged only dependency in terms of the heavy focus on resource extraction and that its model of "debtbook diplomacy" will only further entrench the uneven dynamics of accumulation and dependent relations that have historically characterised the continent. The resource geographies created in these forms of development cooperation clearly require further research, as do the ways in which resources have become a central focus of diplomacy around SSDC. For some observers, if African countries are to avoid a repetition of colonial-era value extraction they require judicious negotiating strategies, improved deliberative capacities and coalitions with local, continental and global civil society and business networks in order to ameliorate their weaker bargaining power and reshape the terms of their engagement with international partners, particularly the (re-)emerging economies (Vickers, 2013). The growing emphasis on the importance of African agency in mediating China–Africa encounters is useful here but the power of recipient states is often confined to incremental bargaining rather than structural change and can be highly circumscribed as African elites have struggled to change the terms of the extraverted and commodity-oriented trading relationship (Carmody and Kragelund, 2016). Again, this underlines the value of assemblage thinking which sees power residing not in actors but in networks, enabling an understanding of Africa and China (and their agency) as inter-constitutive (Carmody and Kragelund, 2016).

In spatial terms, the predominance of extractive industries and natural resources in China–Africa relations has been characterised by the production or reproduction of (often fortified) enclaves – enclosed areas of capital-intensive extraction (e.g. in mining and commercial agriculture) that typically have limited linkages to local communities, firms and institutions while remaining largely self-contained and detached from the national economy. These sub-national and transnational "spaces of post-development" (Sidaway, 2008a) involve subtly reworked articulations between territory, development and sovereignty and are marked by a variety of fractures and boundary practices involving articulations of citizens and subjects and places and spaces of accumulation, exclusion and inclusion. What such spaces illustrate is that contemporary development policy and practice has become strikingly polycentric across multiple scales (local-to-global) and multiple sectors (official, non-official and public–private combinations), simultaneously unfolding both beyond and across states. A focus on the complex and shifting spatialities of development is thus increasingly warranted although this "cannot be assumed a priori" and "needs to be discovered" (Novak, 2016: 503). In Africa, a good place to start would be some of the many SEZs that have emerged in recent years or are planned for the future, inspired by models from China (and others), along with some of the many transboundary infrastructure-related projects funded by (re-)emerging donors.

In this sense it is productive to examine the "borders–development nexus" and to deepen the dialogue between the fields of Border Studies and Development Studies given the recent proliferation of FTAs, SEZs, "growth corridors" and other re-bordering strategies characteristic of contemporary development, whereby territorial jurisdictions are redefined in regional, cross-border and/or sub-national terms (Novak, 2016). The recent "renaissance" of Border Studies has seen a move away from linear and container-like conceptualisations of state-centred cartographies, privileging instead the study of b/ordering processes (across national, transnational, regional and global scales and across both state- and non-state-centred scales of discourse and practice), and this provides a useful focus on the dispersed and fluid ways in which the dichotomies defined by borders unfold in situated and place-specific contexts (Novak, 2016: 494). By creating territorial governance units among, across and within states, these "political technologies" (Elden, 2013) are increasingly seen as important in fostering FDI, regional cooperation and trade-led growth (Novak, 2016).

One key example is China's One Belt, One Road (OBOR) development strategy which aims to build transregional connectivity based on multilateral geoeconomic exchange (Jessop and Sum, 2018). As Blanchard and Flint (2017) note, OBOR has been variously narrated as: (1) a geoeconomic or commercial project to facilitate a cross-continental flow of capital, commodities, labour and resources through infrastructure construction or a "spatial fix" to China's industrial overcapacity problem; (2) as China's strategic move for fulfilling its geopolitical ambition and as a geopolitical-economic strategy not only to counter US imperialist efforts to isolate China but also to promote SSDC and; (3) as a platform to recover the Third World spirit that developed in Maoist China leading to a new Bandung era (Paik, 2016). On their own none of these explanations is particularly satisfactory, however, and it is critical to explore how flows of capital, commodities, labour power and resources impact upon class relations and political formations both within China but also in other concerned countries by paying attention to the role of multiple actors at multiple levels in constructing the components of OBOR, along with the ways in which geopolitics and geoeconomics interconnect in its design, implementation and outcomes. There are of course many definitions of today's "Silk Road" in terms of communities, cultures, communications, ecologies, geographies, hubs, technologies, coexistence etc. and the idea contains a history closely entangled with European imperialism, often skirted around in the contemporary reworking of the concept into OBOR narratives (Sidaway and Woon, 2017). OBOR also recalls the heady days of modernisation with its heavy focus on transport and energy infrastructures (particularly roads, bridges, gas pipelines, ports, railways and power plants) and as a transnational, neoliberal development strategy OBOR has to be understood in the context of China's rise, geopolitical positioning and geoeconomic strategies. The use of such political technologies and (re)b/ordering strategies, the multiple echoes of Third Worldism and SSDC that they involve and the reincarnation of development as an economic and foreign policy strategy to counter the geopolitical ambitions of others, provide promising points of entry for understanding the contemporary nexus between geopolitics and development.

Bibliography

Aaltola, M (2009) *Western Spectacle of Governance and the Emergence of Humanitarian World Politics*, Basingstoke: Palgrave Macmillan.

Abbott, D (2006) "Disrupting the Whiteness of Fieldwork in Geography", *Singapore Journal of Tropical Geography*, 27 (3), 326–341.

Abdenur, A E (2014) "China and the BRICS Development Bank: Legitimacy and Multilateralism in South-South Cooperation", *IDS Bulletin*, 45 (4), 85–101.

Abdenur, A E and Da Fonseca, J M (2013) 'The North's Growing Role in South-South Cooperation: Keeping the Foothold', *Third World Quarterly*, 34 (8), 1475–1491.

Abdenur, A E and Marcondes De Souza Neto, D (2014) "Rising Powers and The Security–Development Nexus: Brazil's Engagement with Guinea-Bissau", *Journal of Peacebuilding and Development*, 9 (2), 1–16.

Abrahamsen, R (2000) *Disciplining Democracy: Development Discourse and Good Governance in Africa*, London: Zed Books.

Abrahamsen, R (2003) "African Studies and the Postcolonial Challenge", *African Affairs*, 102 (407), 189–210.

Abrahamsen, R (2004) "A Breeding Ground for Terrorists? Africa and Britain's 'War on Terrorism'", *Review of African Political Economy*, 31 (102), 677–684.

Abrahamsen, R (2017) "Africa and International Relations: Assembling Africa, Studying the World", Research Note, *African Affairs*, 116 (462), 125–139.

Acharya, A (2014) "Global International Relations (IR) and Regional Worlds", *International Studies Quarterly*, 58 (4), 647–659.

Acharya, A (2016) "The Future of Global Governance: Fragmentation May Be Inevitable and Creative", *Global Governance: A Review of Multilateralism and International Organizations*, 22 (4), 453–460.

Acharya, A and Buzan, B (eds) (2010) *Non-Western International Relations Theory: Perspectives on and beyond Asia*, New York: Routledge.

Acharya, A and Stubbs, R (2006) "Theorizing Southeast Asian Relations: An Introduction", *The Pacific Review*, 19 (2), 125–134.

Acosta, A (2008) "El 'Buen Vivir' para la construcción de alternativas", in Acosta, A (ed) *Entre el quiebre y la realidad: Constitución 2008*. Quito: Abya-Yala: 27–37.

Adas, M (1989) *Machines as the Measure of Men: Science, Technology and Ideologies of Western Dominance*, Ithaca, NY: Cornell University Press.

Adams, H (2012) "Race and the Cuban Revolution: The Impact of Cuba's Intervention in Angola", in Muehlenbeck, P (ed) *Race, Ethnicity, and the Cold War: A Global Perspective*, Nashville, TN: Vanderbilt University Press.

Adebajo, A (2008a) "Hegemony on a Shoestring: Nigeria's Post-Cold War Foreign Policy", in Adebajo, A and Mustapha, A R (eds) *Gulliver's Troubles: Nigeria's Foreign Policy after the Cold War*, Scottsville, University of Kwa-Zulu Natal Press, 1–37.

Adebajo, A (2008b) "An Axis of Evil: China, the United States and France in Africa", in Ampiah, K and Naidu, S (eds) *Crouching Tiger, Hidden Dragon? Africa and China*, Scottsville: KwaZulu-Natal Press, 227–258.

Africa News (2018) "India to Open 18 New Embassies in Africa by 2021", http://www.africanews.com/2018/04/03/india-to-open-18-new-embassies-in-africa-by-2021// (Accessed June 19th 2018).

Africa Report (2011) "Top 500 Companies in Africa", https://paanluelwel.com/2011/08/10/african-economies-up-to-the-global-challenge/ (Accessed January 10th 2018).

AFRICOM (2017) "Fact Sheet: United States Africa Command", https://www.africom.mil/media-room/article/6107/fact-sheet-united-states-africa-command (Accessed August 13th 2017).

Agência Brasileira de Cooperação (ABC) (2011) *Brazilian Technical Cooperation*, Brasília: Agência Brasileira de Cooperação.

Agnew, J (1994) "The Territorial Trap: The Geographical Assumptions of International Relations Theory", *Review of International Political Economy*, 1 (1), 53–80.

Agnew, J (2012) "Looking Back to Looking Forward: Chinese Geopolitical Narratives and China's Past", *Eurasian Geography and Economics*, 53 (3), 301–314.

Agnew, J and Corbridge, S (1995) *Mastering Space: Hegemony, Territory and International Political Economy*, London and New York: Routledge.

Agnew, J and Oslender, U (2013) "Territorialities, Sovereignty in Dispute: Empirical Lessons from Latin America", in Nicholls, W, Miller, B and Beaumont, J (eds) *Spaces of Contention: Spatialities and Social Movements*, Aldershot: Ashgate, 121–140.

Ahluwalia, P (2001) *Politics and Post-Colonial Theory: African Inflections*, London: Routledge.

Ahluwalia, P (2010) "Post-structuralism's Colonial Roots: Michel Foucault", *Social Identities*, 16 (5), 597–606.

Ahmad A (1992) *In Theory: Classes, Nations, Literatures*, London: Verso.

Al Jazeera (2011) "US to Reconsider Aid to Egypt", January 30th, http://www.aljazeera.com/news/middleeast/2011/01/201112915194130323.html (Accessed January 14th 2018).

Alden, C, Alao, A, Zhang, Z and Barber, L (2017) *China and Africa: Building Peace and Security Cooperation on the Continent*, London: Palgrave.

Alden, C and Le Pere, G (2009) "South Africa in Africa: Bound to Lead?", *Politikon*, 36 (1), 145–169.

Alden, C and Verma, R (2016) "India's Pursuit of Investment Opportunities in Africa", in Dubey, A K and Biswas, A (eds) *India and Africa's Partnership: A Vision for a New Future*, New Delhi: Springer.

Alden, C and Vieira, M A (2005) "The New Diplomacy of the South: South Africa, Brazil, India and Trilateralism", *Third World Quarterly*, 36 (7), 1077–1095.

Aldrighi, D and Postali, F (2010) "Business Groups in Brazil", in Colpan, A M and Hikino, J T (eds) *The Oxford Handbook of Business Groups*, Oxford: Oxford University Press, 353–386.

Alessi, C and Xu, B (2015) "China in Africa", CFR Backgrounds, https://www.cfr.org/china/china-africa/p9557 (Accessed August 15th 2016).

Alexander, J (2017) "African Soldiers in the USSR: Oral Histories of ZAPU Intelligence Cadres' Soviet Training, 1964–1979", *Journal of Southern African Studies*, 43 (1), 49–66.

Alexander, P (2010) "Rebellion of the Poor: South Africa's Service Delivery Protests — A Preliminary Analysis", *Review of African Political Economy*, 37 (123), 25–40.

Ali, I (2016) "U.S. Aid to Pakistan Shrinks amid Mounting Frustration over Militants", *Reuters*, http://www.reuters.com/article/us-usa-pakistan-aid-idUSKCN1110AQ (Accessed August 16th 2017).

Ali, K (2012) "Precursors of the Egyptian Revolution", *IDS Bulletin*, 43 (1), 16–25.

Ali, N (2010) "Books vs Bombs? Humanitarian Development and the Narrative of Terror in Northern Pakistan", *Third World Quarterly*, 31 (4), 541–559.

Allison, R (1988) *The Soviet Union and the Strategy of Non-alignment in the Third World*, Cambridge: Cambridge University Press.

Almeida, P R (2007) "O Brasil como ator regional e como emergente global: estratégias de política externa e impacto na nova ordem internacional", *Cena Internacional*, 9 (1), 7–36.

Altmann, P (2015) "Studying Discourse Innovations: The Case of the Indigenous Movement in Ecuador", *Historical Social Research*, 40 (3), 161–184.

Amann, E and Baer, W (2003) "Anchors Away: The Cost and Benefits of Brazil's Devaluation", *World Development*, 31 (6), 1033–1046.

Amanor, K (2013) "South-South Cooperation in Africa: Historical, Geopolitical and Political Economy Dimensions of International Development", *IDS Bulletin*, 44 (4), 20–30.

Amar, P (2012) "Global South to the Rescue: Emerging Humanitarian Superpowers and Globalizing Rescue Industries", *Globalizations*, 9 (1), 1–13.

Ambrose, S and Brinkley, D (1997) *Rise to Globalism: American Foreign Policy since 1938*, 8th edition, New York: Penguin.

Amin, S (1994) *Re-reading the Postwar Period: An Intellectual Itinerary*, New York: Monthly Review Press.

Amin, S (2014) *Samir Amin: Pioneer of the Rise of the South*, London: Springer.

Amnesty International (2014) *China's Trade in Tools of Torture and Repression*, London: Amnesty International.

Amoore, L (2006) "Biometric Borders: Governing Mobilities in the War on Terror", *Political Geography*, 25 (3), 336–351.

Amorim, C (2017) *Acting Globally: Memoirs of Brazil's Assertive Foreign Policy*, Plymouth: Hamilton Books.

Amos, J (2012) "Soviet Diplomacy and Politics on Human Rights, 1945–1977". PhD dissertation, University of Chicago, ProQuest (AAT 3513410).

Amsden, A (1989) *Asia's Next Giant: South Korea and Late Industrialisation*, Oxford: Oxford University Press.

Anderson, B (1998) *The Spectre of Comparisons: Nationalism, Southeast Asia and the World*, London: Verso.

Anderson, P (2010) "Sinomania", *London Review of Books*, 32 (2), 3–6.

Anderson, P (2017) *The H-Word: The Peripeteia of Hegemony*, London: Verso.

Andolina, R, Laurie, N, and Radcliffe, S (2009) *Indigenous Development in the Andes: Culture, Power, and Transnationalism*, Durham, NC: Duke University Press.

Andreasson, S (2005) "Orientalism and African Development Studies: The 'Reductive Repetition' Motif in Theories of African Underdevelopment", *Third World Quarterly*, 26 (6), 971–986.

Andreasson, S (2007) "Thinking beyond Development: The Future of Post-Development Theory in Southern Africa", Prepared for the British International Studies Association annual conference, University of Cambridge, December 17th–19th.

Andreasson, S (2010) *Africa's Development Impasse: Rethinking the Political Economy of Transformation*, London: Zed Books.

Andreasson, S (2017) "Fossil-fuelled Development and the Legacy of Post-Development Theory in Twenty-First Century Africa", *Third World Quarterly*, 38 (12), 2634–2649.

Andrew, C and Mitrokhin, V (2005) *The World Was Going Our Way: The KGB and the Battle for the Third World*, New York: Basic Books.

Andrews-Speed, P (2010) *The Institutions of Energy Governance in China*, Paris: Institut Français des Relations Internationales.

Ansorge, J (2010) "Spirit of War: A Field Manual", *International Political Sociology*, 4 (4), 362–379.

Apaydin, F (2012) "Overseas Development Aid across the Global South: Lessons from the Turkish Experience in Sub-Saharan Africa and Central Asia", *European Journal of Development Research*, 24(2),261–282.

Appadurai, A (1996) *Modernity at Large*, Minneapolis: University of Minnesota Press.

Appadurai, A (2000) "Grassroots Globalization and the Research Imagination", *Public Culture*, 12 (1), 1–19.

Arnold, D (1996) *The Problem of Nature: Environment, Culture and European Expansion*, Oxford: Blackwell.

Arnold, D (2000) "'Illusory Riches': Representations of the Tropical World, 1840-1950", *Singapore Journal of Tropical Geography*, 21 (1), 6–18.

Arnold, D (2005) *The Tropics and the Traveling Gaze: India, Landscape and Science, 1800-1856*, Seattle: Washington University Press.

Arnove, R F (ed) (1982) *Philanthropy and Cultural Imperialism*, Bloomington, IN: Indiana University Press.

Arrighi, G (1991) "World Income Inequalities and the Future of Socialism", *New Left Review*, 189, 39–66.

Arrighi, G (2007) *Adam Smith in Beijing: Lineages of the 21st Century*, London: Verso.

Asad, T (2015) "Reflections on Violence, Law, and Humanitarianism", *Critical Inquiry*, 41 (2), 390–427.

Atkinson, D and Dodds, K (eds) (2000) *Geopolitical Traditions: Critical Histories of a Century of Political Thought*, London: Routledge.

The Atlantic (1962) "Atlantic Report: Afghanistan", October, 26–36.

Attewell, W L (2015) "Ghosts in the Delta: USAID and the Historical Geographies of Vietnam's 'Other' War", *Environment and Planning A: Society and Space*, 27, 2257–2275.

Attewell, W L (2017) "We're Not Counterinsurgents: Development and Security in Afghanistan, 1946-2014", PhD thesis, University of British Columbia.

Atwood, J B, McPherson, M P and Natsios, A (2008) "Arrested Development: Making Aid a More Effective Tool", *Foreign Affairs*, 87 (6), 123–132.

Ayers, A J (2013) "Beyond Myths, Lies and Stereotypes: The Political Economy of a 'New Scramble for Africa'", *New Political Economy*, 18 (2), 227–257.

Ayoob, M (2002) "Inequality and Theorising in International Relations: The Case for Subaltern Realism", *International Studies Review*, 4 (3), 27–48.

Babbit, S A (2014) *José Martí, Ernesto "Che" Guevara and Global Development Ethics: The Battle for Ideas*, New York: St Martin's Press.

Bach, D C (2013) "Africa in International Relations: The Frontier as Concept and Metaphor", *South African Journal of International Affairs*, 20 (1), 1–22.

Bach, D C (2016) *Regionalism in Africa: Genealogies, Institutions and Trans-state Networks*, London: Routledge.

Bachmann, J (2008) "The Danger of Ungoverned Spaces – The 'War on Terror' and its Effects in the Sahel Region", in Eckert, J (ed) *The Social Life of Anti-Terrorism Laws. The War on Terror and the Classifications of the "Dangerous Other"*, Bielefeld: Transcript, 131–162.

Bachmann, J (2010) "Kick down the Door, Clean up the Mess, and Rebuild the House—The Africa Command and the Transformation of the U.S. Military", *Geopolitics*, 15 (3), 564–585.

Bachmann, J (2014) "Policing Africa: The US Military and Visions of Crafting 'Good Order'", *Security Dialogue*, 45(2), 119–136.

Bachmann, J (2017) "Whose Hearts and Minds? A Gift Perspective on the US Military's Aid Projects in Eastern Africa", *Political Geography*, 61, 11–18.

Bachmann, J, Bell, C and Holmqvist, C (eds) (2015) *War, Police and Assemblages of Intervention*, London: Routledge.

Baer, W (2008) *The Brazilian Economy. Growth and Development*, Boulder, CO: Lynne Rienner.

Bakker, I and Silvey, R (2008) *Beyond States and Markets: The Challenges of Social Reproduction*, London: Routledge.

Bakker, K (2010) *Privatizing Water: Governance Failure and the World's Urban Water Crisis*, Ithaca: NY: Cornell University Press.

Bakker, K (2013) "Constructing 'Public' Water: The World Bank, Urban Water Supply, and the Biopolitics of Development", *Environment and Planning D: Society and Space*, 31 (2), 280–300.

Balibar, E (1991) "Is There a 'Neo-Racism'?", in Balibar, E and Wallerstein, I (eds) *Race, Nation, Class: Ambiguous Identities*, London: Verso, 17–28.

Ban, C (2013) "Brazil's Liberal Neo-developmentalism: New Paradigm or Edited Orthodoxy?", *Review of International Political Economy*, 20 (2), 298–331.

Ban, C and Blyth, M (2013) "The BRICs and the Washington Consensus: An Introduction", *Review of International Political Economy*, 20 (2), 241–255.

Bangkok Post (2014) "Chinese Army Shows off Dancing Robots", July 22nd, http://www.bangkokpost.com/news/asia/421784/chinese-army-shows-off-dancing-robots (Accessed August 5th 2014).

Banks, N and Hulme, D (2014) "New Development Alternatives or Business as Usual with a New Face? The Transformative Potential of New Actors and Alliances in Development", *Third World Quarterly*, 35 (1), 181–195.

Baran, Z (2008) "Turkey and the Wider Black Sea Region", in Hamilton, D and Mangott, G (eds) *The Wider Black Sea Region in the 21st Century: Strategic, Economic and Energy Perspectives*, Washington, DC: Center for Transatlantic Relations, 87–102.

Barkawi, T and Laffey, M (2006) "The Postcolonial Moment in Security Studies", *Review of International Studies* 32 (2), 329–352.

Barker, J (2005) "For Whom Sovereignty Matters", in Barker, J (ed) *Sovereignty Matters: Locations of Contestation and Possibility in Indigenous Struggles for Self-determination*, Lincoln: University of Nebraska Press, 1–31.

Barnes, T and Crampton, J (2011) "Mapping Intelligence: American Geographers and the Office of Strategic Services and GHQ/SCAP (Tokyo)", in Kirsch, S and Flint, C (eds) *Reconstructing Conflict: Integrating War and Post-war Geographies*, Farnham: Ashgate, 227–252.

Barney, T (2015) *Mapping the Cold War: Cartography and the Framing of America's International Power*, Chapel Hill, NC: University of North Carolina Press.

Baudrillard, J (1983) *Simulations*, New York: Semiotext(e).

Bayart, J F (1993) *The State in Africa: The Politics of the Belly*, Harlow: Longman.

Bayart, J F (2000) "Africa in the World: A History of Extraversion", *African Affairs*, 99 (395), 217–267.

Bayart, J F (2009) *The State in Africa: The Politics of the Belly*, 2nd edition, Cambridge: Polity.

Beall, J, Goodfellow, T and Putzel, Z (2006) "Introductory Article: On the Discourse of Terrorism, Security and Development", *Journal of International Development*, 18 (1), 51–67.

Bearak, M and Gamio, L (2016) "The US Foreign Aid Budget Visualised", *The Washington Post*, October 18th, https://www.washingtonpost.com/graphics/world/which-countries-get-the-most-foreign-aid/ (Accessed December 30th 2017).

Bebbington, A (ed) (2012) *Social Conflict, Economic Development and Extractive Industry: Evidence from South America*. Abingdon: Routledge.

Bebbington, A, Bornschlegl, T and Johnson, A (2013) "Political Economies of Extractive Industry: From Documenting Complexity to Informing Current Debates. Introduction to *Development and Change* Virtual Issue 2". doi: 10.1111/dech.12057.

Belcher, O (2012) "The Best-laid Schemes: Postcolonialism, Military Social Science and the Making of US Counter-insurgency Doctrine 1947-2009", *Antipode*, 44 (1), 258–263.

Belcher, O (2014) "Staging the Orient: Counterinsurgency Training Sites and the U.S. Military Imagination", *Annals of the Association of American Geographers*, 104 (5), 1012–1029.

Belcher, O (2018) "Anatomy of a Village Razing: Counterinsurgency, Violence, and Securing the Intimate in Afghanistan", *Political Geography*, 62, 94–105.

Bell, C (2015) "The Police Power in Counterinsurgencies: Discretion, Patrolling and Evidence", in Bachmann, J, Bell, C and Holmqvist, C (eds) *War, Police and Assemblages of Intervention*, London: Routledge, 17–35.

Bello, W, Kinley, D and Elinson, E (1982) *Development Debacle: The World Bank in the Philippines*, San Francisco, CA: Institute for Food and Development Policy.

Belloni, R (2007) "The Trouble with Humanitarianism", *Review of International Studies*, 33 (3), 451–474.

Benabdallah, L (2016) "China's Peace and Security Strategies in Africa: Building Capacity is Building Peace?", *African Studies Review*, 16 (3–4), 17–34.

Bender, G J (1978) *Angola under the Portuguese: The Myth and the Reality*, Berkeley, CA: University of California Press.

Bernal-Meza, R (2010) "International Thought in the Lula Era", *Revista Brasileira de Política Internacional*, 53, 193–213.

Berg, L D (2004) "Scaling Knowledge: Towards a Critical Geography of Critical Geographies", *Geoforum*, 35 (5), 553–558.

Berger, M T (1994) "The End of the 'Third World'?", *Third World Quarterly*, 15 (2), 257–275.

Berger, M T (2001) "The Post-Cold War Predicament: A Conclusion", *Third World Quarterly*, 22 (6), 1079–1085.

Berger, M T (2004) *The Battle for Asia*, London: Routledge.

Berger, M and Borer, D (2007) "The Long War: Insurgency, Counterinsurgency, and Collapsing States". *Third World Quarterly*, 28 (2), 197–215.

Beri, R (2003) "India's Africa Policy in the Post-Cold War Era: An Assessment", *Strategic Analysis*, 27 (2), http://www.idsa.in/system/files/strategicanalysis_rberi_0603.pdf (Accessed November 2nd 2018).

Berliner, J S (1958) *Soviet Economic Aid: The New Aid Trade Policy in Underdeveloped Countries*, New York: Praeger.

Berteau, D J, Kiley, G, Lang, H, Zlatnik, M, Callahan, T, Chandler, A and Patterson, T (2010) *Final Report on Lessons Learned: Department of Defense Task Force for Business and Stability Operations*, Center for Strategic and International Studies, June.

Bertelsen, B E (2016) *Violent Becomings: State Formation, Sociality and Power in Mozambique*, New York: Berghahn.

Bervoets, J (2011) "The Soviet Union in Angola: Soviet and African Perspectives on the Failed Socialist Transformation", *estnik: The Journal of Russian and Asian Studies*, Issue 9 (Spring), http://www.sras.org/the_soviet_union_in_angola (Accessed January 10th 2018).

Besada, H (2013) "Assessing China's Relations with Africa", *Africa Development*, 38 (1&2), 81–105.

Beswick, D and Hammerstad, A (2013) "African Agency in a Changing Security Environment: Sources, Opportunities and Challenges", *Conflict, Security and Development*, 13 (5), 471–486.

Beusekom, M (1997) "Colonisation indigène: French Rural Development Ideology at the Office du Niger, 1920–1940", *The International Journal of African Historical Studies*, 30 (2), 299–323.

Bhogal, P (2016) "The Politics of India's Foreign Aid to South Asia", *Global Policy*, November 7th, http://www.globalpolicyjournal.com/blog/07/11/2016/politics-india%E2%80%99s-foreign-aid-south-asia (Accessed September 25th 2017).

Bhungalia, L (2012) "'From the American People': Sketches of the US National Security State in Palestine", *Jaddaliyya*, http://www.jadaliyya.com/pages/index/7412/%E2%80%98from-the-american-people%E2%80%99_sketches-of-the-us-nati (Accessed August 8th 2017).

Bhungalia, L (2015) "Managing Violence: Aid, Counterinsurgency, and the Humanitarian Present in Palestine", *Environment and Planning A*, 47, 2308–2323.

Bhungalia, L (2016) "On USAID, Soft Power and American Globalism", *Dialogues in Human Geography*, 6 (1), 88–90.

Biccum, A R (2009) "Theorising Continuities between Empire and IR", in Duffield, M and Hewitt, V (eds) *Empire, Development and Colonialism: The Past in the Present*, Woodbridge: James Currey, 146–160.

Biersteker, T (1999) "Eroding Boundaries, Contested Terrain", *International Studies Review*, 1 (1), 3–9.

Bilgin, P (2008) "Thinking past 'Western' IR", *Third World Quarterly*, 29 (1), 5–23.

Birtle, A J (2006) *U.S Counterinsurgency and Contingency Operations Doctrine 1942-1976*, Washington, DC: Center of Military History, United States Army.

Bishop, M and Payne, T (2017) "Revisiting the Developmental State 1: Introduction", http://speri.dept.shef.ac.uk/2017/09/26/revisiting-the-developmental-state-1-introduction/ (Accessed November 13th 2017).

Blanchard, J-M F and Flint, C (2017) "The Geopolitics of China's Maritime Silk Road Initiative", *Geopolitics*, 22 (2), 223–245.

Blaney, D L (1996) "Reconceptualizing Autonomy: The Difference Dependency Makes", *Review of International Political Economy* 3 (3), 459–497.

Blaser, M (2009) "Political Ontology. Cultural Studies without 'Cultures'?", *Cultural Studies*, 23 (5–6), 873–896.

Boltanski, L (1999) *Distant Suffering: Morality, Media and Politics*, Cambridge: Cambridge University Press.

Bond, P and Mottiar, S (2013) "Movements, Protests and a Massacre in South Africa", *Journal of Contemporary African Studies*, 31 (2), 283–302.

Bonneuil, C (2000) "Development as Experiment: Science and State Building in Late Colonial and Postcolonial Africa, 1930–1970", *Osiris*, second series, 15, 258–281.

Booker, S and Rickman, A (2018) "The Future is African – and the United States is Not Prepared", *The Washington Post*, June 6th, https://www.washingtonpost.com/news/democracy-post/wp/2018/06/09/the-future-is-african-and-the-united-states-is-not-prepared/?utm_term=.5e6a27a1409f (Accessed June 18th 2018).

Booth, K (1995) "Human Wrongs and International Relations", *International Affairs*, 71 (1), 103–126.

Booysen, S (2011) *The African National Congress and the Regeneration of Political Power*, Johannesburg: Witwatersrand University Press.

Bose, S (1997) "Instruments and Idioms of Colonial and National Development: India's Historical Experience in Comparative Perspective", in Cooper, F and Packard, R (eds) *International Development and the Social Sciences*, Berkeley, CA: University of California Press, 45–63.

Bowd, G P and Clayton, D (2013) "Geographical Warfare in the Tropics: Yves Lacoste and the Vietnam War", *Annals of the Association of American Geographers*, 103 (3), 627–646.

Bowles, D W (1992) "Perestroika and Its Implications for Soviet Foreign Aid", in Lavigne, M (ed) *The Soviet Union and Eastern Europe in the Global Economy*, Cambridge: Cambridge University Press, 66–87.

Bradbury, M and Kleinman, M (2010) *Winning Hearts and Minds. Examining the Relationship between Aid and Security in the Horn of Africa*, Medford, MA: Feinstein International Centre at Tufts University.

Brainard, L (2006) *Security by Other Means: Foreign Assistance, Global Poverty and American Leadership*, Washington, DC: Brookings Institution Press.

Brandes, N and Engels, B (2011) "Social Movements in Africa", *Stichproben Vienna Journal of African Studies*, 11 (20), 1–15.

Brantlinger, P (1988) *Rule of Darkness: British Literature and Imperialism 1830-1914*, Ithaca, NY: Cornell University Press.

Brautigam, D (1998) *Chinese Aid and African Development: Exporting Green Revolution*, Basingstoke: Macmillan;New York: St Martin's.

Brautigam, D (2009) *The Dragon's Gift: The Real Story of China in Africa*, Oxford: Oxford University Press.

Brautigam, D (2010) "China, Africa and the International Aid Architecture", Africa Development Bank, working paper 107, https://www.afdb.org/fileadmin/uploads/afdb/Documents/Publications/WORKING%20107%20%20PDF%20E33.pdf (Accessed November 2nd 2018).

Brenner, N, Peck, J and Theodore, N (2010) "Variegated Neoliberalization: Geographies, Modalities, Pathways", *Global Networks*, 10 (2), 182–222.

Breslin, S (2013a) "China and the Global Order: Signalling Threat or Friendship?", *International Affairs*, 89 (3), 615–634.

Breslin, S (2013b) *China and the Global Political Economy*, London: Palgrave Macmillan.

Breslin, S (2013c) "China and the South: Objectives, Actors and Interactions", *Development and Change*, 44 (6), 1273–1294.

Breslin, S (2017) "Leadership and Followership in Post-Unipolar World: Towards Selective Global Leadership and a New Functionalism?", *Chinese Political Science Review*, 2 (4), 494–511.

Brickell, K (2014) "'The Whole World Is Watching': Intimate Geopolitics of Forced Eviction and Women's Activism in Cambodia", *Annals of the Association of American Geographers*, 104 (6), 1256–1272.

Bridge, G (2010) "Resource Geographies I: Making Carbon Economies, Old and New", *Progress in Human Geography*, 35 (6), 820–834.

Bridge, G (2014) "Resource Geographies II: The Resource-State Nexus", *Progress in Human Geography*, 28 (1), 118–130.

Bridge, G and Le Billon, P (2013) *Oil*, Cambridge: Polity.

Brigg, M (2001) "Empowering NGOs: The Microcredit Movement through Foucault's Notion of Dispositif", *Alternatives*, 26 (3), 233–258.

Brigg, M (2002) "Post-Development, Foucault and the Colonisation Metaphor", *Third World Quarterly*, 23 (3), 421–436.

Brigg, M (2004) "Disciplining the Developmental Subject: Neoliberal Power and Governance through Microcredit" in Fernando, J L (ed) *Microfinance: Perils and Prospects*, London: Routledge, 55–76.

British Broadcasting Corporation (BBC) (2014) "USAid 'Not Plotting to Topple Kenya Government'", February 14th, http://www.bbc.co.uk/news/world-africa-26186188 (Accessed January 2nd 2018).

Brooks, A (2017) *The End of Development: A Global History of Prosperity and Poverty*, London: Zed Books.

Brown, W (1995) *States of Injury: Power and Freedom in Late Modernity*, Princeton, NJ: Princeton University Press.

Brown, W (2006) "Africa and International Relations: A Comment on IR Theory, Anarchy and Statehood", *Review of International Studies*, 32 (1), 119–143.

Brown, W (2012) "A Question of Agency: Africa in International Politics", *Third World Quarterly*, 33 (12), 1889–1908.

Brown, W and Harman, S (eds) (2013) *African Agency in International Politics*, Abingdon: Routledge.

Brun, E and Hersh, J (1990) *Soviet-Third World Relations in a Capitalist World: The Political Economy of Broken Promises*, London: Palgrave.

Bruneau, M (2005) "From a Centred to a Decentred Tropicality: Francophone Colonial and Postcolonial Geography in Monsoon Asia", *Singapore Journal of Tropical Geography*, 26 (3), 304–322.

Brütsch, C and Papa, M (2013) "Deconstructing the BRICS: Bargaining, Coalition, Imagined Community or Geopolitical Fad?", *The Chinese Journal of International Politics*, 6 (3), 299–327.

Bryceson, D F, Gough K V, Rigg, J and Agergaard, J (2009) "Critical Commentary: The World Development Report 2009", *Urban Studies*, 46 (4), 723–738.

Bunnell, T, Ong, C E and Sidaway, J D (2013) "Editorial: Jim Blaut and the Trajectories of Tropical Geography", *Singapore Journal of Tropical Geography*, 34 (3), 285–291.

Burges, S W (2009) *Brazilian Foreign Policy after the Cold War*, Gainesville: University Press of Florida.

Burges, S W (2013) "Brazil as a Bridge between Old and New Powers?", *International Affairs*, 89 (3), 577–594.

Butterfield, S H (2004) *US Development Aid – A Historic First: Achievements and Failures in the Twentieth Century*, London: Praeger.

Buzan, B (1983) *People, States and Fear*, Boulder, CO: Lynne Rienner.

Buzan, B (2006) "Will the 'Global War on Terrorism' Be the New Cold War?", *International Affairs*, 82 (6), 1101–1118.

Cabral, L (2011) "Cooperação Brasil-África para o desenvolvimento: caracterização, tendências e desafios", *Textos Cindes*, 26, Rio de Janeiro: Centro de Estudos de Integração e Desenvolvimento.

Cabral, L and Shankland, A (2013) "Narratives of Brazil-Africa Cooperation for Agricultural Development: New Paradigms? Future Agricultures", CBAA working paper. http://citeseerx.ist.psu.edu/viewdoc/download?doi=10.1.1.434.9512&rep=rep1&type=pdf (Accessed November 2nd 2018).

Cabral, L, Shankland, A, Favareto, A and Vaz, A C (2013) "Brazil–Africa Agricultural Cooperation Encounters: Drivers, Narratives and Imaginaries of Africa and Development", *IDS Bulletin*, 44 (4), 53–68.

Call, C T (2008) "The Fallacy of the 'Failed State'", *Third World Quarterly*, 29 (8), 1491–1507.

Call, C and Abdenur, A E (2017) "A 'Brazilian Way'? Brazil's Approach to Peacebuilding", Brookings Institute, https://www.brookings.edu/wp-content/uploads/2017/03/lai_20170228_brazilian_way_peacebuilding1.pdf (Accessed September 29th 2017).

Callahan, W (2010) *China: The Pessoptimist Nation*, Oxford: Oxford University Press.

Callahan, W (2011) "Tianxia, Empire and the World: Chinese Visions of World Order for the Twenty-First Century", in Callahan, C and Barabantseva, E (eds) *China Orders the World: Normative Soft Power and Foreign Policy*, Washington, DC: Woodrow Wilson Centre Press, 91–117.

Calvert, K (2016) "From 'Energy Geography' to 'Energy Geographies': Perspectives on a Fertile Academic Borderland", *Progress in Human Geography*, 40 (1), 105–125.

Campbell, D (1992) *Writing Security: United States Foreign Policy and the Politics of Identity*, Minneapolis: University of Minnesota Press.

Campbell, D (2007) "Geopolitics and Visuality: Sighting the Darfur Conflict", *Political Geography*, 26 (4), 357–382.

Campbell, D (2008) "Deconstruction, Foundations and the Critique of American Power", *Environment and Planning D: Society and Space*, 26, 181–184.

Campbell, D (2009) "Post-Structuralism", in Dunne, T, Kurki, M and Smith, S (eds) *International Relations Theories*, 2nd edition, Oxford: Oxford University Press, 203–228.

Campbell, D and Power, M (2010) "The Scopic Regime of Africa", in MacDonald, F, Dodds, K and Hughes, R (eds) *Observant States: Geopolitics and Visual Culture*, London: IB Tauris, 167–198.

Cancian, M (2017) "M 3-24-2.0? Why US Counterinsurgency Doctrine Needs an Update", https://mwi.usma.edu/fm-3-24-2-0-us-counterinsurgency-doctrine-needs-update/ (Accessed December 31st 2017).

Cann, J P (1997) *Counterinsurgency in Africa: The Portuguese Way of War, 1961–1974*, London: Greenwood Press.

Carmody, P (2011) *The New Scramble for Africa*, 2nd edition, Cambridge: Polity.

Carmody, P (2012) "Another BRIC in the Wall? South Africa's Developmental Impact and Contradictory Rise in Africa and Beyond", *European Journal of Development Research*, 24 (2), 223–241.

Carmody, P and Kragelund, P (2016) "Who Is in Charge? State Power and Agency in Sino-African Relations", *Cornell International Law Journal*, 49 (1), 1–23.

Carmody, P and Taylor, I (2010) "Flexigemony and Force in China's Resource Diplomacy in Africa: Sudan and Zambia Compared", *Geopolitics*, 15 (3), 496–515.

Carnegie Commission (1997) *Preventing Deadly Conflict: Executive Summary of the Final Report*, Washington, DC: Carnegie Commission on Preventing Deadly Conflict.

Castelo, C (2016) "Reproducing Portuguese Villages in Africa: Agricultural Science, Ideology and Empire", *Journal of Southern African Studies*, 42 (2), 267–281.

Caudill, M (1969) *Helmand-Arghandab Valley: Yesterday, Today, Tomorrow*, Lashkar Gah: USAID.

Central Intelligence Agency (1984) "The USSR and the Third World", NIE 11-10/2-84, https://www.cia.gov/library/readingroom/docs/DOC_0000518056.pdf (Accessed July 18th 2018).

Central Intelligence Agency (1985) "Soviet Military Support to Angola: Intentions and Prospects", SNIE 71/11-85, https://www.cia.gov/library/readingroom/docs/DOC_0000261290.pdf (Accessed July 15th 2017).

Césaire, A (1972) *Discourse on Colonialism*, New York: Monthly Review Press.

Chabal, P and Daloz, J P (1999) *Africa Works: Disorder as a Political Instrument*. Oxford: James Currey.

Chaesung, C (2010) "Why Is There No Non-Western International Relations Theory? Reflections on and from Korea", in Acharya, A and Buzan, B (eds) *Non-Western International Relations Theory*, London: Routledge, 69–91.

Chakrabarty, M (2017) "Indian Investments in Africa: Scale, Trends and Policy Recommendations", ORF Research Paper, May 19th, https://www.orfonline.org/research/indian-investment-africa-scale-trends-and-policy-recommendations/ (Accessed November 3rd 2018).

Chaliand, G (1969) *The Peasants of North Vietnam*. Baltimore, MD: Penguin.

Chaliand, G (1977) *Revolution in the Third World: Myths and Prospects*, Hassocks, UK: Harvester Press.

Chan, S (ed) (2013) *The Morality of China in Africa*, London: Zed Books.

Chan, S (2016) "A Prognosis and Diagnosis for China and the 2016 G20: The Politics of a New Global Economic Geography", IDS evidence report 169, Brighton: IDS.

Chandhoke, N (2010) "Civil Society", in Cornwall, A and Eade, D (eds) *Deconstructing Development Discourse: Buzzwords and Fuzzwords*, Rugby: Practical Action Publishing, 175–184.

Chang, H (2003) *Kicking Away the Ladder: Development Strategy in Historical Perspective*, London: Anthem Press.

Charny, J R (2013) "The US Military's Expanding Role in Foreign Assistance", *InterAction Policy Brief*, https://www.interaction.org/files/FABB%202013_Sec16_NGOAndMilitaryRelations.pdf (Accessed January 7th 2018).

Chatterjee, P (2004) *Politics of the Governed: Reflections on Popular Politics in Most of the World*, New York: Columbia University Press.

Chaturvedi, S (2013) "Development Cooperation: Contours, Evolution and Scope", in Chaturvedi, S, Fues, T and Sidiropoulos, E (eds) *Development Cooperation and Emerging Powers—New Partners or Old Patterns?*, London: Zed Books, 13–36.

Cheesman, N (ed) (2018) *Interpreting Communal Violence in Myanmar*, London: Routledge.

Chege, M (1972) "Systems Management and the Plan Implementation Process in Kenya", Discussion paper 179, Institute of Development Studies, University of Nairobi, https://core.ac.uk/download/pdf/19916331.pdf (Accessed November 3rd 2018).

Chekhutov, A, Ushakova, N and Zevin, L (1991) "U.S.S.R. Chapter 2: Economic Assistance to Developing Countries", in Feinberg, R E and Avakov, R M (eds) *U.S. and Soviet Aid to Developing Countries: From Confrontation to Cooperation?* Washington, DC: Overseas Development Council, 93–122.

Cheng, J Y S and Shi, H (2009) "China's Africa Policy in the Post-Cold War Era", *Journal of Contemporary Asia*, 39 (1), 87–115.

Cheung, C K (2014) "China's Rise and the International Politics of East Asia: The Development of Chinese IR Theory", *China: An International Journal*, 12 (2), 31–45.

Chin, G and Quadir, F (eds) (2012) "Rising States, Rising Donors and the Global Aid Regime", *Cambridge Review of International Affairs*, 25 (4), 493–649.

China Africa Research Institute (CARI) (2017) "Chinese Loans to Africa", http://www.sais-cari.org/data-chinese-loans-and-aid-to-africa (Accessed September 10th 2017).

China Energy Finance database (2018) "China's Global Energy Finance", http://www.bu.edu/cgef/#/intro (Accessed November 2nd 2018).

Chomsky, N (1970) "In North Vietnam", *New York Review of Books*, August 13th, http://www.chomsky.info/articles/19700813.htm (Accessed September 15th 2017).

Chomsky, N (1971) "Foreword", in Limqueco, P, Weiss, P and Coates, K (eds) *Prevent the Crime of Silence: Reports from the Sessions of the International War Crimes Tribunal Founded by Bertrand Russell*, London: Allen Lane/Bertrand Russell Peace Foundation, 9–26.

Chouliaraki, S (2006) *The Spectatorship of Suffering*, London: Sage.

Chow, R (2006) *The Age of the World Target: Self-referentiality in War, Theory, and Comparative Work*, Durham, NC: Duke University Press.

Chow, R (2018) "China's War on Poverty Could Hurt the Poor Most", *Foreign Policy*, January 8th, http://foreignpolicy.com/2018/01/08/chinas-war-on-poverty-could-hurt-the-poor-most/ (Accessed June 15th 2018).

Chowdhry, G and Nair, S (eds) (2003) *Power, Postcolonialism and International Relations*, London: Routledge.

Christian Aid (2004) *The Politics of Poverty: Aid in the New Cold War*, London: Christian Aid.

Christensen, S F (2012) "Brazil's South–South Relations", in Nilsson, M and Gustafsson, J (eds) *Latin American Responses to Globalization in the 21st Century*, Basingstoke: Palgrave Macmillan, 231–252.

Christensen, S F (2013) "Brazil's Foreign Policy Priorities", *Third World Quarterly*, 34 (2), 271–286.

Chronic Poverty Research Centre (2005) *The Chronic Poverty Report 2004–05*, Manchester: Chronic Poverty Research Centre, Institute for Development and Policy Management, University of Manchester.

Chun, Z (2017) "China-Africa Cooperative Partnership for Peace and Security", in Alden, C, Alao, A, Chun, Z and Barber, L *China and Africa: Building Peace and Security Cooperation on the Continent*, London: Palgrave Macmillan, 123–144.

Citino, N J (2008) "The Ottoman Legacy in Cold War Modernization", *International Journal of Middle East Studies*, 40 (4), 579–597.

Citino, N J (2017) *Envisioning the Arab Future: Modernisation in US-Arab Relations 1945-1967*, Cambridge: Cambridge University Press.

Clark, J F (2001) "Realism, Neo-Realism and Africa's International Relations in the Post-Cold War Era", in Dunn, K C and Shaw, T M (eds) (2001) *Africa's Challenge to International Relations Theory*, Basingstoke: Palgrave.

Clarke, J (2008) "Living with/in and without Neo-liberalism", *Focaal*, 51, 135–147.

Clayton, D (2013) "Militant Tropicality: War, Revolution and the Reconfiguration of 'the Tropics' c.1940–c.1975", *Transactions of the Institute of British Geographers*, 38 (1), 180–192.

Clayton, D and Bowd, G (2006) "Geography, Tropicality and Postcolonialism: Anglophone and Francophone Readings of the work of Pierre Gourou", *L'Espace géographique*, 35, 208–221.

Clodfelter, M (2002) *Warfare and Armed Conflicts: A Statistical Reference to Casualty and Other Figures, 1500–2000*, 2nd edition, Jefferson, NC: McFarland.

Cobb, D (2014) "Luanda's Monuments", *The Johannesburg Salon*, 7, 58–66, https://jwtc.org.za/resources/docs/salon-volume-7/13_TheSalon_Vol7.pdf (Accessed November 6th 2018).

Cobbs, E A (1996) "Decolonization, the Cold War and the Foreign Policy of the Peace Corps", *Diplomatic History*, 20 (1), 79–105.

COBRADI (2013) *Cooperação Brasileira para o Desenvolvimento Internacional: 2010*, Brasilia: Instituto de Pesquisa Econômica Aplicada/Agência Brasileira de Cooperação.

Coghe, S (2017) "Reordering Colonial Society: Model Villages and Social Planning in Rural Angola, 1920-45", *Journal of Contemporary History*, 52 (1), 16–44.

Cohen, S (2001) *India: Emerging Power*, Washington, DC: Brookings.

Cole, T (2012) "The White Saviour Industrial Complex", *The Atlantic*, March 21st, https://www.theatlantic.com/international/archive/2012/03/the-white-savior-industrial-complex/254843/ (Accessed April 28th 2017).

Coleman, M (2005) "U.S. Statecraft and the U.S.–Mexico Border as Security/Economy Nexus", *Political Geography*, 24 (2), 185–209.

Collinson, S, Elhawary, S and Muggah, R (2010) "States of Fragility: Stabilisation and its Implications for Humanitarian Action", *Disasters*, 34 (s3), S275–S296.

Comaroff, J (2002) "Governmentality, Materiality, Legality, Modernity: On the Colonial State in Africa", in Deutsch, J-G, Probst, P and Schmidt, H (eds) *Perspectives on African Modernities*, London: James Currey, 107–134.

Comaroff, J and Comaroff, J L (2012) *Theory from the South: Or, How Euro-America Is Evolving toward Africa*, Boulder, CO: Paradigm Publishers.

Connell, R (2007) *Southern Theory: The Global Dynamics of Knowledge in Social Science*, Cambridge: Polity Press.

Connell, R (2009) "Interview with Raewyn Connell", *Southern Perspectives*, February 16th, http://www.southernperspectives.net/region/australia/interview-with-raewyn-Connell (Accessed May 23rd 2017).

Connelly, M (2000) "Taking off the Cold War Lens: Visions of North-South Conflict during the Algerian War for Independence", *American Historical Review*, 105 (3), 739–769.

Connelly, M (2001) "Rethinking the Cold War and Decolonization: The Grand Strategy of the Algerian War for Independence", *International Journal of Middle East Studies*, 33 (2), 221–245.

Connelly, M (2008) *Fatal Misconception: The Struggle to Control World Population*, Cambridge, MA: Belknap.

Constantin, C (2006) "Comprendre la sécurité énergétique en Chine", *Politique et sociétés*, 25 (2–3), 15–45.

Conteh-Morgan, E and Weeks, P (2016) "Is China Playing a Contradictory Role in Africa? Security Implications of its Arms Sales and Peacekeeping", *School of Interdisciplinary*

Global Studies Faculty Publications, 19, http://scholarcommons.usf.edu/sigs_facpub/19 (Accessed June 19th 2018).

Coombes, A E (1997) *Reinventing Africa: Museums, Material Culture and Popular Imagination*, New Haven, CT: Yale University Press.

Cooper, A F and Flemes, D (eds) (2013) "Special Issue: Foreign Policy Strategies of Emerging Powers in a Multipolar World", *Third World Quarterly*, 34 (6), 943–1144.

Cooper, F (1997) "Modernizing Bureaucrats, Backward Africans, and the Development Concept", in Cooper, F and Packard, R (eds) *International Development and the Social Sciences*, Berkeley, CA: University of California Press, 64–92.

Cooper, F (2002) *Africa since 1940: The Past of the Present*, Cambridge: Cambridge University Press.

Cooper, F (2004) "Development, Modernization, and the Social Sciences in the Era of Decolonization: The Examples of British and French Africa", *Revue d'Histoire des Sciences Humaines*, 10, 9–38.

Cooper, F (2005) *Colonialism in Question: Theory, Knowledge, History*, Berkeley; London: University of California Press.

Cooper, F and Packard, R (1997) "Introduction", in Cooper, F and Packard, R (eds) *International Development and the Social Sciences*, Berkeley, CA: University of California Press, 1–41.

Cooper, F and Stoler, A (1989) "Tensions of Empire: Colonial Control and Visions of Rule", *American Ethnologist*, 16 (4), 609–621.

Copley, R (2014) "The Great Africa Switcheroo: U.S. Policy Is Now Ideological, while China's Is Pragmatic", *Background-Brief*, http://www.background-brief.com/the-great-africa-switcheroo-u-s-policy-is-now-ideological-while-chinas-is-pragmatic/ (Accessed August 1st 2017).

Copper, J F (2016) *China's Foreign Aid and Investment Diplomacy, Volume III: Strategy beyond Asia and Challenges to the United States and the International Order*, London: Palgrave.

Corbridge, S (1998) "'Beneath the Pavement Only Soil': The Poverty of Post-Development", *Journal of Development Studies* 34 (6), 138–148.

Corbridge S, Williams G, Srivastava, M and Véron, R (2005) *Seeing the State: Governance and Governmentality in India*, Cambridge: Cambridge University Press.

Cordovez, D and Harrison, S (1995) *Out of Afghanistan*, New York: Oxford University Press.

Corkin, L (2014) *Uncovering African Agency: Angola's Management of China's Credit Lines*, New York: Ashgate.

Cornelissen, S (2009) "Awkward Embraces: Emerging and Established Powers and the Shifting Fortunes of Africa's International Relations in the Twenty-First Century", *Politikon*, 36 (1), 5–26.

Cornwall, A (2010) "Introductory Overview - Buzzwords and Fuzzwords: Deconstructing Development Discourse", in Cornwall, A and Eade, D (eds) *Deconstructing Development Discourse: Buzzwords and Fuzzwords*, Rugby: Practical Action Publishing, 1–18.

Cornwall, A and Brock, K (2005) "What Do Buzzwords Do for Development Policy? A Critical Look at 'Participation', 'Empowerment' and 'Poverty Reduction'", *Third World Quarterly*, 26 (7), 1043–1060.

Costa Vaz, A and Inoue, C (2007) *Emerging Donors in International Development Assistance: The Case of Brazil*, Ottawa: International Development Research Centre.

Cowen, D (2017) "Investigating Infrastructures: A Forum", *Society and Space*, http://societyandspace.org/2017/10/03/investigating-infrastructures-a-forum/ (Accessed November 13th 2017).

Cowen, D and Smith, N (2009) "After Geopolitics? From the Geopolitical Social to Geoeconomics", *Antipode*, 41, 22–48.

Cowen, M and Shenton, R (1995) "The Invention of Development", in Crush, J (ed) *The Power of Development*, London: Routledge.

Cowen, M and Shenton, R (1996) *Doctrines of Development*, London: Routledge.

Craggs, R (2014) "Postcolonial Geographies, Decolonization and the Performance of Geopolitics at the Commonwealth Conference", *Singapore Journal of Tropical Geography*, 35 (1), 39–55.

Craggs, R and Mahony, M (2014) "The Geographies of the Conference: Knowledge, Performance and Protest", *Geography Compass*, 8 (6), 414–430.

Craggs, R and Neate, H (2018) "Post-colonial Careering and the Discipline of Geography: British Geographers in Nigeria and the UK, 1945-1990", *Journal of Historical Geography*, https://doi.org/10.1016/j.jhg.2018.05.014

Craig, D and Porter, D (2006) *Development beyond Neoliberalism: Governance, Poverty Reduction and Political Economy*, London and New York: Routledge.

Cramb, R A, Colfer, C J P, Dressler, W, Laungaramsri, P, Le, Q T, Mulyoutami, E, Peluso, N L and Wadley, R L (2009) "Swidden Transformations and Rural Livelihoods in Southeast Asia", *Human Ecology*, 37 (3), 323–346.

Crush, J (1995) "Imagining Development", in Crush, J (ed) *Power of Development*, London: Routledge, 1–26.

Culcasi, K (2010) "Constructing and Naturalizing the Middle East", *The Geographical Review*, 100 (4), 583–597.

Cull, N J (2008) *The Cold War and the United States Information Agency: American Propaganda and Public Diplomacy, 1945-1989*, Cambridge: Cambridge University Press.

Cullather, N (2000) "Development? Its History", *Diplomatic History*, 24 (4), 641–653.

Cullather, N (2001) "'Fuel for the Good Dragon': The United States and Industrial Policy in Taiwan 1950-1965", in Hahn, P L and Heiss, M A (eds) *Empire and Revolution: The United States and the Third World since 1945*, Columbus: Ohio State University Press, 242–268.

Cullather, N (2002) "Damming Afghanistan: Modernization in a Buffer State", *Journal of American History*, 89 (2), 512–537.

Cullather, N (2009) "The Third Race", *Diplomatic History*, 33 (3), 507–512.

Cummings, B (1997) "Boundary Displacement: Area Studies and International Studies during and after the Cold War", *Bulletin of Concerned Asian Scholars*, 29 (1), 6–26.

Cummings, B (2002) "Boundary Displacement: The State, the Foundations, Area Studies during and after the Cold War", in Masao, M and Harootunian, H D (eds) *Learning Places: The Afterlives of Area Studies*, Durham, NC: Duke University Press, 159–188.

Cummings, B L (2012) "Soviet Cinema and African Filmmaking", http://africasacountry.com/2012/04/soviet-cinema-and-african-filmmaking/ (Accessed July 11th 2017).

Dahlman, C (2011) "Breaking Iraq: Reconstruction as War", in Kirsch, S and Flint, C (eds) *Reconstructing Conflict: Integrating War and Post-war Geographies*, New York: Ashgate, 179–202.

Dalby, S (1990) "American Security Discourse: The Persistence of Geopolitics", *Political Geography Quarterly*, 9 (2), 171–188.

Dalby, S (1991) "Critical Geopolitics: Discourse, Difference, and Dissent", *Environment and Planning D: Society and Space*, 9 (3), 261–283.

Dalby, S (2008) "Imperialism, Domination, Culture: The Continued Relevance of Critical Geopolitics", *Geopolitics*, 13 (3), 414.

Danforth, N (2015) "Malleable Modernity: Rethinking the Role of Ideology in American Policy, Aid Programs, and Propaganda in Fifties' Turkey", *Diplomatic History*, 39 (3), 477–503.

Darby, P (2004) "Pursuing the Political: A Postcolonial Reading of Relations International", *Millennium: Journal of International Studies*, 33 (1), 1–32.

Darracq, V and Neville, D (2014) *South Korea's Engagement in Sub-Saharan Africa: Fortune, Fuel and Frontier Markets, Chatham House Research Paper*, October, https://www.chathamhouse.org/publication/south-korea's-engagement-sub-saharan-africa-fortune-fuel-and-frontier-markets (Accessed June 19th 2018).

Dauvergne, P and Farias, D B L (2012) "The Rise of Brazil as a Global Development Power", *Third World Quarterly*, 33 (5), 903–917.

Dávila, J (2010) *Hotel Trópico: Brazil and the Challenge of African Decolonization, 1950–1980*, Durham, NC: Duke University Press.

David, S R (1989) "Why the Third World Matters", *International Security*, 14 (1), 50–85.

Davies, A (2018) "Milton Santos: The Conceptual Geographer and the Philosophy of Technics" (Review essay), *Progress in Human Geography*, https://doi.org/10.1177/0309132517753809.

Davis, D (1999) "The Power of Distance: Re-Theorizing Social Movements in Latin America", *Theory and Society*, 28 (5), 585–638.

De Castro, J (1952) *Geopolítica da fome* [*Géopolitique de la faim*], Paris: Les éditions Ouvrières.

De Castro, J (1965) *Obras completas. 6, Ensaios de biologia social, 3a. edição*, São Paulo: Editôra Brasiliense.

De Castro, J (1966) *Obras completas. 5, Ensaios de geografia humana, 4a. edição*, São Paulo: Editôra Brasiliense.

de Freitas Barbosa, A, Narciso, T and Biancalana, M (2009) "Brazil in Africa: Another Emerging Power in the Continent?", *Politikon*, 36 (1), 59–86.

De Koninck, R and Déry, S (1997). "Agricultural Expansion as a Tool of Population Redistribution in Southeast Asia", *Journal of Southeast Asian Studies*, 28 (1), 1–26.

De Vries, P (2007) "Don't Compromise Your Desire for Development! A Lacanian/Deleuzian Rethinking of the Anti-politics Machine", *Third World Quarterly*, 28 (1), 25–43.

de Waal, A (2010) "Dollarised", *London Review of Books*, 32 (12), 38–41.

de Waal, A and Ibreck, R (2013) "Hybrid Social Movements in Africa", *Journal of Contemporary African Studies*, 31 (2), 303–324.

Dean, M (1999) *Governmentality: Power and Rule in Modern Society*, London: Sage.

Death, C (2011) "Foucault and Africa: Governmentality, IR Theory, and the Limits of Advanced Liberalism", paper presented at the BISA annual conference, Manchester, April 28th, http://www.open.ac.uk/socialsciences/bisa-africa/conferencepapers.shtml.

Death, C (2013) "Governmentality at the Limits of the International: African Politics and Foucauldian Theory", *Review of International Studies*, 39 (3), 763–787.

Deconde, A, Burns, R D, Logevall, F and Ketz, L B (2001) *Encyclopaedia of American Foreign Policy*, 2nd edition, New York: Charles Scribner's Sons.

Dee, H S (1983) "The Three World Theory and post-Mao China's Global Strategy", *International Affairs*, 29 (2), 239–249.

Deleuze, G (1992) "What Is a Dispositif?", in Armstrong, TJ (ed) *Michel Foucault: Philosopher*, New York: Harvester Wheatsheaf.

Deleuze, G and Guattari, F (2002 [1980]) *A Thousand Plateaus: Capitalism and Schizophrenia*, London: Bloomsbury.

Desai, A and Vahed, G (2015) *The South African Gandhi: Stretcher-Bearer of Empire*, Stanford, CA: Stanford University Press.

Deutschmann, D (1989) *Changing the History of Africa: Angola and Namibia*, Melbourne: Ocean Press.

Dittmer, J (2005) "Captain America's Empire: Reflections on Identity, Popular Culture and Geopolitics", *Annals of the Association of American Geographers*. 95 (3), 626–643.

Dittmer, J (2007) "The Tyranny of the Serial: Popular Geopolitics, the Nation and Comic Book Discourse", *Antipode*, 39 (2), 247–268.

Dittmer, J (2010) *Popular Culture, Geopolitics and Identity*, Lanham, MD: Rowman and Littlefield.

Dittmer, J (2014) "Geopolitical Assemblages and Complexity", *Progress in Human Geography*, 38 (3), 385–401.

Dittmer, J and Dodds, K (2008) "Popular Geopolitics Past and Present: Fandom, Identities and Audiences", *Geopolitics* 13 (3), 437–457.

Dittmer, J and Gray, N (2010) "Popular Geopolitics 2.0: Towards New Methodologies of the Everyday", *Geography Compass*, 4 (11), 1664–1677.

Dodds, K (2007) "Steve Bell's Eye: Cartoons, Popular Geopolitics and the War on Terror", *Security Dialogue*, 38, 157–177.

Dodds, K (2008) "Hollywood and the Popular Geopolitics of the War on Terror", *Third World Quarterly*, 29 (8), 1621–1637.

Dodds, K, Kuus, M and Sharp, J (eds) (2013) *Companion to Critical Geopolitics*, Farnham: Ashgate.

Domosh, M (2015) "Practising Development at Home: Race, Gender, and the 'Development' of the American South", *Antipode*, 47 (4), 915–941.

Domosh, M (2018) "Race, Biopolitics and Liberal Development from the Jim Crow South to Postwar Africa", *Transactions of the Institute of British Geographers*, 43 (2), 312–324.

Doner, R F, Ritchie, B K and Slater, D (2005) "Systemic Vulnerability and the Origins of Developmental States: Northeast and Southeast Asia in Comparative Perspective", *International Organization*, 59 (2), 327–361.

Dorn, C and Ghodsee, K (2012) "The Cold War Politicization of Literacy: Communism, UNESCO and the World Bank", *Diplomatic History*, 36 (2), 373–398.

Doty, L (1996) "Repetition and Variation: Academic Discourses on North-South Relations", in Doty, L, *Imperial Encounters: The Politics and Representation in North-South Relations*, Minneapolis: University of Minnesota Press, 145–162.

Downs, E (2007) "The Fact and Fiction of Sino-African Energy Relations", *China Security*, 3 (3), 42–68.

Drayton, R (2000) *Nature's Government: Science, Imperial Britain and the 'Improvement' of the World*, New Haven, CT: Yale University Press.

Dresch, J, Lacoste, Y, Pinchemel, P Monbeig, P and Moral, P (1967) "Dossier Géographie et développement", *Annales de Géographie*, 418, 641–767.

Dresch, J (1979) *Un géographe au déclin des empires*, Paris: Maspero.

Driessen, M (2015) "The African Bill: Chinese Struggles with Development Assistance", *Anthropology Today*, 31 (1), 3–7.

Driver, F (2004) "Imagining the Tropics: Views and Visions of the Tropical World", *Singapore Journal of Tropical Geography*, 25 (1), 1–17.

Driver, F and Yeoh, B S A (2000) "Constructing the Tropics: Introduction", *Singapore Journal of Tropical Geography*, 21 (1), 1–5.

Dubash, N K and Morgan, B (eds) (2013) *The Rise of the Regulatory State of the South: Infrastructure and Development in Emerging Economies*, Oxford: Oxford University Press.

Dubey, A K (1990) *Indo-African Relations in the Post-Nehru Era (1965–1985)*, Delhi: Kalinga Publications.

Dubey, A K (2010a) "India-Africa Relations: Historical Connections and Recent Trends", in Dubey, A K (ed) *Trends in Indo-African Relations*, New Delhi: Manas Publications, 17–44.

Dubey, A K (2010b) *Indian Diaspora in Africa: A Comparative Perspective*, New Delhi: MD Publications.

Dubey, A K and Biswas, A (2016) "Introduction: A Long-standing Relationship", in Dubey A K and Biswas, A (eds) *India and Africa's Partnership: A Vision for a New Future*, London: Springer, 1–10.

DuBois, M (1991) "The Governance of the Third World: A Foucauldian Perspective on Power Relations in Development", *Alternatives*, 16, 1–30.

Duchâtel, M, Gowan, R and Rapnouil, M L (2016) "Into Africa: China's Global Security Shift", European Council on Foreign Relations, Policy Brief, 179 (June), 1–16.

Duffield, M (1999) "Globalisation and War Economies: Promoting Order or the Return of History?", *Fletcher Forum of World Affairs*, 23 (2), 21–38.

Duffield, M (2001) *Global Governance and the New Wars: The Merger of Development and Security*, London: Zed Books.

Duffield, M (2002) "Social Reconstruction: The Reuniting of Aid and Politics", *Development and Change*, 33 (5), 1049–1071.

Duffield, M (2006a) "Racism, Migration and Development: The Foundations of Planetary Order", *Progress in Development Studies*, 6 (1), 68–78.

Duffield, M (2006b) "Human Security: Linking Development and Security in an Age of Terror", in Klingebiel, S (ed) *New Interfaces between Security and Development: Changing Concepts and Approaches*, Bonn: Deutsches Institut für Entwicklungspolitik, 11–38.

Duffield, M (2007a) *Development, Security and Unending War: Governing the World of Peoples*, Cambridge: Polity Press.

Duffield, M (2007b) "Development, Territories and People: Consolidating the External Sovereign Frontier", *Alternatives: Global, Local, Political*, 32 (2), 225–246.

Duffield, M (2008) "Global Civil War: The Non-Insured, International Containment and Post-Interventionary Society", *Journal of Refugee Studies*, 21 (2), 145–165.

Duffield, M (2009) "Liberal Interventionism and the Fragile State: Linked by Design?", in Duffield, M and Hewitt, V (eds) *Empire, Development and Colonialism: The Past in the Present*, Woodbridge: James Currey, 116–129.

Duffield, M (2010a) "The Liberal Way of Development and the Development-Security Impasse: Exploring the Global Life-Chance Divide", *Security Dialogue*, 41 (1), 53–76.

Duffield, M (2010b) "Risk-Management and the Fortified Aid Compound: Everyday Life in Post-Interventionary Society", *Journal of Intervention and Statebuilding*, 4 (4), 453–474.

Dulles, J F (1955) "Memorandum of Conversation between Dulles and Sir Roger Makins, 7 April 1955", *FRUS, 1955–57*, II, 236, 83–84.

Duncan, C R (2004a) "Legislating Modernity among the Marginalized", in Duncan, C R (ed) *Civilizing the Margins: Southeast Asian Government Policies for the Development of Minorities*, Ithaca, NY; London: Cornell University Press, 1–23.

Duncan, C R (ed) (2004b) *Civilizing the Margins: Southeast Asian Government Policies for the Development of Minorities*, Ithaca, NY; London: Cornell University Press.

Dunn, K C (2001) "Introduction: Africa and International Relations Theory", in Dunn, K C and Shaw, T M (eds) *Africa's Challenge to International Relations Theory*, Basingstoke: Palgrave, 1–8.

Duvall, R and Varadarajan, L (2007) "Traveling in Paradox: Edward Said and Critical International Relations", *Millennium: Journal of International Studies*, 36 (1), 83–99.

Dwyer, M B (2014) "Micro-Geopolitics: Capitalising Security in Laos's Golden Quadrangle", *Geopolitics*, 19 (2), 377–405.

Easterly, W (2002) "The Cartel of Good Intentions: The Problem of Bureaucracy in Foreign Aid", *Journal of Policy Reform*, 5 (4), 223–250.

Eckert, A (2017) "Social Movements in Africa", in Berger, S and Nehring, H (eds) *The History of Social Movements in Global Perspective: A Survey*, Basingstoke: Palgrave, 211–244.

Economic and Social Council (ECOSOC) (1953) Official Records (16th session).

Ekbladh, D (2002) "'Mr. TVA': Grass-roots Development, David Lillienthal, and the Rise and Fall of the Tennessee Valley Authority", *Diplomatic History*, 26 (3), 335–374.

Ekbladh, D (2010) "Meeting the Challenge from Totalitarianism: The Tennessee Valley Authority as a Global Model for Liberal Development, 1933–1945", *International History Review*, 32 (1), 47–67.

Ekbladh, D (2011) *The Great American Mission: Modernization and the Construction of an American World Order*, Princeton, NJ: Princeton University Press.

Elden, S (2013) *The Birth of Territory*, Chicago, IL: University of Chicago Press.

Elwood, S, Bond, P, Novo, C M and Radcliffe, S (2016) "Learning from Postneoliberalisms", *Progress in Human Geography*, 41 (5), 676–695.

Emel, J, Huber, M and Makene, M (2011) "Extracting Sovereignty: Capital, Territory, and Gold Mining in Tanzania", *Political Geography* 30, 70–79.

Engardio, P (2001) "Smart Bombs, so Why No Smart Aid?", *BusinessWeek*, December 24th, 58.

Engerman, D C (2000) "Modernization from the Other Shore: American Observers and the Costs of Soviet Economic Development", *American Historical Review*, 105 (2), 383–416.

Engerman, D C (ed) (2003) *Staging Growth: Modernization, Development, and the Global Cold War*, Amherst, MA: University of Massachusetts Press.

Engerman, D C (2004) "The Romance of Economic Development and New Histories of the Cold War", *Diplomatic History*, 28 (1), 23–54.

Engerman, D C (2010) "Social Science in the Cold War", *Isis*, 101 (2), 393–400.

Engerman, D C and Unger, C R (2009) "Introduction: Towards a Global History of Modernization", *Diplomatic History*, 33 (3), 375–385.

England, K and Ward, K (2007) "Conclusion: Reflections on Neoliberalizations", in England, K and Ward, K (eds) *Neoliberalization: States, Networks, Peoples*, London: Blackwell, 248–262.

Erickson, A S and Strange, A M (2015) *Six Years at Sea... and Counting: Gulf of Aden Anti-Piracy and China's Maritime Commons Presence*, Washington, DC: Brookings Institution Press.

Escobar, A (1984–85) "Discourse and Power in Development: Michel Foucault and the Relevance of His Work to the Third World", *Alternatives*, 10 (3), 377–400.

Escobar, A (1992) "Imagining a Post-Development Era? Critical Thought, Development and Social Movements", *Social Text*, 31/32, 20–56.

Escobar, A (1995) *Encountering Development: The Making and Unmaking of the Third World*, Princeton, NJ: Princeton University Press.

Escobar, A (2000) "Beyond the Search for a Paradigm? Post-development and Beyond", Development 43 (4), 11–14.

Escobar, A (2004) "Beyond the Third World: Imperial Globality, Global Coloniality and Antiglobalization Social Movements", *Third World Quarterly*, 25 (1), 207–230.

Escobar, A (2007a) "The 'Ontological Turn' in Social Theory. A Commentary on 'Human Geography without Scale' by Sallie Marston, John Paul Jones II and Keith Woodward", *Transactions of the Institute of British Geographers*, 32, 106–111.

Escobar, A (2007b) "Worlds and Knowledges Otherwise. The Latin American Modernity/Coloniality Research Program", *Cultural Studies*, 21 (2), 179–210.

Escobar, A (2008) *Territories of Difference: Place, Movements, Life, Redes*, Durham, NC: Duke University Press.

Escobar, A (2010) "Latin America at the Cross-roads: Alternative Modernizations, Postliberalism or Post-Development?", *Cultural Studies*, 24 (1), 1–65.

Escobar, A (2012) *Encountering Development: The Making and Unmaking of the Third World*, 2nd edition, Princeton, NJ: Princeton University Press.

Essex, J (2008) "Deservedness, Development, and the State: Geographic Categorization in the US Agency for International Development's Foreign Assistance Framework", *Geoforum*, 39 (4), 1625–1636.

Essex, J (2013) *Development, Security, and Aid: Geopolitics and Geoeconomics at the U.S. Agency for International Development*, Athens: University of Georgia Press.

Esson, J, Noxolo, P, Baxter, R, Daley, P and Byron, M (2017) "The 2017 RGS-IBG Chair's Theme: Decolonising Geographical Knowledges, or Reproducing Coloniality?", *Area*, 49 (3), 384–388.

Esteva, G (1985) "Beware of Participation" and "Development: Metaphor, Myth, Threat", *Development: Seeds of Change* 3, 77 and 78–79.

Esteva, G (1992) "Development", in Sachs, W (ed) *The Development Dictionary: A Guide to Knowledge as Power*, London: Zed Books, 6–25.

Esteva, G (1999) "The Zapatistas and People's Power", *Capital and Class*, 23 (2), 153–182.

Esteva, G (2009) "Development", in Sachs, W (ed) *The Development Dictionary: A Guide to Knowledge as Power*, 2nd edition, London: Zed Books.

Esteva, G and Escobar, A (2017) "Post-Development @ 25: On 'Being Stuck' and Moving Forward, Sideways, Backward and Otherwise", *Third World Quarterly*, 38 (12), 2559–2572.

European Commission (EC) (1996) *Communication from the Commission to the Council and the European Parliament on: Linking, relief, rehabilitation and development (LRRD)*, Brussels: Commission of the European Communities, April 30th, https://ec.europa.eu/europeaid/sites/devco/files/communication-naturaldisastersandcrisismanagement-com2001153-20010423_en.pdf (Accessed June 19th 2018).

Evans, P (1995) *Embedded Autonomy: States and Industrial Transformation*, Princeton, NJ: Princeton University Press.

Evans, G (2002) "Between the Global and the Local There Are Regions, Culture Areas and National States: A Review Article", *Journal of Southeast Asian Studies*, 33 (1), 147–162.

Evered, K (2010) "The Truman Doctrine in Greece and Turkey: America's Cold War Fusion of Development and Security", *The Arab World Geographer*, 13 (1), 50–66.

Export–Import Bank of China (2006) *Annual Report 2006*, http://english.eximbank.gov.cn/tm/en-AR/index_634_25691.html (Accessed June 18th 2018).

Falk, P S (1988) *The U.S., U.S.S.R., Cuba, and South Africa in Angola, 1974-88*: Pittsburgh, PA: Pew Charitable Trusts.

Fall, I and Koura, B (2017) "New Details Emerge about Attack that Killed US Soldiers in Niger", *VOA News*, October 21st, https://www.voanews.com/a/new-details-emerge-about-attack-that-killed-us-soldiers-in-niger/4080617.html (Accessed January 14th 2017).

Farish, M (2010) *The Contours of America's Cold War*, Minneapolis: University of Minnesota Press.

Farmer, B H (1983) "British Geographers Overseas, 1933-1983", *Transactions of the Institute of British Geographers*, 8, 70–79.

Farrell, M and Lee, J (2015). "Civil-Military Operations in East Africa: Coordinated Approaches", in Piombo, P (ed) *The US Military in Africa. Enhancing Security and Development?*, Boulder, CO: Lynne Rienner, 103–121.

Featherstone, D (2008) *Resistance, Space and Political Identity: The Making of Counter-global Networks*, Oxford: Wiley-Blackwell.

Featherstone, D (2011) "On Assemblage and Articulation", *Area*, 43 (2), 139–142.

Feichtinger, M (2017) "'A Great Reformatory': Social Planning and Strategic Resettlement in Late Colonial Kenya and Algeria, 1952–63", *Journal of Contemporary History*, 52 (1), 45–72.

Fejerskov, A M, Lundsgaarde, E and Cold-Ravnkilde, S M (2016) "Uncovering the Dynamics of Interaction in Development Cooperation: A Review of the 'New Actors in Development' Research Agenda" (No. 2016: 01), DIIS Working Paper.

Ferchen, M (2013) "Whose China Model Is It Anyway? The Contentious Search for Consensus", *Review of International Political Economy*, 20 (2), 390–420.

Ferguson, J (1990) *The AntiPolitics Machine: Development, Depoliticization and Bureaucratic Power in Lesotho*, Cambridge: Cambridge University Press.

Ferguson, J (1999) *Expectations of Modernity: Myths and Meanings of Urban Life on the Zambian Copperbelt*, Berkeley: University of California Press.

Ferguson, J (2005) "Seeing Like an Oil Company: Space, Security, and Global Capital in Neoliberal Africa", *American Anthropologist*, 107 (3), 377–382.

Ferguson, J (2006) *Global Shadows: Africa in the Neoliberal World Order*, Durham, NC: Duke University Press.

Ferguson, J and Gupta, A (eds) (1997) *Culture, Power, Place: Explorations in Critical Anthropology*, Durham, NC: Duke University Press.

Ferguson, J and Gupta, A (2002) "Spatializing States: Toward an Ethnography of Neoliberal Governmentality", *American Ethnologist*, 29 (4), 981–1002.

Ferretti, F (2017) "Tropicality, the Unruly Atlantic and Social Utopias: The French Explorer Henri Coudreau (1859 – 1899)", *Singapore Journal of Tropical Geography*, 38 (3), 332–349.

Ferretti, F (2018) "Geographies of Internationalism: Radical Development and Critical Geopolitics from the Northeast of Brazil", *Political Geography*, 63, 10–19.

Ferretti, F and Pedrosa, B V (2018) "Inventing Critical Development: A Brazilian Geographer and his Northern Networks", *Transactions of the Institute of British Geographers*, 43 (4), 703–717.

Fikes, K and Lemon, A (2002) "African Presence in Former Soviet Spaces", *Annual Review of Anthropology*, 31, 497–524.

Firsing, S and Williams, O (2013) "Increased Chinese and American Defence Involvement in Africa", *Defense and Security Analysis*, 29 (2), 89–103.

Fisher, C T (2006) "The Illusion of Progress", *Pacific Historical Review*, 75 (1), 25–51.

Florentin, B (2016) "Between Policies and Life: The Politics of Buen Vivir in Contemporary Ecuador", *CWiPP working paper No.5*, Centre for Wellbeing in Public Policy, University of Sheffield.

Fluri, J (2009) "'Foreign Passports Only': Geographies of (Post)Conflict work in Kabul, Afghanistan", *Annals of the Association of American Geographers*, 99 (5), 986–994.

Foucault, M (2003 [1975–76]) *Society Must Be Defended: Lectures at the College de France 1975–1976*, London: Alan Lane, The Penguin Press.

Foucault, M (1998 [1976]) *The Will to Knowledge: The History of Sexuality Volume 1*, London: Penguin Books.

Foucault, M (2007 [1977–78]) *Security, Territory, Population: Lectures at the College de France 1977–1978*, Basingstoke: Palgrave Macmillan.

Foucault, M (1979) *Discipline and Punish: The Birth of the Prison*, London: Penguin.

Foucault, M (1980a) "The Confession of the Flesh", in Gordon, C (ed), *Power/Knowledge*, New York: Pantheon, 194–228.

Foucault, M (1980b) "Power and Strategies", in Gordon, C (ed), *Power/Knowledge*, New York: Pantheon, 134–145.

Foucault, M (1980c) "Truth and Power", in Gordon, C (ed), *Power/Knowledge*, New York: Pantheon, 109–133.

Foucault, M (1981) *The History of Sexuality: An Introduction*, London: Penguin.

Foucault, M (1982) "The Subject and Power", in Dreyfus, H L and Rabinow, P (eds) *Michel Foucault: Beyond Structuralism and Hermeneutics*, Chicago, IL: University of Chicago Press, 201–228.

Foucault, M (1991a) "Governmentality", in Burchell, G, Gordon, C and Miller, P (eds) *The Foucault Effect: Studies in Governmentality*, Chicago, IL: University of Chicago Press, 87–104.

Foucault, M (1991b) "Questions of Method", in Burchell, G, Gordon, C and Miller, P (eds) *The Foucault Effect: Studies in Governmentality*, Chicago, IL: University of Chicago Press, 73–86.

Fox, J, Fujita, Y, Ngidang, D, Peluso, N, Potter, L, Sakuntaladewi, N, Sturgeon, J and Thomas, D (2009) "Policies, Political Economy, and Swidden in Southeast Asia", *Human Ecology*, 37 (3), 305–322.

Frank, A G (1966) "The Development of Underdevelopment", *Monthly Review*, 18 (4), 17–31.

Frank, A G (1967) *Capitalism and Underdevelopment in Latin America*, New York: Monthly Review Press.

Frank, A G (1997) "The Cold War and Me", *Bulletin of Concerned Asian Scholars*, 29 (4), http://www.rrojasdatabank.info/agfrank/syfrank.htm (Accessed January 8th 2017).

Frank, M (2014) "Cuba Says U.S Created Other 'Cuban Twitter' Projects", *Reuters*, http://www.reuters.com/article/2014/04/07/us-cuba-usa-idUSBREA350MO20140407 (Accessed August 4th 2014).

Freeman, J (2014) "Raising the Flag over Rio de Janeiro's Favelas: Citizenship and Social Control in the Olympic City", *Journal of Latin American Geography*, 13 (1), 7–38.

Friedman, E (2009) "How Economic Superpower China Could Transform Africa", *Journal of Chinese Political Science*, 14, 1–20.

Friedman, J (2015) *Shadow Cold War: The Sino-Soviet Competition for the Third World*, Chapel Hill, NC: University of North Carolina Press.

Friis, S (2014) "Review of Amer, Ramses; Swain, Ashok; Öjendal, Joakim, The Security-Development Nexus: Peace, Conflict and Development and Spear, Joanna; Williams, Paul D., Security and Development in Global Politics: A Critical Comparison and Tschirgi, Neclâ; Lund, Michael S.;

Mancini, Francesco, Security and Development: Searching for Critical Connections", H-Soz-u-Kult, *H-Net Reviews*, October, http://www.h-net.org/reviews/showrev.php?id=42577 (Accessed January 1st 2017).

Frontera (2016) "How Many Missions Does the US Military Now Carry out in Africa Every Year?", https://fronteranews.com/news/africa/many-military-missions-us-now-carry-africa-every-year/ (Accessed August 13th 2017).

Fujikura, T (2013) *Discourses of Awareness: Social Movements and the Practices of Freedom in Nepal*, Kathmandu: Martin Chautari.

Fulquet, G and Pelfini, A (2015) "Brazil as a New International Cooperation Actor in Sub-Saharan Africa: Biofuels at the Crossroads between Sustainable Development and Natural Resource Exploitation", *Energy Research and Social Science*, 5, 120–129.

Fulquet, G (2015) "A New Cartography of International Cooperation: Emerging Powers in Sub-Saharan Africa – The Case of Biofuels Promotion by Brazil in Senegal", in Delgado-Ramos, G C (ed) *Inequality and Climate Change: Perspectives from the South*, Dakar: CODESRIA, 83–102.

Galbraith, J K (1979) *The Nature of Mass Poverty*, Cambridge, MA: Harvard University Press.

Gallagher, J (2011) "Ruthless Player or Development Partner? Britain's Ambiguous Reaction to China in Africa", *Review of International Studies*, 37 (5), 2293–2310.

Gardner, L (2009) "Walt Whitman Rostow: Hawkeyed Optimist", in Nelson, A K (ed) *The Policy Makers: Shaping American Foreign Policy from 1947 to the Present*, Lanham, MD: Rowman and Littlefield, 59–82.

Garmany, J (2009) "The Embodied State: Governmentality in a Brazilian Favela", *Social and Cultural Geography*, 10 (7), 721–739.

Gay, P (1973) *The Enlightenment: An Interpretation: Volume 2: The Science of Freedom*, London: Wildwood House.

Geertz, C (2010) "What Was the Third Revolution?", in Geertz, C and Inglis, F (eds) *Life among the Anthros and Other Essays*, Princeton, NJ: Princeton University Press, 236–252.

Geidel, M (2015) *Peace Corps Fantasies: How Development Shaped the Global Sixties*, Minneapolis: University of Minnesota Press.

George, E (1999) *Moscow's Gurkhas or the Tail Wagging the Dog? Cuban Internationalism in Angola, 1965-1991*, Bristol: Department of Hispanic, Portuguese and Latin American Studies, University of Bristol.

George, E (2005) *The Cuban Intervention in Angola, 1965-1991: From Che Guevara to Cuito Cuanavale*, London: Frank Cass.

Gibson-Graham, J K (2004) "Area Studies after post-Structuralism", *Environment and Planning A*, 36, 405–419.

Gibson-Graham, J K (2005) "Surplus Possibilities: Postdevelopment and Community Economies", *Singapore Journal of Tropical Geography*, 26 (1), 4–26.

Gibson-Graham, J K (2006) *A Postcapitalist Politics*. Minneapolis: University of Minnesota Press.

Gibson-Graham, J K (2016) "'After' Area Studies? Place-based Knowledge for Our Time", *Environment and Planning D: Society and Space*, 34 (5), 799–806.

Gidwani, V (2002). "The Unbearable Modernity of 'Development'? Canal Irrigation and Development Planning in Western India", *Progress in Planning*, 58 (1), 1–80.

Gilbert, E (2015) "Money as a 'Weapons System' and the Entrepreneurial Way of War", *Critical Military Studies*, 1 (3), 202–219.

Gill, B and Reilly, J (2007) "The Tenuous Hold of China Inc in Africa", *The Washington Quarterly*, (Summer), 37–52.

Gills, B K (2011) "Going South: Capitalist Crisis, Systemic Crisis, Civilisational Crisis", *Third World Quarterly*, 31 (2), 169–184.

Gills, B (2017) "The Future of Development from Global Crises to Global Convergence", *Forum for Development Studies*, 44 (1), 155–161.

Gilman, N (2003) *Mandarins of the Future: Modernization Theory in Cold War America*, Baltimore, MD: Johns Hopkins University Press.

Gilman, N (2016) "The Twin Insurgency: Plutocrats and Criminals Challenge the Westphalian State", in Matfess, H and Miklaucic, M (eds) *Beyond Convergence: World Without Order*, Washington, DC: Centre for Complex Operations, National Defence University, 47–60.

Glassman, J (2018) "Geopolitical Economies of Development and Democratization in East Asia: Themes, Concepts, and Geographies", *Environment and Planning A: Economy and Space*, 50 (2), 407–415.

Glassman, J and Choi, Y-J (2014) "The Chaebol and the US Military–Industrial Complex: Cold War Geopolitical Economy and South Korean Industrialization", *Environment and Planning A*, 46, 1160–1180.

Gleijeses, P (2002) *Conflicting Missions: Havana, Washington, and Africa, 1959-1976*, Chapel Hill, NC: University of North Carolina Press.

Gleijeses, P (2006) "Moscow's Proxy? Cuba and Africa 1975-1988", *Journal of Cold War Studies*, 8 (2), 3–51.

Gleijeses, P (2013) *Piero Gleijeses' International History of the Cold War in Southern Africa*, Omnibus E-Book, Chapel Hill, NC: University of North Carolina Press.

Goldberg, E (2015) "Disgraced 'Three Cups Of Tea' Author To Retire From Charity He Founded", *Huffington Post*, November 20th, https://www.huffingtonpost.com/entry/disgraced-three-cups-of-tea-author-to-retire-from-charity-he-founded_us_564e5d07e4b0d4093a571cee (Accessed January 15th 2018).

Goldman, M I (1967) *Soviet Foreign Aid*, New York: Praeger.

González, R J (2009) *American Counterinsurgency: Human Science and the Human Terrain*, Chicago, IL: Prickly Paradigm Press.

González, R J (2015) "The Rise and Fall of the Human Terrain System", *Counterpunch*, June 29th, https://www.counterpunch.org/2015/06/29/the-rise-and-fall-of-the-human-terrain-system/ (Accessed August 14th 2017).

Gonzalez-Vicente, R (2011) "The Internationalization of the Chinese State", *Political Geography*, 30 (7), 402–411.

Goodman, D S G and Segal, G (eds) (1994) *China Deconstructs: Politics, Trade and Regionalism*, London: Routledge.

Goodman, M K (2010) "The Mirror of Consumption: Celebritization, Developmental Consumption and the Shifting Cultural Politics of Fair Trade", *Geoforum*, 41 (1), 104–116.

Goss, J and Wesley-Smith, T (2010) "Introduction: Remaking Area Studies", in Wesley-Smith, T and Goss, J (eds) *Remaking Area Studies: Teaching and Learning across Asia and the Pacific*, Manoa: University of Hawai'i Press, ix–xxvii.

Goswami, M (2012) "Imaginary Futures and Colonial Internationalisms", *The American Historical Review*, 117 (5), 1461–1485.

Gourou, P (1936) *Les paysans du delta tonkinois, étude de géographie humaine* [The Peasants of the Tonkin Delta: A Study in Human Geography], Paris: L'Ecole française d'Extrême-Orient, Editions d'Art et d'Histoire.

Gourou, P (1947) *Les pays tropicaux: Principes d'une géographie humaine et économique* [The Tropical World: Principles of Human and Economic Geography], Paris: Presses Universitaires de France.

Gourou, P (1953) *The Tropical World*, trans. Laborde, E D, London: Longmans Green and Co.

Gourou, P (1961) "Review of *Oriental Despotism: A Comparative Study of Total Power*", *Annals of the Association of American Geographers*, 51 (4), 401–402.

Gourou, P (1966) *The Tropical World*, trans. Beaver, S H and Laborde, E D, 4th edition, London: Longmans Green and Co.

Graham, S (2009) "Cities as Battlespace: The New Military Urbanism", *City*, 13 (4), 383–402.

Grant, R and Agnew, J (1996) "Representing Africa: The Geography of Africa in World Trade, 1960-1992", *Annals of the Association of American Geographers*, 86 (4), 729–744.

Gray, C S (2015) *Strategic Studies and Public Policy: The American Experience*, Lexington: University Press of Kentucky.

Gray, K and Gills, B (2016). "South–South Cooperation and the Rise of the Global South", *Third World Quarterly*, 37 (4), 557–574.

Gray, K and Murphy, C N (2013) "Introduction: Rising Powers and the Future of Global Governance", *Third World Quarterly*, 34 (2), 183–193.

Greenburg, J (2018) "'Going back to History': Haiti and US Military Humanitarian Knowledge Production", *Critical Military Studies*, 4 (2), 121–139.

Gregory, D (2011) "The Everywhere War", *The Geographical Journal*, 177 (3), 238–250.

Grosfoguel, R (2010) "The Epistemic Decolonial Turn: Beyond Political-Economy Paradigms", in Mignolo, W and Escobar, A (eds) *Globalization and the Decolonial Option*, London: Routledge, 65–77.

Grove, R (1995) *Green Imperialism: Colonial Expansion, Tropical Island Edens and the Origins of Environmentalism, 1600-1860*, Cambridge: Cambridge University Press.

Grovogui, S (1996) *Sovereigns, Quasi Sovereigns, and Africans: Race and Self-determination in International Law*, Minneapolis: University of Minnesota Press.

Grovogui, S N (2006) *Beyond Eurocentrism and Anarchy: Memories of International Order and Institutions*, London: Palgrave.

Grovogui, S N (2011) "A Revolution Nonetheless: The Global South in International Relations", *The Global South*, 5 (1), 175–190.

Gruffydd Jones, B (2005) "Africa and the Poverty of International Relations", *Third World Quarterly*, 26 (6), 987–1003.

Gruffydd Jones, B (ed) (2006) *Decolonizing International Relations*, Lanham, MD: Rowman and Littlefield.

Gruffydd-Jones, B (2014) "'Good Governance' and 'State Failure': The Pseudo-science of Statesmen in Our Times", in Anievas, A, Manchanda, N and Shilliam, R (eds) *Race and Racism in International Relations: Confronting the Global Colour Line*, London: Routledge, 62–80.

Gu, J (2009) "China's Private Enterprises in Africa and the Implications for African Development", *European Journal of Development Research*, 21 (4), 570–587.

Gu, J, Zhang, C, Vaz, A and Mukwereza, L (2016) "Chinese State Capitalism? Rethinking the Role of the State and Business in Chinese Development Cooperation in Africa", *World Development*, 81, 24–34.

Guan-Fu, G (1983) "Soviet Aid to the Third World, an Analysis of Its Strategy", *Soviet Studies*, 35 (1), 71–89.

The Guardian (2011) "US Foreign Aid and the 2012 Budget: Where Will the Axe Fall?", November 7th, http://www.guardian.co.uk/global-development/datablog/2011/nov/07/us-foreign-aid-budget-cuts

The Guardian (2011) "Obama Tells US Officials to Use Overseas Aid to Promote Gay Rights", December 6th, http://www.guardian.co.uk/world/2011/dec/06/us-overseas-aid-human-rights

The Guardian (2014) "US Secretly Created 'Cuban Twitter' to Stir Unrest and Undermine Government", April 3rd, https://www.theguardian.com/world/2014/apr/03/us-cuban-twitter-zunzuneo-stir-unrest (Accessed November 6th 2018).

The Guardian (2016) "Will Trump Honour Pledge to 'Stop Sending Aid to Countries that Hate Us'?", November 13th, https://www.theguardian.com/global-development/2016/nov/13/will-trump-presidency-honour-pledge-stop-sending-foreign-aid-to-countries-that-hate-us-usaid (Accessed August 24th 2017).

The Guardian (2018) "Trump Pans Immigration Proposal as Bringing People from 'Shithole Countries'", January 12th, https://www.theguardian.com/us-news/2018/jan/11/trump-pans-immigration-proposal-as-bringing-people-from-shithole-countries (Accessed July 19th 2018).

The Guardian (2018) "China Rejects Claim It Bugged Headquarters It Built for African Union", January 30th, https://www.theguardian.com/world/2018/jan/30/china-african-union-headquarters-bugging-spying (Accessed June 25th 2018).

Gudynas, E (2011) "Buen Vivir: Today's Tomorrow", *Development*, 54 (4), 441–447.

Guimarães, A (1998) *The Origins of Angolan Civil War*, London: Palgrave Macmillan.

Gupta, A (1995) "Blurred Boundaries: The Discourse of Corruption, the Culture of Politics, and the Imagined State", *American Ethnologist*, 22, 375–402.

Gupta, A (1998) *Postcolonial Developments: Agriculture in the Making of Modern India*. Durham, NC: Duke University Press.

Habermas, J (1987) *The Philosophical Discourse of Modernity*, trans. Lawrence, F G, Cambridge, MA: MIT Press.

Hailey, W M (1957) *An African Survey: A Study of Problems Arising in Africa South of the Sahara*, London: Oxford University Press for the Royal Institute of International Affairs.

Hall, C (2002) *Civilising Subjects: Metropole and Colony in the English Imagination 1830–1867*, Cambridge: Polity Press.

Halper, S (2010) "China Model: This House Believes China Offers a Better Development Model than the West", *The Economist*, August 4th, http://www.economist.com/debate/days/view/553/ (Accessed August 10th 2010).

Halvorsen, S (2017) "Spatial Dialectics and the Geography of Social Movements: The Case of Occupy London", *Transactions of the Institute of British Geographers*, 42 (3), 445–457.

Halvorsen, S (2018) "Cartographies of Epistemic Expropriation: Critical Reflections on Learning from the South", *Geoforum*, 95, 11–20.

Hameiri, S and Jones, L (2016) "Rising Powers and State Transformation: The Case of China", *European Journal of International Relations*, 22 (1), 72–98.

Hammer, J (2016) "Hunting Boko Haram: The US Extends Its Drone War into Africa with Secretive Base", *The Intercept*, February 25th, https://theintercept.com/2016/02/25/us-extends-drone-war-deeper-into-africa-with-secretive-base/ (Accessed January 15th 2018).

Hampson, N (1968) *The Enlightenment*, London: Penguin.

Han, X (2018) "Money, Markets and Hydropower: Chinese Dam Construction in Africa", PhD thesis, School of Geography, University of Melbourne, https://minerva-access.unimelb.edu.au/bitstream/handle/11343/208834/Thesis_hanx_2018Mar.pdf?sequence=1&isAllowed=y (Accessed July 19th 2018).

Hanieh, A (2011) "Egypt's 'Orderly Transition'? International Aid and the Rush to Structural Adjustment", *Jadaliyya*, May 29th, 2011, http://www.jadaliyya.com/pages/index/1711/egypts-%E2%80%98orderly-transition%E2%80%99-international-aid-and- (Accessed November 27th 2017).

Hanieh, A (2013) *Lineages of Revolt: Issues of Contemporary Capitalism in the Middle East*, Chicago, IL: Haymarket Books.

Harding, H (1994) "China's Co-operative Behaviour", in Robinson, T W and Shambaugh, D (eds) *Chinese Foreign Policy: Theory and Practice*, London: Clarendon, 375–400.

Harig, C and Kenkel, K M (2017) "Are Rising Powers Consistent or Ambiguous Foreign Policy Actors? Brazil, Humanitarian Intervention and the 'Graduation Dilemma'", *International Affairs*, 93 (3), 625–641.

Harker, C (2011) "Geopolitics and Family in Palestine", *Geoforum*, 42 (3), 306–315.

Harman, S and Brown, W (2013) "In from the Margins? The Changing Place of Africa in International Relations", *International Affairs*, 89 (1), 69–87.

Harootunian, H D (2002) "Postcoloniality's Unconcious/Area Studies' Desire", in Miyoshi, M and Harootunian, H D (eds) *Learning Places: The Afterlife of Area Studies*, Durham, NC and London: Duke University Press, 150–174.

Harris, L M (2012) "State as Socionatural Effect: Variable and Emergent Geographies of the State in Southeastern Turkey", *Comparative Studies of South Asia, Africa and the Middle East*, 32 (1), 25–39.

Hart, G (2001) "Development Critiques in the 1990s: Culs-de-sac and Promising Paths", *Progress in Human Geography*, 25 (4), 649–658.

Hart, G (2002) *Disabling Globalization: Places of Power in Post-Apartheid South Africa*, Berkeley, CA: University of California Press.

Hart, G (2010) "D/developments after the Meltdown", *Antipode*, 41, 117–141.

Hart, G (2014) *Rethinking the South African Crisis: Nationalism, Populism, Hegemony*, Athens: University of Georgia Press.

Hart, G (2018) "Relational Comparison Revisited: Marxist Postcolonial Geographies in Practice", *Progress in Human Geography*, 42 (3), 371–394.

Hart-Landsberg, M and Burkett, P (2005) *China and Socialism: Market Reforms and Class Struggles*, New York: Monthly Review Press.

Hartman, A (2002) "'The Red Template': US Policy in Soviet-Occupied Afghanistan", *Third World Quarterly*, 23 (3), 467–489.

Harvey, D (2014) *Seventeen Contradictions and the End of Capital*, London: Profile Books.

Haskell, T L (1985a) "Capitalism and the Origins of the Humanitarian Sensibility, Part 1", *American Historical Review*, 90 (2), 339–361.

Haskell, T L (1985b) "Capitalism and the Origins of the Humanitarian Sensibility, Part 2", *American Historical Review*, 90 (3), 547–566.

Hayter, R, Barnes, T J and Bradshaw, M J (2003) "Relocating Resource Peripheries to the Core of Economic Geography's Theorizing: Rationale and Agenda", *Area*, 35 (1), 15–23.

Hayter, T (1971) *Aid as Imperialism*, Harmondsworth: Penguin Books.

Hayter, T (1985) *Aid: Rhetoric and Reality*, London: Pluto.

Hayter, T (2005) "Secret Diplomacy Uncovered: Research on the World Bank in the 1960s and 1980s", in Kothari, U (ed) *Radical History of Development Studies: Individuals, Institutions and Ideologies*, London: Zed Books, 88–108.

Heileg, G K (2006) "'Many' Chinas? The Economic Diversity of China's Provinces", *Population and Development Review*, 32, 147–161.

Hellman, J (1986) *American Myth and the Legacy of Vietnam*, New York: Columbia University Press.

Hettne, B (1995) *Development Theory and the Three Worlds*, London: Longman.

Hettne, B (2010) "Development and Security: Origins and Future", *Security Dialogue*, 41 (1), 31–52.

Hettne, B and Odén, B (2002) *Global Governance in the Twenty-First Century: Alternative Perspectives on World Order*, Stockholm: Almkvist and Wiksell International.

Hickel, J (2017) *The Divide: A Brief Guide to Global Inequality and its Solutions*, London: Windmill Books.

Hill, J (2006) "Beyond the Other? A Postcolonial Critique of the Failed State Thesis", *African Identities*, 3 (2), 139–154.

Hills, S (2006) "Trojan Horses? USAID, Counter-Terrorism and Africa's Police", *Third World Quarterly*, 27 (4), 629–643.

The Hindu (2017) "Bid to Boost India-Africa Trade Ties", May 21st, http://www.thehindu.com/business/Economy/bid-to-boost-india-africa-trade-ties/article18519609.ece (Accessed October 10th 2017).

Hirono, M and Suzuki, S (2014) "Why Do We Need 'Myth-Busting' in the Study of Sino–African Relations?", *Journal of Contemporary China*, 23 (87), 443–461.

Hirsch, P (1989) "The State in the Village: Interpreting Rural Development in Thailand", *Development and Change*, 20 (1), 35–56.

Hirsch, P (2009) "Revisiting Frontiers as Transitional Spaces in Thailand", *The Geographical Journal*, 175 (2), 124–132.

Hirschman, D (1989) "International Relations Studies: Some Questions from the Third World", in Frost, M, Vale, P and Weiner, D (eds) (1989) *International Relations – A Debate on Methodology*, Pretoria: Human Sciences Research Council (HSRC).

Hobson, J and Sharman, J C (2005) "The Enduring Place of Hierarchy in World Politics: Tracing the Social Logics of Hierarchy and Political Change", *European Journal of International Relations*, 11 (1), 63–98.

Hodder, J, Legg, S and Heffernan, M (2015) "Introduction: Historical Geographies of Internationalism, 1900-1950", Political Geography, 49, 1–6.

Hodge, N (2011) *Armed Humanitarians: The Rise of the Nation Builders*, New York: Bloomsbury.

Hoffman, B (2006) "Insurgency and Counterinsurgency in Iraq", *Studies in Conflict and Terrorism*, 29 (2), 103–121.

Hoffmann, S (1977) "An American Social Science: International Relations", *Daedalus*, 3, 41–60.

Holslag, J (2009) "China's New Security Strategy for Africa", *Parameters*, 39 (2), 23–37.

Holsti, K J (2011) "Exceptionalism in American Foreign Policy: Is It Exceptional?", *European Journal of International Relations*, 17 (3), 381–404.

Holston, J (2009) "Insurgent Citizenship in an Era of Global Urban Peripheries", *City and Society*, 21 (2), 245–267.

Hook, S W (1995) *National Interest and Foreign Aid*, Boulder, CO: Lynne Rienner.

Hornby, L (2017) "Communist Party Increasingly Asserts Control over China Inc.", *Financial Times*, October 3rd, https://www.ft.com/content/29ee1750-a42a-11e7-9e4f-7f5e6a7c98a2 (Accessed June 19th 2018).

Horner, R (2016) "A New Economic Geography of Trade and Development? Governing South-South Trade, Value Chains and Production Networks", *Territory, Politics, Governance*, 4 (4), 400–420.

Horner, R and Nadvi, K (2018) "Global Value Chains and the Rise of the Global South: Unpacking Twenty-First Century Polycentric Trade", *Global Networks*, 18 (2), 207–237.

Hornsby, R (2016) "The post-Stalin Komsomol and the Soviet Fight for Third World Youth", *Cold War History*, 16 (1), 83–100.

Horowitz, I L and Suchlicki, J (eds) (2003) *Cuban Communism 1959-2003*, New Brunswick, NJ: Transaction Publishers.

Hosmer, S and Wolfe, T (1983) *Soviet Policy and Practice toward Third World Conflicts*, Lexington, MA: Lexington Books.

Hosmer, S T and Crane, S (eds) (2006) *Counterinsurgency: A Symposium, April 16–20,1962*, Santa Monica, CA: RAND.

Huber, M (2015) "Theorising Energy Geographies", *Geography Compass*, 9 (6), 327–338.

Hughes, J (2013) "State Violence in the Origins of Nationalism: British Counterinsurgency and the Rebirth of Irish Nationalism, 1969-1972", in Hall, J A and Malesevic, S (eds) *Nationalism and War*, Cambridge: Cambridge University Press, 97–123.

Hulme, D (2015) "The SDGs Mark the End of Development as Poverty Reduction", Manchester policy blogs, September 30th, http://blog.policy.manchester.ac.uk/posts/2015/09/the-sdgs-mark-the-end-of-development-as-poverty-reduction/ (Accessed November 5th 2018).

Hulme, D and Edwards, M (eds) (1997) *NGOs, States and Donors: Too Close for Comfort?*, Houndmills, Basingstoke; London: Macmillan.

Human Security Centre (HSC) (2005) *Human Security Report 2005: War and Peace in the 21st Century*, Oxford: Oxford University Press.

Humphrey, J (2007) "Forty Years of Development Research: Transformations and Reformations", *IDS Bulletin*, 38, 14–19.

Hunt, M H (1987) *Ideology and U.S. Foreign Policy*, Princeton, NJ: Princeton University Press.

Hunt, M H (2007) *The American Ascendancy*, Chapel Hill: University of North Carolina Press.

Hunt, M H (2009) *Ideology and US Foreign Policy*, New Haven, CT: Yale University Press.

Hwang, J, Brautigam, D and Eom, J (2016) "How Chinese Money is Transforming Africa: It's Not What You Think", *CARI Policy Brief 11*, April, John Hopkins University.

Hyndman, J (2004) "Mind the Gap: Bridging Feminist and Political Geography through Geopolitics", *Political Geography*, 23 (3), 307–322.

Hyndman, J (2007) "The Securitization of Fear in post-Tsunami Sri Lanka", *Annals of the Association of American Geographers*, 97 (2), 361–372.

Hyndman, J (2009) "Acts of Aid: Neoliberalism in a War Zone", *Antipode*, 41 (5), 867–889.

Hyndman, J (2010) "Forging the political", in Jones, L and Sage, D "New Directions in Critical Geopolitics: An Introduction. With Contributions of: Gearoid Ó Tuathail, Jennifer Hyndman, Fraser Mac-Donald, Emily Gilbert and Virginie Mamadouh", *GeoJournal*, 75 (4), 315–325.

Inayatullah, N and Blaney, D L (2004) *International Relations and the Problem of Difference*, London: Routledge.

Institute of Applied Economic Research (IPEA) and Brazilian Agency for Cooperation (ABC) (2013) *Brazilian Cooperation for International Development 2010*, Brasilia: IPEA/ABC.

Institute of Security Studies (ISS) (2016) *Cooperation between African States and the Democratic People's Republic of Korea*, DuPre, A, Kasprzyk, N and Stott, N, https://issafrica.s3.amazonaws.com/site/uploads/research-report-dprk.pdf (Accessed April 17th 2017)

Integrated Regional Information Networks (IRIN) (2010) "Afghanistan: Money Well Spent?", http://www.irinnews.org/report.aspx?reportid=88502 (Accessed January 10th 2018).

International Institute for Strategic Studies (IISS) (2017) *The Military Balance*, http://www.iiss.org/en/publications/military%20balance

Ireland, P and McKinnon, K (2013) "Strategic Localism for an Uncertain World: A Postdevelopment Approach to Climate Change Adaptation", *Geoforum*, 47, 158–166.

IRIN (2011) "Who's Who among the 'New' Aid Donors?", *The Guardian*, https://www.theguardian.com/global-development/2011/oct/25/whos-who-new-aid-donors (Accessed November 2nd 2018).

IRIN(2014) "NGOs against MONUSCO Drones for Humanitarian Work", http://www.irinnews.org/report/100391/ngos-against-monusco-drones-for-humanitarian-work (Accessed June 18th 2018).

Ismael, T (1971) "The People's Republic of China and Africa", *The Journal of Modern African Studies*, 9 (4), 507–529.

Ismail, Y (2018) "On US-Africa Relations: Interview with David Shinn", *Dehai Eritrea online*, http://www.dehai.org/dehai/dehai/256713 (Accessed June 4th 2018).

Jabri, V (2007) "Michel Foucault's Analytics of War: The Social, the International, and the Racial", *International Political Sociology*, 1 (1), 67–81.

Jackson, R H and Rosberg, C G (1982) "Why Africa's Weak States Persist: The Empirical and the Juridical in Statehood", *World Politics*, 35 (1), 1–24.

Jackson, R H and Rosberg, C G (1986) "Sovereignty and Underdevelopment: Juridical Statehood in the African Crisis", *Journal of Modern African Studies*, 24 (1), 1–31.

Jackson, R H (1993) *Quasi-states: Sovereignty, International Relations and the Third World*, Cambridge: Cambridge University Press.

Jackson, S F (1995) "China's Third World Foreign Policy: The Case of Angola and Mozambique, 1961-93", *China Quarterly*, 142, 388–422.

Jazeel, T (2014) "Subaltern Geographies: Geographical Knowledge and Postcolonial Strategy", *Singapore Journal of Tropical Geography*, 35 (1), 88–103.

Jazeel, T (2016) "Between Area and Discipline: Progress, Knowledge Production and the Geographies of Geography", *Progress in Human Geography*, 40 (5), 649–667.

Jessop, B (2009) "Avoiding Traps, Rescaling States, Governing Europe", in Keil, R and Mahon, R (eds) *Leivathan Undone? Towards a Political Economy of Scale*, Vancouver: UBC Press, 87–104.

Jessop, B, Brenner, N and Jones, M (2008) "Theorizing Sociospatial Relations", *Environment and Planning D: Society and Space*, 26 (3), 389–401.

Jessop, B and Sum, N-L (2018) "Geopolitics: Putting Geopolitics in its Place in Cultural Political Economy", *Environment and Planning A*, 50 (2), 474–478.

Johnson, C (1999) "The Developmental State: Odyssey of a Concept", in Woo-Cumings, M (ed) *The Developmental State*, Ithaca, NY: Cornell University Press, 32–60.

Johnston, A I (2003) "Is China a Status Quo Power?", *International Security*, 27 (4), 5–56.

Jolly, R (2004) *UN Contributions to Development Thinking and Practice*, Bloomington, IN: Indiana University Press.

Jones, L and Sage, D (2010) "New Directions in Critical Geopolitics: An Introduction. With Contributions of: Gearoid Ó Tuathail, Jennifer Hyndman, Fraser MacDonald, Emily Gilbert and Virginie Mamadouh", *GeoJournal*, 75 (4), 315–325.

Jones, M (1965) *Two Ears of Corn: Oxfam in Action*, London: Hodder and Stoughton.

Jones, M (2005) "A 'Segregated' Asia?: Race, the Bandung Conference, and Pan-Asianist Fears in American Thought and Policy, 1954–1955", *Diplomatic History*, 29 (5), 841–868.

Jones, S G and Johnston, P B (2013) "The Future of Insurgency", *Studies in Conflict and Terrorism*, 36 (1), 1–25.

Joseph, J (2010) "The Limits of Governmentality: Social Theory and the International", *European Journal of International Relations*, 16 (2), 223–246.

Kahler, M (2013) "Rising Powers and Global Governance: Negotiating Change in a Resilient Status Quo", *International Affairs*, 89 (3), 711–729.

Kalinowski, T and Cho, H (2012) "Korea's Search for a Global Role between Hard Economic Interests and Soft Power", *European Journal of Development Research*, 24 (2), 242–260.

Kanet, R E (1967) "Soviet Economic Policy in Sub-Saharan Africa", *Canadian Slavic Studies*, I, 566–586.

Kanet, R E (1974) "Soviet Attitudes toward Developing Nations since Stalin", in Kanet, R E (ed) *The Soviet Union and the Developing Nations*, Baltimore, MD: Johns Hopkins University Press.

Kanet, R E (2006) "The Superpower Quest for Empire: The Cold War and Soviet Support for 'Wars of National Liberation'", *Cold War History*, 6 (3), 331–352.

Kanet, R E (2010) "Four Decades of Soviet Economic Assistance: Superpower Economic Competition in the Developing World", ACDIS Occasional paper, July 2010, http://acdis.illinois.edu/assets/docs/551/FourDecadesofSovietEconomicAssistanceSuperpowerEconomicCompetitionintheDevelopingWorld.pdf (Accessed April 23rd 2017).

Kanet, R E, Miner, D N and Resler, T J (1992) (eds) *Soviet Foreign Policy in Transition*, Cambridge: Cambridge University Press.

Kapoor, I (2012) *Celebrity Humanitarianism: The Ideology of Global Charity*, London: Routledge.

Kapoor, I (2014a) "Psychoanalysis and Development: An Introduction", *Third World Quarterly*, 35 (7), 1117–1119.

Kapoor, I (2014b) "Psychoanalysis and Development: Contributions, Examples, Limits", *Third World Quarterly*, 35 (7), 1120–1143.

Kapoor, I (2017) "Cold Critique, Faint Passion, Bleak Future: Post-Development's Surrender to Global Capitalism", *Third World Quarterly*, 38 (12), 2664–2683.

Kapur, D, Lewis, J and Webb, R (1997) *The World Bank: Its First Half Century*, Washington, DC: Brookings Institution.

Karabell, Z (1999) *Architects of Intervention*, Baton Rouge: Louisiana State University Press.

Keen, D J (2008) *Complex Emergencies*, London: Polity.

Keenan, J (2009) *The Dark Sahara: America's War on Terror in Africa*, London: Pluto Press.

Kelsall, T and Booth, D (2010) "Developmental Patrimonialism? Questioning the Orthodoxy on Political Governance and Economic Progress in Africa", APPP working paper 9, July, https://assets.publishing.service.gov.uk/media/57a08b1bed915d3cfd000b30/20100708-appp-working-paper-9-kelsall-and-booth-developmental-patrimonialism-july-2010.pdf (Accessed November 7th 2017).

Kennan, G (1932) "Memorandum for the Minister", in Robert Skinner to the Secretary of State, 19 August 1932, State Department Decimal File (Record Group 59, U.S. National Archives), 861.5017, Living Conditions/510.

Kenny, M and Pearce, N (2018) *Shadows of Empire: The Anglosphere in British Politics*, Cambridge: Polity.

Kessler, S (1990) "Cuba's Involvement in Angola and Ethiopia: A Question of Autonomy in Cuba's Relationship with the Soviet Union", Thesis (M.A.), Monterey, CA: Naval Postgraduate School.

Keyes, C (1992) "A Conference at Wingspread and Rethinking Southeast Asian Studies", in Hirschman, C, Keyes, C and Hutterer, K (eds) *Southeast Asian Studies in the Balance: Reflections from America*, Ann Arbor, MI: Association of Asian Studies, 9–24.

Khalili, L (2010) "The Location of Palestine in Global Counterinsurgencies", *International Journal of Middle East Studies*, 42 (3), 413–433.

Khalili, L (2012) *Time in the Shadows: Confinement in Counterinsurgencies*, Stanford, CA: Stanford University Press.

Kickbusch, I (2011) "Global Health Diplomacy: How Foreign Policy Can Influence Global Health", *British Medical Journal*, 342, June 10th.

Kiely, R (1999) "The Last Refuge of the Noble Savage? A Critical Assessment of Post-Development Theory", *The European Journal of Development Research* 11 (1), 30–55.

Kilcullen, D (2006a) "Counter-Insurgency *Redux*", *Survival*, 48 (4), 111–130.

Kilcullen, D (2006b) "Twenty-Eight Articles: Fundamentals of Company-Level Counterinsurgency", *Military Review* (May–June), 29–35.

Kilcullen, D (2011) *The Accidental Guerilla: Fighting Small Wars in the Midst of a Big One*, London: Hurst and Co.

Kim, S (1980) "Mao Zedong and China's Changing World View", in Hsiung, J C and Kim, S S (eds) *China in the Global Community*, New York: Praeger, 16–39.

Kim, S-M (2015) "The Domestic Politics of International Development in South Korea: Stakeholders and Competing Policy Discourses", *The Pacific Review*, 29 (1), 67–91.

Kirby, W (2006) "China's Internationalization in the Early People's Republic: Dreams of a Socialist World Economy", *China Quarterly*, 188, 870–890.

Kirsch, S and Flint, C (eds) (2011) *Reconstructing Conflict: Integrating War and Post-war Geographies*, Farnham: Ashgate.

Kirshner, J and Power, M (2015) "Mining and Extractive Urbanism: Postdevelopment in a Mozambican Boomtown", *Geoforum*, 61, 67–78.

Kitchin, R (2005) "Commentary: Disrupting and Destabilizing Anglo-American and English-Language Hegemony in Geography", *Social and Cultural Geography*, 6 (1), 1–15.

Klein, C (2003) "Musicals and Modernization: Rodgers and Hammerstein's *The King and I*", in Engerman, D, Gilman, N, Haefele, M H and Latham, M E (eds) *Staging Growth: Modernization, Development, and the Global Cold War*, Amherst: University of Massachusetts Press, 129–162.

Kofman, E (1994) "Unfinished Agendas: Acting upon Minority Voices of the Past Decade", *Geoforum*, 25 (4), 429–445.

Kofman, E, Peake, L and Staeheli, L (eds) (2004) *Mapping Women, Making Politics: Feminist Perspectives on Political Geography*, London: Routledge.

Kolko, G (1988) *Confronting the Third World: United States Foreign Policy, 1945–1980*, New York: Pantheon Books.

Kolluoglu-Kirli, B (2003) "From Orientalism to Area Studies", *The New Centennial Review* 3 (3), 93–111.

Konyndyk, J (2017) "Trump's Aid Budget is Breathtakingly Cruel – Cuts Like These Will Kill People", *The Guardian*, May 31st, https://www.theguardian.com/global-development-professionals-network/2017/may/31/trumps-aid-budget-is-breathtakingly-cruel-cuts-like-these-will-kill-people?CMP=share_btn_tw (Accessed July 18th 2018).

Korean International Cooperation Agency (KOICA) (2017) "KOICA's Smart Saemaul Undong of the 21st Century", http://webzine.koica.go.kr/201411/eng/sub1_1.php (Accessed November 2nd 2018).

Korbonski, A and Fukuyama, D (eds) (1987), *The Soviet Union and the Third World: The Last Three Decades*, Ithaca, NY: Cornell University Press.

Kraske, J (1996) *Bankers with a Mission: The Presidents of the World Bank, 1946–91*, Washington, DC: World Bank.

Kratoska, P H, Raben, R and Nordholt, H S (eds) (2005) *Locating Southeast Asia*. Singapore: National University of Singapore Press.

Kulp, E M (1970) *Rural Development Planning: Systems Analysis and Working Method*, New York: Praeger.

Kurtz, L R (2010) "The Anti-Apartheid Struggle in South Africa (1912-1992)", International Center on Nonviolent Conflict, https://www.nonviolent-conflict.org/the-anti-apartheid-struggle-in-south-africa-1912-1992/ (Accessed June 18th 2018).

Kuziemo, I and Werker, E (2006) "How Much is a Seat on the Security Council Worth? Foreign Aid and Bribery at the United Nations", *Journal of Political Economy*, 114 (5), 905–930.

Kuzmarov, J (2009) "Modernizing Repression: Police Training, Political Violence, and Nation-Building in the American Century", *Diplomatic History*, 33 (2), 191–221.

Kuzmarov, J (2017) "American Military Assistance Programmes since 1945", *Oxford Research Encyclopaedia*, American History, http://americanhistory.oxfordre.com/view/10.1093/acrefore/9780199329175.001.0001/acrefore-9780199329175-e-346 (Accessed July 17th 2017).

La-Orngplew, W (2012) "Living under the Rubber Boom: Market Integration and Agrarian Transformations in the Lao Uplands", Durham theses, Durham University, http://etheses.dur.ac.uk/6372/

Lacoste, Y (1972) "The Problems Raised by Bombing of Red River Dikes", in *Proceedings of the 22nd Pugwash Conference on Science and World Affairs*, Oxford, UK: Pugwash, 565–573.

Lacoste, Y (1973) "An Illustration of Geographical Warfare: Bombing the Dikes of the Red River, North Vietnam", *Antipode* 5 (2), 1–13.

Laïdi, Z (1988) "Introduction: What Use Is the Soviet Union?", in Laïdi, Z (ed) *The Third World and the Soviet Union*, London: Zed Books, 1–23.

Lancaster, C (2008) *George Bush's Foreign Aid: Transformation or Chaos?*, Washington, DC: Centre for Global Development.

Landau, P and Kaspin, D (2002) *Images and Empires: Visuality in Colonial and Postcolonial Africa*, Berkeley, CA: University of California Press.

Lanteigne, M (2009) *Chinese Foreign Policy: An Introduction*, London; New York: Routledge.

Large, D (2008) "Beyond 'Dragon in the Bush': The Study of China-Africa Relations", *African Affairs*, 107 (426), 45–61.

Larkin, B (1971) *China and Africa, 1949-70: The Foreign Policy of the People's Republic of China*, Berkeley and Los Angeles: University of California Press.

Larmer, M (2010) "Social Movement Struggles in Africa", *Review of African Political Economy*, 37 (125), 251–262.

Latham, M E (2000) *Modernization as Ideology: Social Science and "Nation-Building" in the Kennedy Era*, Chapel Hill: University of North Carolina Press.

Latham, M E (2011) *The Right Kind of Revolution: Modernization, Development, and U.S. Foreign Policy from the Cold War to the Present*, Ithaca, NY: Cornell University Press.

Lee, C J (2010) "Between a Moment and an Era: The Origins and Afterlives of Bandung", In Lee, C J (ed) *Making a World after Empire: The Bandung Moment and Its Political Afterlives*, Athens: Ohio University Press, 1–42.

Lee, S-O, Wainwright, J and Glassman, J (2018) "Geopolitical Economy and the Production of Territory: The Case of US–China Geopolitical-Economic Competition in Asia", *Environment and Planning A: Economy and Space*, 50 (2), 416–436.

Leffler, M (1992) *A Preponderance of Power: National Security, the Truman Administration, and the Cold War*, Stanford, CA: Stanford University Press.

Legg, S (2002) "Assemblage/apparatus: Using Deleuze and Foucault", *Area*, 43 (2), 128–133.

Legg, S (2007) *Spaces of Colonialism: Delhi's Urban Governmentalities*, Oxford: Blackwell.

Legg, S (2008) "Ambivalent Improvements: Biography, Biopolitics, and Colonial Delhi", *Environment and Planning A*, 40 (1), 37–56.

Legg, S (2011) "Assemblage/apparatus: Using Deleuze and Foucault", *Area*, 43 (2), 128–133.

Leitner, H, Sheppard, E and Sziarto, M (2008) "The Spatialities of Contentious Politics", *Transactions of the Institute of British Geographers*, 33 (2), 157–172.

Lemke, D (2011) "Intra-national IR in Africa", *Review of International Studies*, 37 (1), 49–70.

Lester, A (2002) "Obtaining the 'Due Observance of Justice': The Geographies of Colonial Humanitarianism", *Environment and Planning D: Society and Space*, 20 (3), 277–293.

Levander, C and Mignolo, W D (2011) "Introduction: The Global South and World Dis/Order", *The Global South*, 5 (1), 1–11.

Levi, W (1954) "Political Rivalries in Nepal", *Far Eastern Survey*, 23 (7), 102–107.

Lévy, J (ed) (2007) *Milton Santos: philosophe du mondial, citoyen du local*, Lausanne: Presses Polytechniques et Universitaires Romandes.

Lewis, M W (2015) "Martin Lewis: Metageographies, Postmodernism and Fallacy of Unit Comparability", interview by Leonhardt van Efferink, http://www.exploringgeopolitics.org/interview_lewis_martin_metageographies_postmodernism_fallacy_of_unit_comparability_historical_spatial_ideological_development_constructs_taxonomy_ideas_systems/ (Accessed May 23rd 2017).

Lewis, M W, and Wigen, K E (1997) *The Myth of Continents: A Critique of Metageography*, Berkeley: University of California Press.

Li, G Y (2003) "The 21st Century Strategy for the Yellow River Harnessing with the 'Three Yellow Rivers' Construction Project", in Hongqi, S (ed) *Proceedings of the 1st International Yellow River*

Forum on River Basin Management, Volume 1, Zhengzhou, China: Yellow River Conservancy Commission, 3–6.

Li, T (1999) "Compromising Power: Development, Culture, and Rule in Indonesia", *Cultural Anthropology*, 14 (3), 295–322.

Li, T (2007) *The Will to Improve: Governmentality, Development, and the Practice of Politics*, Durham, NC: Duke University Press.

Lieberthal, K G (1992) "Introduction: The 'Fragmented Authoritarianism' Model and Its Limitations", in Lieberthal, K G and Lampton, D M (eds) *Bureaucracy, Politics, and Decision Making in Post-Mao China*, Berkeley, CA: University of California Press, 1–30.

Liew, L H (2005) "China's Engagement with Neo-liberalism: Path Dependency, Geography and Party Self-reinvention", *Journal of Development Studies*, 41 (2), 331–352.

Lim, K F (2010) "On China's Growing Geo-economic Influence and the Evolution of Variegated Capitalism", *Geoforum*, 41 (5), 677–688.

Lima, M R S and Hirst, H (2006) "Brazil as an Intermediate State and Regional Power: Action, Choice and Responsibilities", *International Affairs*, 82 (1), 21–23.

Lin, Z (1989) "China's Third World Policy", in Hao, Y and Huan, G (eds) *The Chinese View of the World*, New York: Pantheon Books, 225–259.

Liou, C (2009) "Bureaucratic Politics and Overseas Investment by Chinese State-Owned Oil Companies: Illusory Champions", *Asian Survey*, 49 (4), 670–690.

Liska, G (1960) *The New Statecraft: Foreign Aid in American Foreign Policy*, Chicago, IL: University of Chicago Press.

Livingston, D (1993) *The Geographical Tradition: Episodes in a Contested Enterprise*, London: Blackwell.

Livsey, T (2017) *Nigeria's University Age: Reframing Decolonisation and Development*, London: Palgrave.

Low, D A and Lonsdale, J M (1976) "Introduction: Towards the New Order, 1945–1963", in Low, D A and Smith, A (eds) *History of East Africa*, vol. 3, Oxford: Clarendon Press, 1–64.

Lubeck, P (2000) "The Islamic Revival: Antinomies of Islamic Movements under Globalization", in Cohen, R and Rai, S (eds) *Global Social Movements*, London: Continuum, 146–164.

Luckham, R (2009) "Development and Security: A Shotgun Marriage?", IDS working paper 8, http://www.ids.ac.uk/idspublication/democracy-and-security-a-shotgun-marriage (Accessed January 10th 2018).

Luckham, R and Kirk, T (2013) "The Two Faces of Security in Hybrid Political Orders: A Framework for Analysis and Research", *Stability: International Journal of Security and Development*, 2 (2), 1–30.

Ludden, D (2000) "Area Studies in the Age of Globalization", *FRONTIERS: The Interdisciplinary Journal of Area Studies Abroad*, 6, 1–22.

Ludden, D (2003) "Why Area Studies?", in Mirsepassi, A, Basu, A and Weaver, F (eds) *Localizing Knowledge in a Globalizing World: Recasting the Area Studies Debate*, New York: Syracuse University Press, 131–137.

Luk, T C (2008) "Regulating China? Regulating Globalisation?", in Guerrero, D-G and Manji, F (eds) *China's New Role in Africa and the South: A Search for a New Perspective*, Cape Town: Fahamu, 13–16.

Luke, T W (1991) "The Discourse of Development: A Genealogy of 'Developing Nations' and the Discipline of Modernity", *Current Perspectives in Social Theory*, 11, 271–293.

Luke, T W (1996) "Governmentality and Contragovernmentality: Rethinking Sovereignty and Territoriality after the Cold War", *Political Geography*, 15 (6/7), 491–507.

Lula da Silva, L I (2004) Statement opening the general debate of the 59th session of the General Assembly of the United Nations, September 21st, http://www.un.org/webcast/ga/59/statements/braeng040921.pdf (Accessed June 20th 2018).

Lulu, J (2018) "UN with Chinese Characteristics", *China Digital Times*, June 24th, https://chinadigitaltimes.net/2018/06/sinoposis-united-nations-with-chinese-characteristics/ (Accessed July 17th 2018).

Lummis, C D (1996) *Radical Democracy*, Ithaca, NY: Cornell University Press.

Lumsdaine, D H (1993) *Moral Vision in International Politics: The Foreign Aid Regime (1949–1989)*, Princeton, NJ: Princeton University Press.

Luttwak, E W (2012) *The Rise of China vs. the Logic of Sovereignty*, Cambridge, MA: Harvard University Press.

Lyman, P and Morrison, J (2004) "The Terrorist Threat in Africa", *Foreign Affairs*, 83, 75–86.

Mabogunje, A (1980) *The Development Process: A Spatial Perspective*, New York: Holmes and Meier.

Mabogunje, A (1981) "Geography and the Dilemma of Rural Development in Africa", *Geographiska Annaler*, 63 B, 73–86.

Mabogunje, A (1984) "The Poor Shall Inherit the Earth: Issues of Environmental Quality and Third World Development", *Geoforum*, 15 (3), 295–306.

Macey, D (2009) "Rethinking Biopolitics, Race and Power in the Wake of Foucault", *Theory, Culture and Society*, 26 (6), 186–205.

Magrin, G and Perrier-Buslé, L (2011) "New Geographies of Resource Extraction", *Echo Géo*, 17 (June–August), 1–11.

Magubane, Z (2008) "The (Product) Red Man's Burden: Charity, Celebrity, and the Contradictions of Coevalness", *Journal of Pan African Studies*, 2 (6), 134–149.

Mailer, N (1957) "The White Negro", *Dissent*, June 20th, https://www.dissentmagazine.org/online_articles/the-white-negro-fall-1957 (Accessed December 21st 2017).

Majavu, M (2014) "'Failed States' and 'Ungoverned Spaces': The Thinking behind US Foreign Policy in Africa'", *Pambazuka News*, August, http://pambazuka.org/en/category/comment/92810 (Accessed July 16th 2018).

Malaquias, A (2001) "Reformulating International Relations Theory: African Insights and Challenges", in Dunn, K C and Shaw, T M (eds) *Africa's Challenge to International Relations Theory*, Basingstoke: Palgrave, 12–15.

Mallavarapu, S (2005) "Introduction", in Bajpai, K and Mallavarapu, S (eds) *International Relations in India: Bringing Theory Back Home*, Hyderabad: Orient Longman.

Mallin, J (1987) *Cuba in Angola*, Miami: University of Miami Press.

Malm, J (2016) "How Does China Challenge the IMF's Power in Africa?", http://eba.se/wp-content/uploads/2016/12/DDB_2016_9_Malm_webb.pdf (Accessed October 2nd 2017).

Malone, D M (2000) "Eyes on the Prize: The Quest for Nonpermanent Seats on the U.N. Security Council", *Global Governance*, 6 (1), 3–24.

Malone, D M (2011) *Does the Elephant Dance?*, Oxford: Oxford University Press.

Mamadouh, V (2003) "Some Notes on the Politics of Political Geography", *Political Geography*, 22 (6), 663–676.

Mamadouh, V (2010) "Critical Geopolitics at a (Not so) Critical Junction", in Jones, L and Sage, D "New Directions in Critical Geopolitics: An Introduction. With Contributions of: Gearoid Ó Tuathail, Jennifer Hyndman, Fraser MacDonald, Emily Gilbert and Virginie Mamadouh", *GeoJournal*, 75 (4), 315–325.

Mamadouh, V, Meijer, A, Sidaway, J D and van der Wustern, H (2015) "Toward an Urban Geography of Diplomacy: Lessons from The Hague", *The Professional Geographer*, 67 (4), 564–574.

Mamdani, M (2009) *Saviours and Survivors: Darfur, Politics and the War on Terror*, London: Verso.

Manchanda, N (2017) "Rendering Afghanistan Legible: Borders, Frontiers and the 'State' of Afghanistan", *Politics*, 37 (4), 386–401.

Manji, M and Marks, S (eds) (2007) *African Perspectives on China in Africa*, Cape Town, Nairobi and Oxford: Fahamu.

Mao, Z (1961) "Talk with the American Correspondent Anna Louise Strong August 1946", in *Selected Works of Mao Tse-fung*, Vol. IV, Peking: Foreign Languages Press, 99–101.

Mao, Z (1965) *Selected Works*, Volume 2, *On New Democracy*, Beijing: Foreign Languages Press, 339–394.

Mao, Z (1969) *Mao Zedong sixiang wansui* [Long Live Mao Zedong Thought], Beijing.

Marais, H (2011) *South Africa Pushed to the Limit: The Political Economy of Change*, Claremont, CA: UCT Press.

Marcum, J A (1969) *The Angolan Revolution: The Anatomy of an Explosion (1950–1962)*, Cambridge, MA: MIT Press.

Marcum, J A (1976) "Lessons of Angola", *Foreign Affairs*, 54 (3), 407–425.

Marcum, J A (1978) *The Angolan Revolution: Exile Politics and Guerrilla Warfare (1962–1976)*, Cambridge, MA: MIT Press.

Marquez, G G (1977a) "Cubans in Angola", *Australian Left Review*, 1 (60), 27–32. http://ro.uow.edu.au/alr/vol1/iss60/6 (Accessed January 11th 2018).

Marquez, G G (1977b) "Operation Carlota", *New Left Review I*, 101–102 (January–April),123–137.

Marshall, K (2008) *The World Bank: From Reconstruction to Development to Equity*, New York: Routledge.

Marx, A and Soares, J (2013) "South Korea's Transition from Recipient to DAC Donor: Assessing Korea's Development Cooperation Policy", *International Development Policy*, 4 (2), 107–142, https://poldev.revues.org/1535 (Accessed July 10th 2018).

Mason, E S and Asher, R E (1973) *The World Bank since Bretton Woods*, Washington, DC: Brookings Institution Press.

Mason, M (1997) *Development and Disorder: A History of the Third World since 1945*, Hanover, NH: University of New England Press.

Massey, D (1993) "Power-geometry and a Progressive Sense of Place", in Bird, J, Curtis, B, Putman, T, Robertson, G and Tickner, L (eds) *Mapping the Futures: Local Cultures, Global Change*, London: Routledge, 59–69.

Mathers, K (2012) "Mr. Kristof, I Presume? Saving Africa in the Footsteps of Nicholas Kristof", *Transitions*, 107, 14–31.

Matin, K (2010) "Decoding Political Islam: Uneven and Combined Development and Ali Shariati's Political Thought", in Shilliam, R (ed) *International Relations and Non-Western Thought: Imperialism, Colonialism and Investigations of Global Modernity*, London: Routledge, 108–124.

Matthews, S (2004) "Post-Development Theory and the Question of Alternatives: A View from Africa", *Third World Quarterly*, 25 (2), 373–384.

Matthews, S (2017) "Colonised Minds? Post-Development Theory and the Desirability of Development in Africa", *Third World Quarterly*, 38 (12), 2650–2663.

Maul, D (2009) "'Help Them Move the ILO Way': The International Labour Organization and the Modernization Discourse in the Era of Decolonization and the Cold War", *Diplomatic History*, 33 (3), 387–404.

Mawdsley, E (2008) "Fu Manchu versus Dr Livingstone in the Dark Continent? Representing China, Africa and the West in British Broadsheet Newspapers", *Political Geography*, 27 (5), 509–529.

Mawdsley, E (2013) *From Recipients to Donors: The Emerging Powers and the Changing Development Landscape*, London: Zed Books.

Mawdsley, E (2017) "Development geography 1: Cooperation, Competition and Convergence between North and South", *Progress in Human Geography*, 41 (1), 108–117.

Mawdsley, E (2018) "Development geography II: Financialization", *Progress in Human Geography*, 42 (2), 264–274.

Mayer, R (2002) *Artificial Africas: Colonial Images in the Times of Globalization*, Hanover, NH: University Press of New England.

Mazrui, A A (1980) *The African Condition – A Political Diagnosis*, London: Heinemann.

Mbembe, A and Nuttall, S (2004) "Writing the World from an African Metropolis", *Public Culture*, 16 (3), 347–372.

McConnell, F, Moreau, T and Dittmer, J (2012) "Mimicking State Diplomacy: The Legitimizing Strategies of Unofficial Diplomacies", *Geoforum*, 43 (4), 804–814.

McCoy, A (2009) *Policing America's Empire: The United States, the Philippines, and the Rise of the Surveillance State*, Madison: University of Wisconsin Press.

McElwee, P (2004) "Becoming Socialist or Becoming Kinh? Government Policies for Ethnic Minorities in the Socialist Republic of Vietnam", in Duncan, C R (ed) *Civilizing the Margins: Southeast Asian Government Policies for the Development of Minorities*, Ithaca, NY; London: Cornell University Press, 182–213.

McEwan, C (2008) "Subaltern", in Kitchen, R and Thrift, N (eds) (2009) *International Encyclopaedia of Human Geography*, London: Elsevier, 59–64.

McEwan, C (2009) *Postcolonialism and Development*, London: Routledge.

McEwan, C and Mawdsley, E (2012) "Trilateral Development Cooperation: Power and Politics in Emerging Aid Relationships", *Development and Change*, 43 (6), 1185–1209.

McFarlane, C and Rutherford, J (2008) "Political Infrastructures: Governing and Experiencing the Fabric of the City", *International Journal of Urban and Regional Research*, 32 (2), 363–374.

McFaul, M (1990) "The Demise of the World Revolutionary Process: Soviet-Angolan Relations under Gorbachev", *Journal of Southern African Studies*, 16 (1), 165–189.

McGregor, A (2007) "Development, Foreign Aid and Post-Development in Timor-Leste", *Third World Quarterly*, 28 (1), 155–170.

McLin, J (1979) "Surrogate International Organization and the Case of World Food Security, 1949-1969", *International Organization*, 33 (1), 35–55.

McLure, J A (1994) *Late Imperial Romance*, London: Verso.

McMahon, R J (2001) "Introduction: The Challenge of the Third World", in Hahn, P L and Heiss, M A (eds) *Empire and Revolution: The United States and the Third World since 1945*, Columbus: Ohio State University Press, 1–16.

McMahon, R (2006) "Q&A: Transforming U.S. Foreign Aid", *New York Times*, March 17th, https://archive.nytimes.com/www.nytimes.com/cfr/international/slot1_031706.html (Accessed November 7th 2018).

McMichael, P (2008) *Development and Social Change. A Global Perspective*, 4th edition, Thousand Oaks, CA: Pine Forge/Sage.

McNally, C A (2013) "The Challenge of Refurbished State Capitalism: Implications for the Global Political Economic Order", *DMS – Der moderne Staat*, 6 (1), 33–48.

McNally, D (2011) *Global Slump: The Economics and Politics of Crisis and Resistance*, Winnipeg: Fernwood.

McVety, A K (2008) "Pursuing Progress: Point Four in Ethiopia", *Diplomatic History*, 32 (3), 371–403.

McVety, A K (2012) *Enlightened Aid: U.S. Development as Foreign Policy in Ethiopia*, New York: Oxford University Press.

McVety, A K (2015) "JFK and Modernization Theory", in Hoberek, A (ed) *The Cambridge Companion to John F Kennedy*, Cambridge: Cambridge University Press, 103–117.

Meehan, K M (2014) "Tool-power: Water Infrastructure as Well Springs of State Power", *Geoforum*, 57, 215–224.

Megoran, N (2006) "For Ethnography in Political Geography: Experiencing and Re-imagining Ferghana Valley Boundary Closures", *Political Geography*, 25 (6), 622–640.

Mehta, U S (1999) *Liberalism and Empire*, Chicago, IL: University of Chicago Press.

Meisler, S (2011) *When the World Calls: The Inside Story of the Peace Corps and Its First 50 Years*, Boston, MA: Beacon Press.

Mello, P C (2015) "Brasil recua e reduz projetos de cooperação e doações para a África", *Folha da S.Paulo*, December 13th, http://www1.folha.uol.com.br/mundo/2015/03/1606466-brasil-recua-e-reduz-projetos-de-cooperacao-e-doacoes-para-a-africa.shtml (Accessed October 10th 2017).

Mercer, C, Mohan, G and Power, M (2003) "Towards a Critical Political Geography of African Development", *Geoforum*, 34 (4), 419–436.

Mercille, J (2008) "The Radical Geopolitics of US Foreign Policy: Geopolitical and Geoeconomic Logics of Power", *Political Geography*, 27 (5), 570–586.

Michel, S, Beuret, M and Woods, P (2009) *China Safari: On the Trail of Beijing's Expansion in Africa*, New York: Nation Books.

Mies, M and Shiva, V (1993) *Ecofeminism*, London: Fernwood Publications.

Mignolo, W (2000) *Local Histories/Global Designs: Coloniality, Subaltern Knowledges and Border Thinking*, Princeton, NJ: Princeton University Press.

Mignolo, W D (2005) *The Idea of Latin America*, London: Blackwell.

Mihaly, E B (1965) *Foreign Aid and Politics in Nepal*, Oxford: Oxford University Press.

Miller, B A (2000) *Geography and Social Movements: Comparing Antinuclear Activism in the Boston Area*, Minneapolis: University of Minnesota Press.

Millikan, M and Rostow, W W (1954) "Notes on Foreign Economic Policy", May 21st, in Simpson, C (ed) (1998) *Universities and Empire: Money and Politics in the Social Sciences during the Cold War*, New York: New Press, 41.

Millikan, M F and Rostow, W W (1957) *A Proposal: Key to an Effective Foreign Policy*, New York: Harper and Brothers.

Milne, D (2008) *America's Rasputin: Walt Rostow and the Vietnam War*, New York: Hill and Wang.

Minca, C (2003) "Empire Goes to War: Or, the Ontological Shift in the Transatlantic Divide", *ACME: An Electronic Journal of Critical Geographies*, 2, 227–235.

Ministry of Defence (MOD) (2009) *Annual Report and Accounts: Volume One: 2008–2009*, London: MOD.

Ministry of Foreign Affairs (MFA) China (2000) "Premier Zhou Enlai's Tour of Three Asian and African Countries", http://www.fmprc.gov.cn/mfa_eng/ziliao_665539/3602_665543/3604_665547/t18001.shtml (Accessed July 22nd 2018).

Mirsepassi, A, Basu, A and Weaver, F (2003) "Introduction: Knowledge, Power and Culture", in Mirsepassi, A, Basu, A and Weaver, F (eds) *Localizing Knowledge in a Globalising World: Recasting the Area Studies Debate*, New York: Syracuse University Press, 1–21.

Mitchell, K (2017) "Education, Race and Empire: A Genealogy of Humanitarian Governance in the United States", *Transactions of the Institute of British Geographers*, 42 (3), 349–362.

Mitchell, T (1991a) "America's Egypt: Discourse of the Development Industry", *Middle East Report*, 21 (March/April), 18–34+36.

Mitchell, T (1991b) "The Limits of the State: Beyond Statist Approaches and Their Critics", *American Political Science Review*, 85 (1), 77–96.

Mitchell, T (1995) "The Object of Development: America's Egypt", in Crush, J (ed), *Power of Development*, London: Routledge, 129–157.

Mitchell, T (2002) *Rule of Experts: Egypt, Techno-Politics, Modernity*, Berkeley: University of California Press.

Mitchell, T (2003) "Deterritorialization and the Crisis of Social Science", in Mirsepassi, A, Basu, A and Weaver, F (eds) *Localizing Knowledge in a Globalising World: Recasting the Area Studies Debate*, New York: Syracuse University Press, 148–170.

Mitchell, T (2011) *Carbon Democracy: Political Power in the Age of Oil*, London: Verso Books.

Mohan, G and Lampert, B (2012) "Negotiating China: Reinserting African Agency into China-African Relations", *African Affairs*, 112 (446), 92–110.

Moisio, S (2016) "Geopolitics/Critical Geopolitics", in Agnew, J, Mamadouh, V, Secor, A J and Sharp, J (eds) *The Wiley Blackwell Companion to Political Geography*, Chichester, UK: John Wiley and Sons, 220–234.

Monson, J (2009) *Africa's Freedom Railway: How a Chinese Development Project Changed Lives and Livelihoods in Tanzania*, Bloomington: Indiana University Press.

Moore, A and Walker, J (2016) "Tracing the US Military's Presence in Africa", *Geopolitics*, 21 (3), 686–716.

Moore, A (2017) "US Military Logistics Outsourcing and the Everywhere of War", *Territory, Politics, Governance*, 5 (1), 5–27.

Moore, D S (1998) "Subaltern Struggles and the Politics of Place: Remapping Resistance in Zimbabwe's Eastern Highlands", *Cultural Anthropology*, 13 (3), 1–38.

Moore, D S (1999). "The Crucible of Cultural Politics: Reworking 'Development' in Zimbabwe's Eastern Highlands", *American Ethnologist*, 26 (3), 654–689.

Moore, D S (2005) *Suffering for Territory: Race, Place, and Power in Zimbabwe*, Durham, NC and London: Duke University Press.

Moran, M (2014) *Private Foundations and Development Partnerships: American Philanthropy and Global Development Agendas*, London: Routledge.

Morgenthau, H (1973) *Politics among Nations: The Struggle for Peace and Power*, 5th edition, New York: Knopf.

Morikawa, J (2005) "Japan and Africa after the Cold War", *African and Asian Studies*, 4 (4), 485–508.

Morrissey, J (2015). "Securitizing Instability: The US Military and Full Spectrum Operations", *Environment and Planning D: Society and Space*, 33 (4), 609–625.

Moss, T, Roodham, D and Standley, S (2005) "The Global War on Terror and U.S. Development Assistance: USAID Allocation by Country, 1998–2005", Washington, DC: Center for Global Development, working paper number 62, http://www.cgdev.org/content/publications/detail/2863/ (Accessed January 11th 2018).

Mosse, D (2005) *Cultivating Development: An Ethnography of Aid Policy and Practice*, London; Ann Arbor, MI: Pluto Press.

Mosse, D (2007) "Notes on the Ethnography of Expertise and Professionals in International Development". Ethnografeast III: "Ethnography and the Public Sphere". Lisbon, June 20th–23rd.

Mosse, D (ed) (2011) *Adventures in Aidland: The Anthropology of Professionals in International Development*, Oxford: Berghahn.

Motta, S C and Nilsen, A G (2011) "Social Movements and/in the Postcolonial: Dispossession, Development and Resistance in the Global South", in Motta, S C and Nilsen, A G (eds) *Social Movements in the Global South: Dispossession, Development and Resistance*, London: Palgrave Macmillan, 1–31.

Mottiar, S (2013) "From 'Popcorn' to 'Occupy': Protest in Durban, South Africa", *Development and Change*, 44 (3), 603–619.

Moyo, D (2009) *Dead Aid: Why Aid Is Not Working and How There Is a Better Way for Africa*, London: Allen Lane.

Muehlenbeck, P (2016) *Czechoslovakia in Africa, 1945-1968*, London: Palgrave.

Muggah, R (2017) "The Trouble with Brazil's Expanding Arms Trade", *DefenseOne*, April 18th, http://www.defenseone.com/ideas/2017/04/trouble-brazils-expanding-arms-trade/137123/ (Accessed October 2nd 2017).

Müller, M (2008) "Reconsidering the Concept of Discourse for the Field of Critical Geopolitics: Towards Discourse as Language *and* Practice", *Political Geography*, 27 (3), 322–338.

Munck, R (1999) "Deconstructing Development Discourse: Of Impasses, Alternatives and Politics", in Munck, R and O'Hearn, D (eds) *Critical Development Theory: Contributions to a New Paradigm*, London: Zed Books, 198–210.

Muppidi, H (2012) *The Colonial Signs of International Relations*, London: Hurst.

Murphy, C (2007) "The Promise of Critical IR, Partially Kept", *Review of International Studies*, 33, 117–133.

Myers, G (2014) "Toward Expanding Links between Political Geography and African Studies", *Geography Compass*, 8 (2), 125–136.

Myrdal, G (1966) "FAO—The Imperative of Altruism", *The Nation*, 203 (21), December 19th, 666–670.

Myrdal, G (1968) *Asian Drama: An Inquiry into the Poverty of Nations*, New York: Pantheon Books.

Nadelmann, E (1993) *Cops across Borders: The Internationalization of U.S. Criminal Law Enforcement*, University Park: Pennsylvania State University Press.

Naft, T (1993) *Paths to the Middle East*, Albany: SUNY Press.

Nagar, R, Lawson, V, McDowell, L and Hanson, S (2002) "Locating Globalization: Feminist (Re) readings of the Subjects of Spaces of Globalization", *Economic Geography*, 78 (3), 257–284.

Nagl, J A (2002) *Counterinsurgency Lessons from Malaya to Vietnam: Learning to Eat Soup with a Knife*, London: Praeger.

Nakano, Y (2007) "Serge Latouche's *Destruktion* of Development and the Possibility of Emancipation", in Ziai, A (ed) *Exploring Post-Development: Theory and Practice, Problems and Perspectives*, New York: Routledge, 63–80.

Nally, D and Taylor, S (2015) "The Politics of Self-Help: The Rockefeller Foundation, Philanthropy and the 'Long' Green Revolution", *Political Geography*, 49, 51–63.

Nandy, A (1987) *Traditions, Tyranny and Utopia*, New Delhi: Oxford University Press.

Narlikar, A (2013) "India Rising: Responsible to Whom?", *International Affairs*, 89 (3), 595–614.

Nash, A (2003) "Third Worldism", *African Sociological Review*, 7 (1), 94–116.

Nash, M (ed) (2016) *Red Africa: Affective Communities and the Cold War*, London: Black Dog Publishing.

Nashel, J (2000) "The Road to Vietnam: Modernization Theory in Fact and Fiction", in Appy, C G (ed) *Cold War Constructions: The Political Culture of United States Imperialism 1945-1966*, Amherst: University of Massachusetts Press, 132–156.

Nation, C R and Kauppi, M (1984) *The Soviet Impact in Africa*, Lexington, MA: Lexington Books.

National Intelligence Council (1982) "The US-Soviet Competition for Influence in the Third World: How the LDCs Play It, Memorandum 82-10005", http://www.foia.cia.gov/docs/DOC_0000273391/DOC_0000273391.pdf (Accessed July 17th 2017).

National Security Council (1993) "NSC-68", in May, E R (ed) *American Cold War Strategy: Interpreting NSC-68*, New York: Bedford Books of St. Martin's Press, 40–41.

Natsios, A (2010) *The Clash of the Counter-Bureaucracy and Development*, Washington, DC: Center for Global Development, http://www.cgdev.org/content/publications/detail/1424271 (Accessed August 4th 2017).

Nehru, J (1941) *Toward Freedom: The Autobiography of Jawaharlal Nehru*, London: John Day.

Nehru, J (1949 [1929]), *Soviet Russia: Some Random Sketches and Impressions*, Bombay: Chetana.

Network of Concerned Anthropologists (2009) *The Counter-Counterinsurgency Manual*, Chicago, IL: Prickly Paradigm Press.

Neuhauser, C (1968) *Third World Politics: China and the Afro-Asian People's Solidarity Organisation*, Cambridge, MA: Harvard University Press.

Neuman, S G (ed) (1998) *International Relations Theory and the Third World*, London: Macmillan.

Neumann, R P (2004) "Nature-State-Territory: Toward a Critical Theorisation of Conservation Enclosures", in Peet, R and Watts, M (eds) *Liberation Ecologies: Environment, Development, Social Movements*, London: Routledge.

Nicholls, W, Miller, B and Beaumont, J (eds) (2013) *Spaces of Contention: Spatialities and Social Movements*, Farnham: Ashgate.

Nilsen, A G (2010) *Dispossession and Resistance in India: The River and the Rage*, London: Routledge.

Nilsen, A G (2015) "Postcolonial Social Movements", in Ness, I and Cope, Z (eds) *The Palgrave Encyclopaedia of Imperialism and Anti-imperialism*, New York: Palgrave, 932–938.

Nilsen, A G (2016) "Power, Resistance and Development in the Global South: Notes toward a Critical Research Agenda", *International Journal of Politics, Culture and Society*, 29 (3), 269–287.

Nkiwane, T C (2001) "Africa and International Relations: Regional Lessons for a Global Discourse", *International Political Science Review*, 22 (3), 279–290.

Nolutshungu, S (1985) "Soviet Involvement in Southern Africa", *Annals of the American Academy of Political and Social Science*, 481, 138–146.

Nonini, D M (2008) "Is China Becoming Neoliberal?", *Critique of Anthropology*, 28 (2), 145–176.

Norris, J (2012) "Hired Gun Fight", *Foreign Policy*, July 18th, http://foreignpolicy.com/2012/07/18/hired-gun-fight/ (Accessed July 17th 2018).

Norris, J (2014) "Kennedy, Johnson and the Early Years", https://www.devex.com/news/kennedy-johnson-and-the-early-years-83339 (Accessed November 2nd 2018).

Novak, J (2017) "After Tax Reform Victory, Trump Team's Next Big Win Will Be Cutting Foreign Aid", *CNBC News*, December 29th, https://www.cnbc.com/2017/12/29/trump-should-slash-foreign-aid-budget-for-next-big-win-commentary.html (Accessed December 30th 2017).

Novak, P (2016) "Placing Borders in Development", *Geopolitics*, 21 (3), 483–512.

Nunan, T (2014) "The Sino-Soviet Split and the Left as Global History: An Interview with Jeremy Friedman", *Toynbee Prize Foundation*, http://toynbeeprize.org/global-history-forum/the-sino-soviet-split-and-the-left-as-global-history-an-interview-with-jeremy-friedman/ (Accessed March 20th 2017).

Nunan, T (2016) *Humanitarian Invasion: Global Development in Cold War Afghanistan*, Cambridge: Cambridge University Press.

Nustad, K G (2007) "Development: The Devil We Know?", in Ziai, A (ed) *Exploring Post-Development: Theory and Practice, Problems and Perspectives*, New York: Routledge, 35–46.

O'Loughlin, J, Raento, P and Sidaway, J D (2008) "Editorial: Where the Tradition Meets Its Challengers", *Political Geography*, 27 (1), 1–4.

O'Neill, J (2011) *The Growth Map: Economic Opportunity in the BRICs and Beyond*, London: Penguin.

O'Neill, J (2012) "SA's Brics Score Not All Doom and Gloom", *Mail and Guardian*, March 30th, http://www.mg.co.za/article/2012-03-30-sas-bric-score-not-all-doom-and-gloom (Accessed April 9th 2013).

Odgaard, L and Nielsen, T G (2014) "China's Counterinsurgency Strategy in Tibet and Xinjiang", *Journal of Contemporary China*, 23 (87), 535–555.

OECD (1985) *Twenty-Five Years of Development Co-operation: A Review*, Paris: OECD.

OECD (1998) *Development Co-operation Guideline Series, Conflict, Peace and Development Co-operation on the Threshold of the 21st Century*, Paris: Organisation for Economic Co-operation and Development.

OECD (2017) "Development Aid Rises Again in 2016", https://www.oecd.org/dac/financing-sustainable-development/development-finance-data/ODA-2016-detailed-summary.pdf (Accessed July 11th 2017).

OECD DAC (2003) *A Development Co-operation Lens on Terrorism Prevention: Key Entry Points for Action*, Paris: OECD Development Assistance Committee.

Office of Public Safety (OPS) (1966) *The Public Safety Program Vietnam*, Saigon, Vietnam: USAID.

Offiler, B (2015) *US Foreign Policy and the Modernization of Iran: Kennedy, Johnson, Nixon and the Shah*, London: Palgrave.

Ogunbadejo, O (1981) "Angola: Ideology and Pragmatism in Foreign Policy", *International Affairs*, 57 (2), 254–269.

Oliva, A R (2009) "A África não está em nós: a História Africana no imaginário de estudantes do Recôncavo Baiano", *Fronteiras* 11 (20): 73–91.

Ong, A (1999) *Flexible Citizenship: The Cultural Logics of Transnationality*, Durham, NC: Duke University Press.

Ong, A (2000) "Graduated Sovereignty in South-East Asia", *Theory, Culture and Society*, 17 (4), 55–75.

Ong, A (2006) *Neoliberalism as Exception: Mutations in Citizenship and Sovereignty*, Durham, NC: Duke University Press.

Oonk, G (2017) "India and Indian Diaspora in East Africa: Past Experiences and Future Challenges", Presentation given to a conference of the Organisation for Diaspora Initiatives (ODI), December 2015, https://in.boell.org/2017/04/13/india-and-indian-diaspora-east-africa-past-experiences-and-future-challenges-lecture (Accessed June 19th 2018).

Osborne, M E (1965) *Strategic Hamlets in South Vietnam: A Survey and Comparison*, Ithaca, NY: Cornell University.

Ó Tuathail, G (1986) "The Language and Nature of the 'New Geopolitics' – The Case of US-El Salvador Relations", *Political Geography Quarterly*, 5 (1), 73–85.

Ó Tuathail, G (1994) "Critical Geopolitics and Development Theory: Intensifying the Dialogue", *Transactions of the Institute of British Geographers*, 19 (2), 228–233.

Ó Tuathail, G (1996a) *Critical Geopolitics: The Politics of Writing Global Space*, Minneapolis: University of Minnesota Press.

Ó Tuathail, G (1996b) "An Anti-geopolitical Eye: Maggie O'Kane in Bosnia 1992–1993", *Gender, Place and Culture*, 3 (2), 171–185.

Ó Tuathail, G (2003) "Re-asserting the Regional: Political Geography and Geopolitics in World Thinly Known", *Political Geography*, 22 (6), 653–655.

Ó Tuathail, G (2010a) "Opening Remarks", in Jones, L and Sage, D "New Directions in Critical Geopolitics: An Introduction. With Contributions of: Gearoid Ó Tuathail, Jennifer Hyndman, Fraser MacDonald, Emily Gilbert and Virginie Mamadouh", *GeoJournal*, 75 (4), 315–325.

Ó Tuathail, G (2010b) "Localizing Geopolitics: Disaggregating Violence and Return in Conflict Regions", *Political Geography*, 29 (5), 256–265.

Ó Tuathail, G and Agnew, J (1992) "Geopolitics and Discourse: Practical Geopolitical Reasoning in American Foreign Policy", *Political Geography*, 11 (2), 190–204.

Ó Tuathail, G and Dalby, S (1998) "Introduction: Rethinking Geopolitics: Towards a Critical Geopolitics", in Ó Tuathail, G and Dalby, S (eds) *Rethinking Geopolitics*. London; New York: Routledge, 1–16.

Ovadia, J and Woolf, C (2018) "Studying the Developmental State: Theory and Method in Research on Industrial Policy and State-led Development in Africa", *Third World Quarterly*, 39 (6), 1056–1076.

Oxfam (2005) *Paying the Price: Why Rich Countries Must Invest Now in a War on Poverty*, Oxford, UK: Oxfam.

Paasi, A (2000) "Rethinking Geopolitics", *Environment and Planning D: Society and Space*, 18 (2), 282–284.

Paasi, A (2006) "Texts and Contexts in the Globalizing Academic Marketplace: Comments on the Debate on Geopolitical Remote Sensing", *Eurasian Geography and Economics*, 47 (2), 216–220.

Packenham, R (1973) *Liberal America and the Third World: Political Development Ideas and Social Science*, Princeton, NJ: Princeton University Press.

Paik, W (2016) "The 60th Anniversary of the Bandung Conference and Asia", *Inter-Asia Cultural Studies* 17 (1), 148–157.

Pain, R and Staeheli, L (2014) "Introduction: Intimacy-Geopolitics and Violence", *Area*, 46 (4), 344–347.

Painter, J (2006) "Prosaic Geographies of Stateness", *Political Geography*, 25 (7), 752–774.

Palat, R A (2008) "A New Bandung? Economic Growth vs. Distributive Justice among Emerging Powers", *Futures*, 40 (8), 721–734.

Parenti, C (2011) *Tropic of Chaos: Climate Change and the New Geography of Violence*, New York: Nation.

Parker, S and Chefitz, G (2018) "China's Debtbook Diplomacy: How China is Turning Bad Loans into Strategic Investments", *The Diplomat*, May 30th, https://thediplomat.com/2018/06/chinas-debtbook-diplomacy-how-china-is-turning-bad-loans-into-strategic-investments/ (Accessed July 22nd 2018).

Patey, L (2016) "China's New Crisis Diplomacy in Africa and the Middle East", Danish Institute for International Studies, January 26th, https://www.ethz.ch/content/specialinterest/gess/cis/center-for-securities-studies/en/services/digital-library/articles/article.html/195713 (Accessed October 4th 2017).

Patrick, S and Brown, K (2007) "The Pentagon and Global Development: Making Sense of the DoD's Expanding Role" (Working Paper Number 131, November 2007), Washington, DC: Center for Global Development.

Paudel, D (2016) "The Double Life of Development: Empowerment, USAID and the Maoist Uprising in Nepal", *Development and Change*, 47 (5), 1025–1050.

Pearce, R D (1982) *The Turning Point in Africa: British Colonial Policy, 1938–48*, London: Frank Cass.

Pearson, J L (2017) "Defending Empire at the United Nations: The Politics of International Colonial Oversight in the Era of Decolonisation", *Journal of Imperial and Commonwealth History*, 45 (3), 525–549.

Pechota, V (1981) "Czechoslovakia and the Third World", in Radu, M (ed), *Eastern Europe and the Third World: East vs. South*, New York: Praeger, 77–105.

Peck, J and Tickell, A (2002) "Neoliberalizing Space", *Antipode*, 34 (3), 380–404.

Peluso, N (1995) "Whose Woods Are These? Counter-Mapping Forest Territories in Kalimantan, Indonesia", *Antipode*, 27 (4), 383–406.

People's Daily (2018) "China-Africa Development Fund Reaches 10 Bln USD", http://en.people.cn/n3/2018/0902/c90000-9496355.html (Accessed November 2nd 2018).

Peralta, E (2014) "USAID Says Building of 'Cuban Twitter' was Part of Public Record", *NPR*, https://www.npr.org/sections/thetwo-way/2014/04/07/300285197/usaid-says-building-of-cuban-twitter-was-part-of-public-record (Accessed January 16th 2017).

Perkins, J H (1997) *Geopolitics and the Green Revolution: Wheat, Genes, and the Cold War*, New York; Oxford: Oxford University Press.

Perry, P (1987) "Editorial Comment. Political Geography Quarterly: A Content (but Discontented) Review", *Political Geography Quarterly*, 6 (1), 5–6.

Petras, J and Veltmeyer, H (2002) "Age of Reverse Aid: Neo-liberalism and Catalyst of Regression", *Development and Change*, 33 (2), 281–293.

Phillips, R (2008) *Why Vietnam Matters: An Eyewitness Account of Lessons Not Learned*, Annapolis, MD: Naval Institute Press.

Pieterse, E (2011) "Introduction: Rogue Urbanisms", *Social Dynamics*, 37 (1), 1–4.

Pieterse, J N (1995) *White on Black: Images of Africa and Blacks in Western Popular Culture*, London: Yale University Press.

Pieterse, J N (1998) "My Paradigm or Yours?: Alternative Development, Post-Development, Reflexive Development", *Development and Change*, 29 (2), 342–373.

Pieterse, J N (2001) *Development Theory: Deconstructions/Reconstructions*, London: Sage Publications.

Pieterse, J N (2011) "Global Rebalancing: Crisis and the East–South Turn", *Development and Change*, 42 (1), 22–48.

Pieterse, J N and Parekh, B (eds) (1995) *Decolonization of the Imagination: Culture, Knowledge and Power*, New York: St Martin's Press.

Pietz, W (1988) "The 'Post-Colonialism' of Cold War Discourse", Social *Text*, 19/20, 55–75.

Pilling, D (2018) *The Growth Delusion: The Wealth and Well-Being of Nations*, London: Bloomsbury.

Pimlott, J (1985) "The British Army", in Beckett, I and Pimlott, J (eds) *Armed Forces and Modern Counter-Insurgency*, New York: St. Martin's Press, 16–17.

Pirie, I (2008) *The Korean Developmental State: From Dirigisme to Neo-liberalism*, London: Routledge.

Pithouse, R (2009) "The Thoroughly Democratic Logic of Refusing to Vote", *The South African Civil Society Information Service* (Sacsis), April 2nd, http://sacsis.org.za/site/article/258.1 (Accessed October 24th 2018).

Pithouse, R (2011) "The Service Delivery Myth", *The South African Civil Society Information Service* (Sacsis), January 26th, http://sacsis.org.za/site/article/610.1 (Accessed July 17th 2018).

Pletsch, C (1981) "The Three Worlds or the Division of Social Scientific Labour 1950-1975", *Comparative Studies in Society and History*, 23 (4), 565–590.

Polet, F (ed) (2007) *The State of Resistance: Popular Struggles in the Global South*, New York: Zed Books.

Pope Francis (2017) "Pope Francis's Urbi et Orbi 2017 Christmas Message: Full Text", *Catholic Herald*, December 25th, http://catholicherald.co.uk/news/2017/12/25/pope-franciss-urbi-et-orbi-2017-christmas-message-full-text/ (Accessed January 10th 2018).

Popke, J (1994) "Recasting Geopolitics: The Discursive Scripting of the International Monetary Fund", *Political Geography*, 13 (3), 255–269.

Porter, B D (1984) *The USSR in Third World Conflicts: Soviet Arms and Diplomacy in Local Wars 1945-1980*, Cambridge: Cambridge University Press.

Porter, P (2009) *Military Orientalism: Eastern War through Western Eyes*, Oxford: Oxford University Press.

Porto-Gonçalves, C W (2012) *A Reinvenção dos territórios na América Latina/Abya Yala*, Universidad Autónoma de México.

Poster, A (2012) "The Gentle War: Famine Relief, Politics and Privatization in Ethiopia, 1983-1986", *Diplomatic History*, 26 (2), 399–425.

Potter, D M (2002) "Japan's Official Development Assistance", http://office.nanzan-u.ac.jp/ncia/about-cia/item/pdf_12/kenkyu_02.pdf (Accessed November 6th 2018).

Potter, D M (2009) *Modes of Asian Development Assistance*, The Bulletin of the Center for International Education, Nanzan University, 10, 19–34.

Power, M (2003) *Rethinking Development Geographies*, London: Routledge.

Power, M (2009) "Tropical Geography", in Kitchin, R and Thrift, N (eds) *International Encyclopaedia of Human Geography*, Volume 11, Oxford: Elsevier, 493–498.

Power, M (2010) "Geopolitics and Development: An Introduction", *Geopolitics*, 15 (3), 433–440.

Power, M (2011) "Angola 2025: The Future of the 'World's Richest Poor Country' as Seen through a Chinese Rear-view Mirror", *Antipode*, 44 (3), 993–1014.

Power, M (2015) "The Rise of the BRICS", in Agnew, J, Mamadouh, V, Secor, A J and Sharp, J (eds) *The Wiley Blackwell Companion to Political Geography*, Chichester, UK: John Wiley and Sons, 379–392.

Power, M and Campbell, D (2010) "The State of Critical Geopolitics", *Political Geography*, 29 (5), 243–246.

Power, M and Kirshner, J (2018) "Powering the State: The Political Economy of Electrification in Mozambique", *Environment and Planning C*, http://journals.sagepub.com/doi/abs/10.1177/2399654418784598.

Power, M and Mohan, G (2010) "Towards a Critical Geopolitics of China's Engagement with African Development", *Geopolitics*, 15 (3), 462–495.

Power, M, Mohan, G and Tan-Mullins, M (2012) *China's Resource Diplomacy and Africa's Future: Powering Development?*, London: Palgrave.

Power, M, Newell, P, Baker, L, Bulkeley, H, Kirshner, J and Smith, A (2016) "The Political Economy of Energy Transitions in Mozambique and South Africa: The Role of the Rising Powers", *Energy Research & Social Science*, 17, 10–19.

Power, M and Sidaway, J D (2004) "The Degeneration of Tropical Geography", *Annals of the Association of American Geographers*, 94 (3), 585–601.

Prakash, G (1999) *Another Reason: Science and the Imagination of Modern India*, Princeton, NJ: Princeton University Press.

Prashad, V (2007) *The Darker Nations: A People's History of the Third World*, New York: The New Press.

Presley, C A (1988) "The Mau Rebellion, Kikuyu Women, and Social Change", *Canadian Journal of African Studies/Revue Canadienne des Études Africaines*, 22 (3), 502–527.

Price, G (2011) *For the Global Good: India's Developing International Role*, London: Chatham House.

Rabe, S G (1988) *Eisenhower and Latin America: The Foreign Policy of Anti-Communism*, Chapel Hill, NC: University of North Carolina Press.

Rabe, S G (1999) *The Most Dangerous Area in the World: John F Kennedy Confronts Communist Revolution in Latin America*, Chapel Hill, NC: University of North Carolina Press.

Radcliffe, S A (2012) "Development for a Postneoliberal Era? Sumak kawsay, Living Well and the Limits to Decolonisation in Ecuador", *Geoforum*, 43 (2), 240–249.

Radcliffe, S A (2015) "Development Alternatives", *Development and Change*, 46 (4), 855–874.

Radcliffe, S A (2017) "Decolonizing Geographical Knowledges", *Transactions of the Institute of British Geographers*, 42 (3), 329–348.

Rafael, V L (1994) "The Cultures of Area Studies in the United States", *Social Text*, 41, 91–111.

Raghuram, P and Madge, C (2006) "Towards a Method for Postcolonial Development Geography? Possibilities and Challenges", *Singapore Journal of Tropical Geography*, 27 (3), 270–288.

Rahnema, M (1997a) "Introduction", in Rahnema, M and Bawtree, V (eds) *The Post-Development Reader*, London: Zed Books, ix–xix.

Rahnema, M (1997b) "Towards Post-Development: Searching for Signposts, a New Language and New Paradigms", in Rahnema, M and Bawtree, V (eds) *The Post-Development Reader*, London: Zed Books, 377–403.

Ramamurthy, A (2003) *Imperial Persuaders: Images of Africa and Asia in British Advertising*, Manchester: Manchester University Press.

Rangan, H (2000) *Of Myths and Movements: Rewriting Chipko into Himalayan History*, Delhi: Oxford University Press.

Ranjan, S (2016) "India–Africa Defence Cooperation against the Backdrop of the 'Make in India' Initiative", *African Security Review*, 25 (4), 407–419.

Raposo, P A (ed) (2014a) *Japan's Foreign Aid Policy in Africa: Evaluating the TICAD Process*, New York: Palgrave Macmillan.

Raposo, P A (ed) (2014b) *Japan's Foreign Aid to Africa: Angola and Mozambique within the TICAD Process*, London: Routledge.

Reid, R (2014) "Horror, Hubris and Humanity: The International Engagement with Africa 1914–2014", *International Affairs*, 90 (1), 143–165.

Reilly, J (2013) "China and Japan in Myanmar: Aid, Natural Resources and Influence", *Asian Studies Review*, 37 (2), 141–157.

Renfrew, B (2003) "France Battles U.S. to Line up U.N. Votes", *Associated Press*, March 1st.

Repo, J and Yrjölä, R (2011) "The Gender Politics of Celebrity Humanitarianism in Africa", *International Feminist Journal of Politics*, 13 (1), 44–62.

Reynolds, G (2008) *Apostles of Modernity: American Writers in the Age of Development*, Lincoln: University of Nebraska Press.

Rice, G T (1985) *The Bold Experiment: JFK's Peace Corps*, Notre Dame, IN: University of Notre Dame Press.

Riddell, R (1998), *Aid in the 21st Century*, New York: United Nations Development Programme, Office of Development Studies.

Riddell, R (2007) *Does Foreign Aid Really Work?*, Oxford: Oxford University Press.

Riech, B (1996) "The United States and Israel: The Nature of A Special Relationship", in Lesch, D W (ed) *The Middle East and the United States: A Historical and Political Reassessment*, Boulder, CO: Westview Press, 233–248.

Rigg, J (2007) *An Everyday Geography of the Global South*, London: Routledge.

Rist, G (1997) *History of Development: From Western Origins to Global Faith*, London: Routledge.

Rist, G (2010) "Development as a Buzzword", in Cornwall, A and Eade, D (eds) *Deconstructing Development Discourse: Buzzwords and Fuzzwords*, Rugby: Practical Action Publishing, 19–28.

Rivera Cusicanqui, S R (2012) "*Ch'ixinakax utxiwa*: A Reflection on the Practices and Discourses of Decolonization", *South Atlantic Quarterly*, 111 (1), 95–109.

Robbins, C A (1979) *Looking for Another Angola: Cuban Policy Dilemmas in Africa*, Washington, DC: Latin American Program, Wilson Center.

Roberts, S, Secor, A and Sparke, M (2003) "Neoliberal Geopolitics", *Antipode*, 35 (5), 886–897.

Roberts, S M (2012) "Worlds Apart? Economic Geography and Questions of 'Development'", in Barnes, T, Peck, J and Sheppard, E (eds) *The Wiley-Blackwell Companion to Economic Geography*, London: Wiley, 552–566.

Roberts, S M (2014) "Development Capital: USAID and the Rise of Development Contractors", *Annals of the Association of American Geographers*, 104 (5), 1030–1051.

Robinson, H (2007) *Russians in Hollywood, Hollywood's Russians: Biography of an Image*, London: University of New England Press.

Robinson, J (2003a) "Postcolonialising Geography: Tactics and Pitfalls", *Singapore Journal of Tropical Geography*, 24 (3), 273–289.

Robinson, J (2003b) "Political Geography in a Postcolonial Context", *Political Geography*, 22 (6), 647–651.

Robinson, J (2016a) "Thinking Cities through Elsewhere: Comparative Tactics for a More Global Urban Studies", *Progress in Human Geography*, 40 (1), 3–29.

Robinson, J (2016b) "Comparative Urbanism: New Geographies and Cultures of Theorizing the Urban", *International Journal of Urban and Regional Research*, 40 (1), 187–199.

Robinson, P and Dixon, J (2010) "Soviet Development Theory and Economic and Technical Assistance to Afghanistan, 1954-1991", *The Historian*, 72 (3), 599–623.

Robinson, P and Dixon, J (2013) *Aiding Afghanistan: A History of Soviet Assistance to a Developing Country*, London: Hurst.

Roe, A (2015) "Riverine Environments", in McNeil, J R and Mauldin, E S (eds) *A Companion to Global Environmental History*, London: Blackwell, 297–318.

Roennfeldt, C (2011) "Productive War: A Re-conceptualization of War", *Journal of Strategic Studies*, 34 (1), 39–62.

Rose, N (1999) *Powers of Freedom: Reframing Political Thought*, Cambridge: Cambridge University Press.

Rostow, W W (in collaboration with Levin, A) (1953) *The Dynamics of Soviet Society*, New York: Norton and Co.

Rostow, W W (1960) *The Stages of Economic Growth: A Non-Communist Manifesto*, Cambridge: Cambridge University Press.

Rostow, W W (1961a) "Guerrilla Warfare in the Underdeveloped Areas", Department of State Bulletin, August 7th.

Rostow, W W (1961b) Memorandum: Walt Rostow to Kennedy, March 2nd 1961, Presidents Office Files Staff Memos, box 64a, JFK Presidential Library, Boston, MA.

Rostow, W W (1968) *The Economics of Take-off into Sustained Growth*, New York: St Martin's Press.

Rostow, W W (1985) *Eisenhower, Kennedy, and Foreign Aid*, Austin: University of Texas Press.

Rothkopf, D (2009) "The BRICS and What the BRICS Would Be without China", *Foreign Policy*, June 15.

Routledge, P (1993) *Terrains of Resistance: Nonviolent Social Movements and the Contestation of Place in India*, Westport, CT: Praeger.

Routledge, P (1996) "Critical Geopolitics and Terrains of Resistance", *Political Geography*, 12 (6/7), 509–552.

Routledge, P (1998) "Introduction", in Ó Tuathail, G, Dalby, S and Routledge, P (eds) *The Geopolitics Reader*, London: Routledge, 245–255.

Routledge, P (2002) "Travelling East as Walter Kurtz: Identity, Performance and Collaboration in Goa, India", *Environment and Planning D: Society and Space*, 20 (4), 477–498.

Routledge, P (2003) "Anti-geopolitics", in Agnew, J, Mitchell, K and Ó Tuathail, G (eds) *A Companion to Political Geography*, London: Blackwell, 236–248.

Routledge, P (2015) "Territorialising Movement: The Politics of Land Occupation in Bangladesh", *Transactions of the Institute of British Geographers*, 40 (4), 445–463.

Routledge, P (2017) *Space Invaders: Radical Geographies of Protest*, London: Pluto.

Roy, A N (1999) *The Third World in the Age of Globalism: Requiem or New Agenda?*, London: Zed Books.

Roy, A (2009) "The 21st-Century Metropolis: New Geographies of Theory", *Regional Studies*, 43 (6), 819–830.

Roy, A (2014a) "Worlding the Global South: Toward a Postcolonial Urban Theory", in Parnell, S and Oldfield, S (eds) *The Routledge Handbook on Cities of the Global South*, London: Routledge, 9–20.

Roy, A (2014b) "The NGOization of Resistance", http://massalijn.nl/new/the-ngo-ization-of-resistance/ (Accessed November 1st 2017).

Roy, A (2015) "What Is Urban about Critical Urban Theory?", *Urban Geography*, 37 (6), 810–823.

Roy, A (2016a) "Divesting from Whiteness: The University in the Age of Trumpism", *Society and Space*, November 28th, http://societyandspace.org/2016/11/28/divesting-from-whiteness-the-university-in-the-age-of-trumpism/ (Accessed July 31st 2018).

Roy, A (2016b) "Who's Afraid of Postcolonial Theory?", *International Journal of Urban and Regional Research*, 40 (1), 200–209.

Roy, A and Crane, E (2015) *Territories of Poverty: Rethinking North and South*, Athens, GA: University of Georgia Press.

Roy, A, Schrader, S and Crane, E S (2015) "Gray Areas: The War on Poverty at Home and Abroad", in Roy, A and Crane, E S (eds) *Territories of Poverty: Rethinking North and South*, Athens, GA: University of Georgia Press, 289–314.

Rudner, M (1996) "East European Aid to Asian Developing Countries: The Legacy of the Communist Era", *Modern Asian Studies*, 30 (1), 1–28.

Rupiya, M and Southall, R (2009) "The Militarisation of the New Scramble in Africa", in Southall, R and Melber, H (eds) *A New Scramble for Africa? Imperialism, Investment and Development*, Scottsville: University of KwaZulu-Natal Press.

Ryan, J R (1997) *Picturing Empire: Photography and the Visualization of the British Empire*, Chicago, IL: University of Chicago Press.

Ryan, M (2011) "'War in Countries We Are Not at War With': The 'War on Terror' on the Periphery from Bush to Obama", *International Politics*, 48 (2–3), 364–389.

Sachs, W (1992) "Introduction", in Sachs, W (ed), *The Development Dictionary: A Guide to Knowledge as Power*, 1st edition, London: Zed.

Sachs, W (1999) "The Archaeology of the Development Idea", in Sachs, W (ed) *Planet Dialectics: Explorations in Environment and Development*, London: Zed, 1–23.

Sachs, W (2010) "Introduction", in Sachs, W (ed), *The Development Dictionary: A Guide to Knowledge as Power*, 2nd edition, London: Zed.

Sachs, W (2017) "The Sustainable Development Goals and *Laudato si'*: Varieties of Post-Development?", *Third World Quarterly*, 38 (12), 2573–2587.

Sacks, B (2012) "'Othering' Tropical Environments", *Geography Directions*, https://blog.geographydirections.com/tag/cuba/ (Accessed July 3rd 2018).

Said, E (1978) *Orientalism*, New York: Pantheon.

Said, E (1983) "Opponents, Audiences, Constituencies and Community", in Foster, H (ed) *Post-modern Culture*, London: Pluto, 135–159.

Saldinger, A and Igoe, M (2018) "Set of Congressional Budget Hearings Lay out US Aid Funding", *Devex*, June 22nd, https://www.devex.com/news/set-of-congressional-budget-hearings-lay-out-us-aid-funding-92986 (Accessed July 17th 2018).

Santicola, R (2014) "China's Consistently Inconsistent South China Sea Policy", *The Diplomat*, May 24th.

Santos, B S (2014) *Epistemologies of the South: Justice against Epistemicide*, London: Routledge.

Santos, M (1969) *Aspects de la géographie et de l'économie urbaines des pays sous-développés*, Paris: Centre de Documentation Universitaire.

Santos, M (1970) *Dix essais sur les villes des pays sous-développés*, Paris: Éditions Ophrys.

Santos, M (1971) *Le Métier de géographe en pays sous-développé, un essai méthodologique*, Paris: Éditions Ophrys.

Santos, M (1972) "Les Villes incomplètes des pays sous-développés", *Annals de Géographie*, 445, 316–323.

Santos, M (1979) *The Shared Space: The Two Circuits of the Urban Economy in Underdeveloped Countries*, London: Methuen.

Saraiva, J (2012) *África parceira do Brasil Atlantico: Relações internacionais do Brasil e da África*, Belo Horizonte: Fino Traço.

Saurin, J (2006) "International Relations as the Imperial Illusion: Or, the Need to Decolonize IR", in Gruffyd Jones, B (ed) *Decolonizing International Relations*, Lanham, MD: Rowman and Littlefield, 23–42.

Sautman, B and Hairong, Y (2016) "The Discourse of Racialization of Labour and Chinese Enterprises in Africa", *Ethnic and Racial Studies*, 39 (12), 2149–2168.

Sauvy, A (1952) "Trois Mondes, Une Planète", *L'Observateur*, August 14th, n°118, 14.

Schlesinger, A M (2002) *A Thousand Days: John F Kennedy in the White House*, Wilmington, MA: Mariner Books.

Schuurman F J (2001) "Globalization and Development Studies: Introducing the Challenges", in Schuurman, F J (ed) *Globalization and Development: Challenges for the 21st Century*, London: Sage, 3–16.

Scott, C G (2009) "Swedish Vietnam Criticism Reconsidered: Social Democratic Vietnam Policy a Manifestation of Swedish *Ostpolitik*?", *Cold War History*, 9 (2), 243–266.

Scott, D (1999) *Refashioning Futures: Criticism after Postcoloniality*, Princeton, NJ: Princeton University Press.

Scott, D (1995) "Colonial Governmentality", *Social Text*, 43, 191–220.

Scott, F (2016) *Outlaw Territories: Environments of Insecurity/Architectures of Counterinsurgency*, New York: Zone Books.

Scott, J (1985) *Weapons of the Weak: Everyday Forms of Peasant Resistance*, Hartford, CT: Yale University Press.

Scott, J (1998) *Seeing Like a State: How Certain Schemes to Improve the Human Condition Have Failed*, New Haven, CT: Yale University Press.

Seabra, P (2014) "A Harder Edge: Reframing Brazil's Power Relation with Africa", *Revista Brasileira de Política Internacional*, 57 (1), 77–97.

Selby, J (2013) "The Myth of Liberal Peace-building", *Conflict, Security and Development*, 13 (1), 57–86.

Sengupta, K (2013) "Special Report: The New Model Armies – Why Are Western Forces Being Deployed across Africa?", *The Independent*, November 12th, http://www.independent.co.uk/news/world/politics/special-report-the-new-model-armies-why-are-western-forces-being-deployed-across-africa-8935715.html (Accessed August 11th 2017).

Setzekorn, E (2017) "Eisenhower's Mutual Security Program and Congress: Defense and Economic Assistance for Cold War Asia", *Federal History Journal*, 9, 7–25, http://www.shfg.org/resources/Documents/03%20Eisenhower%20web.pdf (Accessed July 18th 2017).

Sewell, B (2010) "Early Modernisation Theory? The Eisenhower Administration and the Foreign Policy of Development in Brazil", *English Historical Review*, CXXV (517), 1449–1480.

Shakya, Y B and Rankin, K (2008) "The Politics of Subversion in Development Practice: An Exploration of Microfinance in Nepal and Vietnam", *Journal of Development Studies*, 44 (8), 1214–1235.

Shani, G (2010) "De-colonizing Foucault", *International Political Sociology*, 4 (2), 210–212.

Shapley, D (1993) *Promise and Power: The Life and Times of Robert McNamara*, Boston, MA: Little, Brown and Co.

Shariati, A (1979) *Civilization and Modernization*, Aligarh: Iranian Students Islamic Association.

Sharp, J (2007) "Geography and Gender: Finding Feminist Political Geographies", *Progress in Human Geography*, 31 (3), 381–387.

Sharp, J (2011) "Subaltern Geopolitics: Introduction", *Geoforum*, 42 (3), 271–273.

Sharp, J (2013) "Geopolitics at the Margins? Reconsidering Genealogies of Critical Geopolitics", *Political Geography*, 37, 20–29.

Sharp, J M (2010) "U.S Foreign Assistance to the Middle East: Historical Background, Recent Trends and the FY2011 Request", Congressional Research Service, Federation of American Scientists, http://www.fas.org/sgp/crs/mideast/RL32260.pdf (Accessed January 10th 2018).

Shaw, T M (2014) "Conclusion: The BRICS and beyond: New Global Order, Reorder and/or Disorder. Insights from 'Global Governance'", in Xing, L (ed) *The BRICS and beyond: The International Political Economy of the Emergence of a New World Order*, London: Routledge, 201–218.

Shaw, T M, Cooper, A F and Chin, G T (2009) "Emerging Powers and Africa: Implications for/from Global Governance?", *Politikon: South African Journal of Political Studies*, 36 (1), 27–44.

Shaxson, N (2012) *Treasure Islands: Uncovering the Damage of Offshore Banking and Tax Havens*, New York: Palgrave Macmillan.

Shen, W and Power, M (2017) "Africa and the Export of China's Clean Energy Revolution", *Third World Quarterly*, 38 (3), 678–697.

Sheppard, E (2013) "Thinking through the Pilbara", *Australian Geographer*, 44 (3), 265–282.

Sheppard, E, Porter, P, Faust, D R and Nagar, R (eds) (2009) *A World of Difference: Encountering and Contesting Development*, 2nd edition, New York: Guilford.

Shigetomi, S (2009) "Rethinking Theories on Social Movements and Development", in Shigetomi, S and Makino, K (eds) *Protest and Social Movements in the Developing World*, Cheltenham: Edward Edgar, 1–16.

Shimazu, N (2012) "Places in Diplomacy: Guest Editorial", *Political Geography*, 31 (6), 335–336.

Shimazu, N (2014) "Diplomacy as Theatre: Staging the Bandung Conference of 1955", *Modern Asian Studies*, 48 (1), 225–252.

Shinn, D (2015) *Turkey's Engagement in Sub-Saharan Africa: Shifting Alliances and Strategic Diversification*, Chatham House Research Paper, September, London: Chatham House.

Shinn, D H and Eisenman, J (2012) *China and Africa: A Century of Engagement*, Philadelphia: University of Pennsylvania Press.

Shivji, I (2007) *Silences in NGO Discourse: The Role and Future of NGOs in Africa*, Oxford: Fahamu.

Shohat, E and Stam, R (1994) *Unthinking Eurocentrism: Multiculturalism and the Media*, London: Routledge.

Short, A (1975) *The Communist Insurrection in Malaya*, London: Frederick Muller.

Shubin, V (2008) *The Hot 'Cold War': The USSR in Southern Africa*, London: Pluto Press.

Sidaway, J D (2000) "Postcolonial Geographies: An Exploratory Essay", *Progress in Human Geography*, 24 (4), 573–594.

Sidaway, J D (2003) "Sovereign Excesses? Portraying Postcolonial Sovereigntyscapes", *Political Geography*, 22 (1), 157–178.

Sidaway, J D (2007) "Enclave Space: A New Metageography of Development?", *Area*, 39 (3), 331–339.

Sidaway, J D (2008a) "The Geography of Political Geography", in Cox, K, Low, M and Robinson, J (eds) *The Handbook of Political Geography*, London: Sage, 41–55.

Sidaway, J D (2008b) "Spaces of Post(development)", *Progress in Human Geography*, 31 (3), 345–361.

Sidaway, J D (2011) "The Ends of Development Geography", in Simon, D (ed) "Symposium: Geographers and/in Development", *Environment and Planning A: Economy and Space*, 43 (12), 2791–2792.

Sidaway, J D (2012) "Geography, Globalization and the Problematic of Area Studies", *Annals of the Association of American Geographers*, 103 (4), 984–1002.

Sidaway, J D (2013) "Geographies of Development: New Maps, New Visions?", *Professional Geographer*, 64 (1), 49–62.

Sidaway, J D (2017) "Foreword: Third Wave Area Studies", in Mielke, K and Hornridge, A K (eds) *Area Studies at the Crossroads: Knowledge Production after the Mobility Turn*, New York, NY: Palgrave Macmillan, v–vii.

Sidaway, J D, Ho, E L, Rigg, J D and Woon, C Y (2016) "Area Studies and Geography: Trajectories and Manifesto", *Environment and Planning D: Society and Space*, 34 (5), 777–790.

Sidaway, J D, Mamadouh, V and Power, M (2013) "Reappraising Geopolitical Traditions", in Dodds, K, Kuus, M and Sharp, J (eds) *The Ashgate Research Companion to Critical Geopolitics*, Farnham: Ashgate, 165–187.

Sidaway, J D and Woon, C Y (2017) "Chinese Narratives on 'One Belt, One Road' (一带一路) in Geopolitical and Imperial Contexts", *The Professional Geographer*, 69 (4), 591–603.

Sidaway, J D, Woon, C Y and Jacobs, J M (2014) "Planetary Postcolonialism", *Singapore Journal of Tropical Geography*, 25 (1), 4–21.

Sider, G (1987) "When Patriots Learn to Talk and Why They Can't: Domination, Deception and Self-Deception in Indian-White Relations", *Comparative Studies in Society and History*, 29 (1), 3–23.

Silver, B and Slater, E (1999) "The Social Origins of World Hegemonies", in Arrighi, G and Silver, B (eds) *Chaos and Governance in the World System*, Minneapolis: University of Minnesota Press, 151–216.

Silvey, R and Rankin, K (2010) "Development Geography: Critical Development Studies and Political Geographic Imaginaries", *Progress in Human Geography*, 35 (5), 696–704.

Simon, D (2007) "Beyond Anti-development: Discourses, Convergences, Practices", *Singapore Journal of Tropical Geography*, 28 (2), 205–218.

Simone, A M (2004a) "People as Infrastructure: Intersecting Fragments in Johannesburg", *Public Culture*, 16 (3), 407–429.

Simone, A M (2004b) *For the City Yet to Come: Changing African Life in Four Cities*, London: Duke University Press.

Simone, A M (2016) "Cities that Are just Cities", in Lancione, M (ed) *Rethinking Life at the Margins: The Assemblage of Contexts, Subjects and Politics*, Abingdon: Routledge, 76–88.

Simpson, B (2008) *Economists with Guns: Authoritarian Development and U.S.-Indonesian Relations, 1960-1968*, Stanford, CA: Stanford University Press.

Sioh, M (2010) "Anxious Enactments: Postcolonial Anxieties and the Performance of Territorialization", *Environment and Planning D: Society and Space*, 28 (3), 467–486.

Siu, H F and McGovern, M (2017) "China–Africa Encounters: Historical Legacies and Contemporary Realities", *Annual Review of Anthropology*, 46, 337–355.

Six, C (2009) "The Rise of Postcolonial States as Donors: A Challenge to the Development Paradigm?", Third *World Quarterly*, 30 (6), 1103–1121.

Slater, D (1976) "Anglo-Saxon Geography and the Study of Underdevelopment", *Antipode*, 8 (3), 88–93.

Slater, D (1977) "Geography and Underdevelopment - Part 2", *Antipode*, 9 (3), 1–31.

Slater, D (1989) *Territory and State Power in Latin America: The Peruvian Case*, London: Macmillan Press.

Slater, D (1993) "The Geopolitical Imagination and Enframing of Development Theory", *Transactions of the Institute of British Geographers*, 18 (4), 419–437.

Slater, D (1994) "Reimagining the Geopolitics of Development: Continuing the Dialogue", *Transactions of the Institute of British Geographers*, 19 (2), 233–239.

Slater, D (1997) "Geopolitical Imaginations across the North-South Divide: Issues of Different Development and Power", *Political Geography*, 16 (8), 631–653.

Slater, D (2004) *Geopolitics and the Post-colonial: Rethinking North-South Relations*, Oxford: Blackwell.

Slater, M B (2008) "Imagining Numbers: Risk, Quantification and the Aviation Industry", *Security Dialogue* 39 (2–3), 243–266.

Slim, H (2004) "With or Against? Humanitarian Agencies and Coalition Counter-Insurgency", *Refugee Survey Quarterly*, 23 (4), 34–47.

Slotkin, R (1992) *Gunfighter Nation: The Myth of the Frontier in Twentieth Century America*, New York: Harper Perennial.

Smith, B H (1990) *More than Altruism: The Politics of Private Foreign Aid*, Princeton, NJ: Princeton University Press.

Smith, N (2010) "Remaking Area Knowledge: Beyond Global/Local", in Wesley-Smith, T and Goss, J (eds) *Remaking Area Studies: Teaching and Learning across Asia and the Pacific*, Manoa: University of Hawai'i Press, 24–40.

Smith, N and Cowen, D (2009) "After Geopolitics? From the Geopolitical Social to Geoeconomics", *Antipode*, 41 (1), 22–48.

Smyth, R (2004) "The Roots of Community Development in Colonial Office Policy and Practice in Africa", *Social Policy and Administration*, 38 (4), 418–436.

Sneddon, C (2012) "The 'Sinew of Development': Cold War Geopolitics, Technical Expertise, and Water Resource Development in Southeast Asia, 1954-1975", *Social Studies of Science*, 42 (4), 564–590.

Sneddon, C (2015) *Concrete Revolution: Large Dams, Cold War Geopolitics, and the US Bureau of Reclamation*, Chicago, IL: The University of Chicago Press.

Sneddon, C and Fox, C (2011) "The Cold War, the US Bureau of Reclamation, and the Technopolitics of River Basin Development, 1950-1970", *Political Geography*, 30 (8), 450–460.

Snow, P (1988) *The Star Raft: China's Encounter with Africa*, London: The Bath Press.

Snow, P (1994) "China and Africa: Consensus and Camouflage", in Robinson, T W and Shambaugh, D (eds) *Chinese Foreign Policy: Theory and Practice*, London: Clarendon, 283–321.

So, A (2005) "Beyond the Logic of Capital and the Polarization Model: The State, Market Reforms, and the Plurality of Class Conflict in China", *Critical Asian Studies*, 37 (3), 481–494.

Social Science Research Council (SSRC) (1957) *Social Science Research Council Annual Report 1956-1957*, Washington, DC: SSRC.

Sogge, D (2002) *Give and Take: What's the Matter with Foreign Aid?*, London: Zed Books.

Solovey, M (2001) "Project Camelot and the 1960s Epistemological Revolution: Rethinking the Politics-Patronage-Social Science Nexus", *Social Studies of Science*, 31 (2), 171–206.

Sorre, M (1952), "La géographie de l'alimentation", *Annales de Géographie*, 61 (325), 184–199.

Soske, J (2017) *Internal Frontiers: African Nationalism and the Indian Diaspora in Twentieth-Century South Africa*, Athens: Ohio University Press.

Southall, R and Comninos, A (2009) "The Scramble for Africa and the Marginalization of African Capitalism", in Southall, R and Melber, H (eds) *A New Scramble for Africa? Imperialism, Investment and Development*, Durban: University of Kwa-Zulu Natal Press, 357–385.

Sparke, M (2005) *In the Space of Theory: Postfoundational Geographies of the Nation-state*, Minneapolis, MN: University of Minnesota Press.

Sparke, M (2006) "A Neoliberal Nexus: Economy, Security and the Biopolitics of Citizenship on the Border", *Political Geography*, 25 (2), 151–180.

Sparke, M (2007) "Everywhere but Always Somewhere: Critical Geographies of the Global South", *The Global South*, 1 (1), 117–126.

Sparke, M (2008) "Political Geographies of Globalization (3): Resistance", *Progress in Human Geography*, 32 (1), 1–18.

Sparke, M (2016) "On the Overlaps of Geopolitics, Geoeconomics and USAID", *Dialogues of Human Geography*, 6 (1), 95–98.

Sparke, M, Brown, E, Corva, D, Day, H, Faria, C, Sparks, T, et al. (2005) "The World Social Forum and the Lessons for Economic Geography", *Economic Geography*, 81 (4), 359–380.

Spiegel, S J and Le Billon, P (2009) "China's Weapons Trade: From Ships of Shame to the Ethics of Global Resistance", *International Affairs*, 85 (2), 323–346.

Spillane, J F and Wolcott, D B (2012) *A History of Modern American Criminal Justice*, Thousand Oaks, CA: Sage.

Spivak, G C (2003) *Death of a Discipline*, New York: Columbia University Press.

Staar, R (1991) *Foreign Policies of the Soviet Union*, Stanford, CA: Hoover Institute Press.

Staples, A (2006) *The Birth of Development: How the World Bank, Food and Agriculture Organization, and World Health Organization Have Changed the World 1945–1965*, Ohio: Kent State University Press.

Steinecke, T (2016) "Malaysia: Africa's Silent Partner?", *The Diplomat*, May 30th, https://thediplomat.com/2016/05/malaysia-africas-silent-partner/ (Accessed October 11th 2017).

Steinmetz, G (2008) "The Colonial State as a Social Field: Ethnographic Capital and Native Policy in the German Overseas Empire before 1914", *American Sociological Review*, 73 (4), 589–612.

Stepan, N L (2001) *Picturing Tropical Nature*, Chicago, IL: University of Chicago Press.

Stephen, M D (2014) "Rising Powers, Global Capitalism and Liberal Global Governance: A Historical Materialist Account of the BRICs Challenge", *European Journal of International Relations*, 20 (4), 912–938.

Stevis-Gridneff, M (2018) "Middle East Power Struggle Plays out on New Stage", *Wall Street Journal*, June 1st, https://www.wsj.com/articles/global-powers-race-for-position-in-horn-of-africa-1527861768 (Accessed June 4th 2018).

Stewart, G C (2017) *Vietnam's Lost Revolution: Ngo Dinh Diem's Failure to Build an Independent Nation, 1955–1963*, Cambridge: Cambridge University Press.

Stiglitz, J (2002) *Globalization and Its Discontents*, New York: Norton.

Stockholm International Peace Research Institute (SIPRI) (2017) "Trends in International Arms Transfers, 2016", https://www.sipri.org/sites/default/files/Trends-in-international-arms-transfers-2016.pdf (Accessed August 23rd 2017).

Stoler, A L (1995) *Race and the Education of Desire: Foucault's History of Sexuality and the Colonial Order of Things*, Durham, NC: Duke University Press.

Stoler, A L (2016) *Duress: Imperial Durabilities in Our Times*, Durham, NC: Duke University Press.

Stolte, C (2015) *Brazil's Africa Strategy: Role Conception and the Drive for International Status*, London: Palgrave.

Strauss, J (2009) "The Past in the Present: Historical and Rhetorical Lineages in China's Relations with Africa", *The China Quarterly*, 199, 777–795.

Strüver, A (2007) "The Production of Geopolitical and Gendered Images through Global Aid Organisations", *Geopolitics*, 12 (4), 680–703.

Stubbs, R (1999) "War and Economic Development: Export-oriented Industrialization in East and Southeast Asia", *Comparative Politics*, 31 (3), 337–355.

Stubbs, R (2005) *Rethinking Asia's Economic Miracle: The Political Economy of War, Prosperity and Crisis*, New York: Palgrave Macmillan.

Stuenkel, O (2016) *Post-Western World: How Emerging Powers Are Remaking Global Order*, Cambridge: Polity.

Sukarno (1955) Opening Address by President Sukarno to Bandung Conference, April 18th, http://international.ucla.edu/institute/article/18432 (Accessed November 1st 2018).

Sutter, R (2008) *Chinese Foreign Relations: Power and Policy since the Cold War*, Plymouth, UK: Rowman and Littlefield Publishers.

Sylvester, C (2011) "Development and Postcolonial Takes on Biopolitics and Economy", in Pollard, J, McEwan, C and Hughes, A (eds) (2011) *Postcolonial Economies*, London: Zed, 185–205.

Szamosszegi, A and Kyle, C (2011) *An Analysis of State-Owned Enterprises and State Capitalism in China*, Washington, DC: Capital Trade Inc.

Taffet, J F (2007) *Foreign Aid as Foreign Policy: The Alliance for Progress in Latin America*, New York: Routledge.

Tan-Mullins, M, Mohan, G and Power, M (2010) "Redefining Aid in the China-Africa Context", *Development and Change*, 41 (5), 857–881.

Tarnoff, C (2009) "Iraq: Reconstruction Assistance", U.S. Congressional Research Service, https://fas.org/sgp/crs/mideast/RL31833.pdf (Accessed July 18th 2018).

Taylor, I (2004), "The 'All-Weather Friend'? Sino-African Interaction in the Twenty-First Century", in Taylor, I and Williams, P (eds) *Africa in International Politics: External Involvement in Africa*, London: Routledge, 83–101.

Taylor, I (2006) "China's Oil Diplomacy in Africa", *International Affairs*, 82 (5), 937–959.

Taylor, I (2010) *The International Relations of Sub-Saharan Africa*, London: Continuum.

Taylor, I (2014) *Africa Rising? BRICS – Diversifying Dependency*, Woodbridge: James Currey.

Taylor, I (2016) "Dependency Redux: Why Africa Is Not Rising", *Review of African Political Economy*, 43 (147), 8–25.

Telepneva, N (2017) "Mediators of Liberation: Eastern-Bloc Officials, Mozambican Diplomacy and the Origins of Soviet Support for Frelimo, 1958–1965", *Journal of Southern African Studies*, 43 (1), 67–81.

Teng, E J (2004) *Taiwan's Imagined Geography: Chinese Colonial Travel Writing and Pictures 1683-1895*, Cambridge, MA: Harvard University Asia Center.

Tharoor, S (2016) "From Aid-Taker to Donor: India Is Now Global Rule-Maker", *The Quint*, October 19th, https://www.thequint.com/voices/opinion/being-aid-donor-establishes-india-as-the-globes-fulcrum-tharoor-g-20-president-obama-africa-india-summit (Accessed October 11th 2017).

Thérien, J P (2002) "Debating Foreign Aid: Right versus Left", *Third World Quarterly*, 23 (3), 449–466.

Thomas, C and Wilkin, P (2004) "Still Waiting after All These Years: 'The Third World' on the Periphery of International Relations", *British Journal of Politics and International Relations*, 6 (2), 241–258.

Thomas, D, Ekasingh, B, Ekasingh, M, Lebel, L, Ha, H, Ediger, L, Thongmanivong, S, Xu, J, Sangchyoswat, C and Nyberg, Y (2008) *Comparative Assessment of Resource and Market Access of the Poor in Upland Zones of the Greater Mekong Region*, Chiang Mai: World Agroforestry Center.

Thompson, R (1966) *Defeating Communist Insurgency: Experiences from Malaya and Vietnam*, New York: Frederick A Praeger.

Thrift, N (2000) "It's the Little Things", in Dodds, K and Atkinson, D (eds) *Geopolitical Traditions: A Century of Geopolitical Thought*, London: Routledge, 380–387.

Tickner, A (2003a) "Seeing IR Differently: Notes from the Third World", *Millennium: Journal of International Studies*, 32 (2), 295–324.

Tickner, A (2003b) "Hearing Latin American Voices in International Relations Studies", *International Studies Perspectives*, 4 (4), 325–350.

Tickner, A B (2007) "A Political Economy of Dialogic IR Scholarship", paper presented at the 48th Annual International Studies Association Convention, Chicago, IL, February 28th–March 3rd.

Toye, J (2010) "Poverty Reduction", in Cornwall, A and Eade, D (eds) *Deconstructing Development Discourse: Buzzwords and Fuzzwords*, Rugby: Practical Action Publishing, 45–52.

Truman, H (1949) "Remarks at the Women's National Democratic Club (Excerpts)", in Geselbracht, R H (2015) *Foreign Aid and the Legacy of Harry S Truman*, Kirksville, MO: Truman State University Press, 291–292.

Tsui, S, Wong, E, Lau, K C and Tiejun, W (2016) "The Rhetoric and Reality of the Trans-Pacific Partnership", *Monthly Review*, 68 (7), 24–35.

Tuck, E and Yang, K W (2012) "Decolonization Is Not a Metaphor", *Decolonization: Indigeneity, Education and Society*, 1 (1), 1–40.

Turse, N (2014a) "Why Is the U.S. Military Averaging More than a Mission a Day in Africa", *The Nation*, March 27th, https://www.thenation.com/article/why-us-military-averaging-more-mission-day-africa/ (Accessed July 5th 2018).

Turse, N (2014b) "Tomgram: Nick Turse, American Monuments to Failure in Africa?", *TomDispatch*, September 7th, http://www.tomdispatch.com/post/175891/tomgram%3A_nick_turse,_american_monuments_to_failure_in_africa/ (Accessed August 7th 2018).

Turse, N (2014c) "A New Cold War in Africa?", *The Investigative Fund*, July 31st, https://www.theinvestigativefund.org/investigation/2014/07/31/new-cold-war-africa/ (Accessed August 7th 2018).

Turse, N (2015) *Tomorrow's Battlefield: U.S. Proxy Wars and Secret Ops in Africa*, Chicago, IL: Haymarket Books.

Turse, N (2016) "In Africa the US Military Sees Enemies Everywhere", *The Intercept*, July 11th, https://theintercept.com/2016/07/11/in-africa-u-s-military-sees-enemies-everywhere/ (Accessed July 16th 2016).

Turse, N (2017) "The US Military Is Conducting Secret Missions All over Africa", *Vice News*, https://news.vice.com/en_us/article/ywn5yy/us-military-secret-missions-africa (Accessed June 18th 2018).

Tyner, J (2004) "Territoriality, Social Justice and Gendered Revolutions in the Speeches of Malcolm X", *Transactions of the Institute of British Geographers*, 29 (3), 330–343.

Tyner, J A (2007) *America's Strategy in Southeast Asia: From the Cold War to the War on Terror*, Lanham, MD: Rowman and Littlefield.

Uhlig, H (1988) "Spontaneous and Planned Settlement in South-East Asia", in Manshard, W and Morgan, W B (eds) *Agricultural Expansion and Pioneer Settlements in the Humid Tropics*, Tokyo: The United Nations University, 7–43.

United Nations Conference on Trade and Development (UNCTAD) (2004) "Follow up to UNCTAD XI: New Developments in International Economic Relations", http://unctad.org/en/Docs/tdb51d6_en.pdf (Accessed October 12th 2017).

United Nations Conference on Trade and Development (UNCTAD) (2017) *World Investment Report 2017*, Geneva: UNCTAD.

United Nations Development Programme (UNDP) (1994) Draft, Position Paper of the Working Group on Operational Aspects of the Relief to Development Continuum, January 12th, New York, NY: UNDP.

United Nations Development Programme (UNDP) (2013) *Human Development Report 2013: The Rise of the South: Human Progress in a Diverse World*, New York, NY: UNDP.

United Nations Development Programme (UNDP)/China Development Bank (CDB) (2017) *The Economic Development along the Belt and Road*, http://www.cn.undp.org/content/china/en/home/library/south-south-cooperation/the-economic-development-along-the-belt-and-road-2017.html (Accessed January 1st 2018).

United Nations General Assembly (2009) "Report of the Secretary-General—Promotion of South–South Cooperation for Development: A Thirty-Year Perspective", Doc A/64/504.

United Nations Security Council (UNSC) (2017) "Note by the President of the Security Council", http://www.un.org/ga/search/view_doc.asp?symbol=S/2017/150 (Accessed October 10th 2017).

United States (2002) *The National Security Strategy of the United States of America*, Washington, DC: The White House.

United States (2006) *The National Security Strategy of the United States of America*, Washington, DC: The White House.

United States (2017) *The National Security Strategy of the United States of America*, Washington, DC: The White House.

United States Agency for International Development (USAID) (2004) *US Foreign Aid: Meeting the Challenges of the Twenty-first Century*, Washington, DC: USAID.

United States Agency for International Development (USAID) (2005) *Fragile States Strategy*, Washington, DC: USAID.

United States Agency for International Development (USAID) (2017) "Budget", https://www.usaid.gov/results-and-data/budget-spending (Accessed August 17th 2017).

United States Army (2003) *Stability Operations and Support Operations*, FM 3-07, Washington, DC: Department of the Army.

United States Army (2006) *Civil Affairs Operations*, FM 3-05.40, Washington, DC: Department of the Army.

United States Army (2008) *Stability Operations*, FM 3-07, Washington, DC: Department of the Army.

United States Army (2014) *Insurgencies and Countering Insurgencies*, FM 3-24/MCWP 3-33.5, Washington, DC: US Army, https://fas.org/irp/doddir/army/fm3-24.pdf (Accessed July 31st 2018).

United States Army/Marine Corps (2006) *Counterinsurgency Field Manual* FM 3-24, Chicago, IL: University of Chicago Press.

United States Census Bureau (2012a) "The 2012 Statistical Abstract: Table 1299 - U.S. Foreign Economic and Military Aid by Major Recipient Country", https://www.census.gov/prod/2011pubs/12statab/foreign.pdf (Accessed January 14th 2018).

United States Census Bureau (2012b) "The 2012 Statistical Abstract: Table 1297 - U.S. Government Foreign Grants and Credits by Type and Country", https://www.census.gov/prod/2011pubs/12statab/defense.pdf (Accessed January 14th 2018).

United States Census Bureau (2012c) "The 2012 Statistical Abstract: Table 1298 - U.S. Foreign Economic and Military Aid Programs", https://www.census.gov/prod/2011pubs/12statab/foreign.pdf (Accessed January 14th 2018).

United States Congress (1973) *Congressional Record – House*, Washington, DC: U.S. Government Printing Office, January 18th.

United States Department of Defense (2006), FM3-24/FMFM3-24, *Counterinsurgency*, Final Draft, June 16th, http://www.fas.org/irp/doddir/army/fm3-24fd.pdf (Accessed January 8th 2018).

United States Department of Defense (2016) *Annual Report to Congress: Military and Security Developments Involving the People's Republic of China 2016*, US Department of Defense, http://www.defense.gov/Portals/1/Documents/pubs/2016%20China%20Military%20Power%20Report.pdf (Accessed January 6th 2018).

United States Department of State (1959) *The Situation in Laos*, Washington, DC: US Department of State.

United States Department of State (2003) *United States Participation in the United Nations: A Report by the Secretary of State to the Congress for the Year 2002*, Washington, DC: U.S. Government Printing Office.

United States Department of State (2017) "Appendix 1: Foreign Assistance Framework", https://www.state.gov/s/d/rm/rls/dosstrat/2007/html/82981.htm (Accessed July 17th 2018).

United States Department of State (2018) "US Government Foreign Assistance 2017", https://foreignassistance.gov/explore (Accessed July 17th 2018).

United States Global Leadership Coalition (2017) Letter to Congress, http://www.usglc.org/downloads/2017/02/FY18_International_Affairs_Budget_House_Senate.pdf (Accessed August 17th 2017).

Urban, F, Siciliano, G and Nordensvard, J (2018) "China's Dam-builders: Their Role in Transboundary River Management in South-East Asia", *International Journal of Water Resources Development*, 34 (5), 747–770.

Uvin, P (1999) *The Influence of Aid in Situations of Violent Conflict*, Paris: Development Assistance Committee, Informal Task Force on Conflict, Peace and Development Co-operation, OECD.

Valkenier, E K (1983) *The Soviet Union and the Third World: An Economic Bind*, New York: Praeger.

Van Mead, N (2018) "China in Africa: Win-Win Development or a New Form of Colonialism?", *The Guardian*, July 31st, https://www.theguardian.com/cities/2018/jul/31/china-in-africa-win-win-development-or-a-new-colonialism?CMP=share_btn_tw (Accessed July 31st 2018).

Van Wyck, J (2016) "Africa in International Relations: Agent, Bystander or Victim?", in Bischoff, P, Aning, K and Acharya, A (eds) *Africa in Global International Relations: Emerging Approaches to Theory and Practice*, London: Routledge.

Vanolo, A (2010) "The Border between Core and Periphery: Geographical Representations of the World System", *Tijdschrift voor Economische en Sociale Geogragrafie*, 101 (1), 26–36.

Varin, C (2015) "Why Nigeria Is Turning to South African Mercenaries to Help Fight Boko Haram", *The Conversation*, March 20th, http://theconversation.com/why-nigeria-is-turning-to-south-african-mercenaries-to-help-fight-boko-haram-38948 (Accessed June 19th 2018).

Vasudevan, A (2015) *Metropolitan Preoccupations: The Spatial Politics of Squatting in Berlin*, Oxford: Wiley.

Vasudevan, A, McFarlane, C and Jeffrey, A (2008) "Spaces of Enclosure", *Geoforum*, 39 (5), 1641–1646.

Vergara-Camus, L (2014) *Land and Freedom: The MST, the Zapatistas and Peasant Alternatives to Neoliberalism*, London: Zed Books.

Verschaeve, J and Orbie, J (2015) "The DAC Is Dead, Long Live the DCF? A Comparative Analysis of the OECD Development Assistance Committee and the UN Development Cooperation Forum", *European Journal of Development Research*, 28 (4), 1–17.

Vezirgiannidou, S-E (2013) "The United States and Rising Powers in a Post-hegemonic World Order", *International Affairs*, 83 (3), 635–651.

Vickers, B (2013) "Africa and the Rising Powers: Bargaining for the 'Marginalized Many'", *International Affairs*, 89 (3), 673–693.

Vieira, M and Menezes, H (2016) "Brazil is Breaking with Its South-South Focus. What It Means for BRICS", *The Conversation*, November 20th, http://theconversation.com/brazil-is-breaking-with-its-south-south-focus-what-it-means-for-brics-69008 (Accessed August 30th 2017).

Villalba, U (2013) "*Buen Vivir* vs Development: A Paradigm Shift in the Andes?", *Third World Quarterly*, 34 (8), 1427–1442.

Vine, D (2015) *Base Nation: How U.S. Military Bases Abroad Harm America and the World*, New York: Henry Holt and Company.

Vines, A and Campos, I (2010) "China and India in Angola", in Cheru, F and Obi, C (eds) *The Rise of China and India in Africa: Challenges, Opportunities and Critical Interventions*, London: Zed, 193–207.

Vivek, V (2017) "India Gives Most Foreign Aid to Bhutan, Not Its New Priorities Afghanistan and Africa", https://scroll.in/article/835481/india-gives-most-foreign-aid-to-bhutan-not-its-new-priorities-afghanistan-and-africa (Accessed September 25th 2017).

Voice of America (VOA) (2018) "Tillerson: China's Approach to Africa Encourages Dependency", March 7th, https://www.voanews.com/a/tillerson-china-approach-to-africa-encourages-dependency/4282809.html (Accessed April 11th 2018).

Wade, R (1990) *Governing the Market: Economic Theory and the Role of Government in East Asian Industrialisation*, Princeton, NJ: Princeton University Press.

Wainwright, J (2008) *Decolonizing Development: Colonial Power and the Maya*, Oxford: Blackwell.

Wainwright, J (2016) "The U.S. Military and Human Geography: Reflections on Our Conjuncture", *Annals of the American Association of Geographers*, 106 (3), 513–520.

Wainwright, J and Bryan, J (2009) "Cartography, Territory, Property: Postcolonial Reflections on Indigenous Counter-mapping in Nicaragua and Belize", *Cultural Geographies*, 16 (2), 153–178.

Wallerstein, I (1988) "Development: Lodestar or Illusion?", *Economic and Political Weekly*, 23 (39), 2017–2019 + 2021–2023.

Wallerstein, I (1997) "The Unintended Consequences of Cold War Area Studies", in Chomsky, N (ed) The *Cold War and the University: Toward an Intellectual History of the Postwar Years*, New York: The New Press, 195–231.

Walton, J and Seddon, D (1994) *Free Markets and Food Riots: The Politics of Global Adjustment*, Oxford: Blackwell.

Waltz, K (1979) *Theory of International Politics*, New York: Random House.

Wansleben, L (2013) "Dreaming with BRICs", *Journal of Cultural Economy*, 6 (4), 453–471.

Ward, K and England, K (2007) "Introduction: Reading Neoliberalization", in England, K and Ward, K (eds) *Neoliberalization: States, Networks, Peoples*, London: Blackwell, 1–22.

Watnick, M (1952–53) "The Appeal of Communism to the Peoples of Underdeveloped Areas", *Economic Development and Cultural Change*, 1 (1), 22–36.

Watts, M (1995) "'A New Deal in Emotions': Theory and Practice and the Crisis of Development", in Crush, J (ed) *Power of Development*, London: Routledge, 44–62.

Watts, M (2003) "Development and Governmentality", *Singapore Journal of Tropical Geography*, 24 (1), 6–34.

Wayne, M I (2008) *China's War on Terrorism: Counterinsurgency, Politics, and Internal Security*, New York: Routledge.

Weheliye, A G (2014) *Habeas Viscus: Racializing Assemblages, Biopolitics, and Black Feminist Theories of the Human*, London: Duke University Press.

Weinberger, N J (1986) *Syrian Intervention in Lebanon: The 1975-76 Civil War*, Oxford: Oxford University Press.

Weizman, E (2011) *The Least of All Possible Evils: Humanitarian Violence from Arendt to Gaza*, London: Verso.

Westad, O A (2006) *The Global Cold War: Third World Interventions and the Making of Our Times*, Cambridge: Cambridge University Press.

White, B (1999) "Nucleus and Plasma: Contract Farming and the Exercise of Power in Upland West Java", in Murray Li, T (ed) *Transforming the Indonesian Uplands: Marginality, Power, and Production*, Amsterdam: Harwood Academic, 231–256.

White, E (2003) "Kwame Nkrumah: Cold War Modernity, Pan-African Ideology and the Geopolitics of Development", *Geopolitics*, 8 (2), 99–124.

White, J (1974) *The Politics of Foreign Aid*, New York: St Martin's Press.

White, L (2010), "Understanding Brazil's New Drive for Africa", *South African Journal of International Affairs*, 17 (2), 221–242.

White, S (2002) "Thinking Race, Thinking Development", *Third World Quarterly*, 23 (3), 407–419.

Whitlock, C (2012) "U.S Expands Secret Intelligence Operations in Africa", *The Washington Post*, June 13th, https://www.washingtonpost.com/world/national-security/us-expands-secret-intelligence-operations-in-africa/2012/06/13/gJQAHyvAbV_story.html (Accessed January 15th 2018).

Wilder, G (1999) "Practising Citizenship in Imperial Paris", in Comaroff, J and Comaroff, J (eds) *Civil Society and the Political Imagination in Africa*, Chicago, IL: Chicago University Press.

Williams, A (1997) "The Postcolonial Flaneur and Other Fellow-Travellers: Conceits for a Narrative of Redemption", *Third World Quarterly*, 18 (5), 821–841.

Williams, M, Kean, J, Jenkins, C, Feldman, J and Fisher-Harris, P (1988) "Retrospective Review of US Assistance to Afghanistan, 1950-1979", Bethesda, MD: Devres Inc. for USAID contract no. PDC-0085-I-00-6095, October 31st.

Williams, R (1976) *Keywords: A Vocabulary of Culture and Society*, London: Fontana.

Williams, R (1983) *Toward 2000*, London: Penguin.

Williams, S (2010) *Climbing the Bookshelves: The Autobiography of Shirley Williams*, London: Virago.

Wilson, H (1953) *The War on World Poverty*, London: Victor Gollancz.

Wilson, J D (2015) "Resource Powers? Minerals, Energy and the Rise of the BRICS", *Third World Quarterly*, 36 (2), 223–239.

Winrow, G M (2009) *The Foreign Policy of the GDR in Africa*, Cambridge: Cambridge University Press.

Winther, T (2008) *The Impact of Electricity: Development, Desires and Dilemmas*, Oxford: Berghahn Books.

Wisner, B (1986) "Geography: War or Peace Studies?", *Antipode* 18 (2), 212–217.

Wittfogel, K A (1957) *Oriental Despotism: A Comparative Study of Total Power*, New Haven, CT: Yale University Press.

Wolfe, A J (2013) *Competing with the Soviets: Science, Technology, and the State in Cold War America*, Baltimore, MD: Johns Hopkins University Press.

Woll, J (2004) "The Russian Connection: Soviet Cinema and the Cinema of Francophone Africa", in Pfaff, F (ed) *Focus on African Films*, Indianapolis: Indiana University Press, 223–240.

Wolvers, A, Tappe, O, Salverda, T and Schwarz, T (2015) "Concepts of the Global South - Voices from around the World", Global South Studies Center, University of Cologne, Germany, http://gssc.uni-koeln.de/node/452 (Accessed January 2nd 2018).

Woo-Cumings, M (1999) "Introduction: Chalmers Johnson and the Politics of Nationalism and Development", in Woo-Cumings, M (ed) *The Developmental State*, Ithaca, NY: Cornell University Press.

Woods, N (2005) "The Shifting Politics of Foreign Aid", *International Affairs*, 81 (2), 393–409.

Woods, N (2008) "Whose Aid? Whose Influence? China, Emerging Donors and the Silent Revolution in Development Assistance", *International Affairs*, 84 (6), 1205–1221.

Woon, C Y (2011) "Undoing Violence, Unbounding Precarity: Beyond the Frames of Terror in the Philippines", *Geoforum*, 42 (3), 285–296.

Worby, E (2000) "'Discipline without Oppression': Sequence, Timing and Marginality in Southern Rhodesia's Post-war Development Regime", *The Journal of African History*, 41 (1), 101–125.

World Bank (1998) *Assessing Aid: What Works, What Doesn't and Why*, New York: Oxford University Press.

World Friends Korea (WFK) (2017) "About World Friends Korea", http://www.worldfriendskorea.or.kr/eng/ (Accessed September 10th 2017).

Wright, R (1956) *The Color Curtain: A Report on the Bandung Conference*, Cleveland, OH: The World Publishing Company.

Xinhua (2017) "China Focus: Chinese Fund Helps Development in Africa", December 24th, http://www.xinhuanet.com/english/2017-12/24/c_136848691.htm (Accessed July 18th 2018).

Yacobi, H (2015) *Israel and Africa: A Genealogy of Moral Geography*, London: Routledge.

Yamada, S (2015) "From Humanitarianism to Trade Promotion: The Changing Emphasis of Japanese Development Co-operation to Africa", *African-East Asian Affairs: The China Monitor*, 1–2 (June), 28–49.

Yao, Y (2010a) "The End of the Beijing Consensus: Can China's Model of Authoritarian Growth Survive?", *Foreign Affairs*, February 2nd, http://www.foreignaffairs.com/articles/65947/the-end-of-the-beijing-consensus/ (Accessed July 1st 2018).

Yao, Y (2010b) "The China Model and Its Future", in Garnaut, R, Golley, J and Song, L (eds) *China: The Next Twenty Years of Reform and Development*, Canberra: The Australian National University Press, 39–52.

Yengde, S (2015) "Caste among Indian Diaspora in Africa", *Economic and Political Weekly*, 50 (37), https://www.epw.in/journal/2015/37/notes/caste-among-indian-diaspora-africa.html (Accessed June 14th 2018).

Yeung, H W (2017) "Rethinking the East Asian Developmental State in Its Historical Context: Finance, Geopolitics and Democracy", *Area Development and Policy*, 2 (1), 1–23.

Yi-Chong, X (2014) "Chinese State-owned Enterprises in Africa: Ambassadors or Freebooters?", *Journal of Contemporary China*, 23 (89), 822–840.

Young, J (2006) "Review of *The Global Cold War: Third World Interventions and the Making of Our Times*", *Reviews in History*, http://www.history.ac.uk/reviews/review/534 (Accessed July 31st 2018).

Young, R (2005) "Postcolonialism from Bandung to the Tricontinental", *Historein*, 5, 11–21.

Zakaria, R (2018) "Who Benefits from Pakistan's Loss of US Aid?", *Al Jazeera*, January 10th, https://www.aljazeera.com/indepth/opinion/aid-tirade-pakistan-story-180108095929965.html (Accessed June 16th 2018).

Zezela, P T (2006) *The Study of Africa (Volume 1): Disciplinary and Interdisciplinary Encounters*, Dakar: Codesria.

Zezela, P T (2007) *The Study of Africa (Volume 2): Transnational and Global Engagements*, Dakar: Codesria.

Zhang, C (2015) *The Domestic Dynamics of China's Energy Diplomacy*, Singapore: World Scientific.

Zhang, J and Peck, J (2016) "Variegated Capitalism, Chinese Style: Regional Models, Multi-scalar Constructions", *Regional Studies*, 50 (1), 52–78.

Zhao, J and Wu, B (2010) *Lun Zhongguo moshi* [Discussing the China Model], Beijing: China Social Sciences Press.

Zhao, S (2008) "The Making of Chinese Foreign Policy: Actors and Institutions", in Ampiah, K and Naidu, S (eds) *Crouching Tiger, Hidden Dragon? Africa and China*, South Africa: University of KwaZulu-Natal Press, 39–52.

Zhao, T Y (2005) *Tianxia Tixi: Shijie Zhidu Zhexue Daolun* [The Tianxia System: A Philosophy for the World Institution], Nanjing: Jiangsu Jiaoyu Chubanshe.

Ziai, A (2004) "The Ambivalence of Post-Development: Between Reactionary Populism and Radical Democracy", *Third World Quarterly*, 25 (6), 1045–1060.

Ziai, A (ed) (2007) *Exploring Post-Development: Theory and Practice, Problems and Perspectives*, New York: Routledge.

Ziai, A (2017) "'I Am Not a Post-Developmentalist, but…' The influence of Post-Development on Development Studies", *Third World Quarterly*, 38 (12), 2719–2734.

Zibechi, R (2012) *Territories in Resistance: A Cartography of Latin American Social Movements*, Oakland, CA: AK Press.

Zimmerman, R F (1993) *Dollars, Diplomacy and Dependency*, Boulder, CO: Lynne Rienner.

Žižek, S (2012) *Organs without Bodies: On Deleuze and Consequences*, London: Routledge.

Index

Entries in *italic* denote figures.